OXFORD IB STUDY GUIDES

Garrett Nagle
Briony Cooke

Geography

FOR THE IB DIPLOMA

OXFORD
UNIVERSITY PRESS

OXFORD
UNIVERSITY PRESS

Great Clarendon Street, Oxford OX2 6DP

Oxford University Press is a department of the University of Oxford. It furthers the University's objective of excellence in research, scholarship, and education by publishing worldwide in

Oxford New York

Auckland Cape Town Dar es Salaam Hong Kong Karachi Kuala Lumpur Madrid Melbourne Mexico City Nairobi New Delhi Shanghai Taipei Toronto

With offices in

Argentina Austria Brazil Chile Czech Republic France Greece Guatemala Hungary Italy Japan Poland Portugal Singapore South Korea Switzerland Thailand Turkey Ukraine Vietnam

Oxford is a registered trade mark of Oxford University Press in the UK and in certain other countries

© Garret Nagle 2012

The moral rights of the author have been asserted

Database right Oxford University Press (maker)

First published 2009

All rights reserved. No part of this publication may be reproduced, stored in a retrieval system, or transmitted, in any form or by any means, without the prior permission in writing of Oxford University Press, or as expressly permitted by law, or under terms agreed with the appropriate reprographics rights organization. Enquiries concerning reproduction outside the scope of the above should be sent to the Rights Department, Oxford University Press, at the address above

You must not circulate this book in any other binding or cover and you must impose this same condition on any acquirer

British Library Cataloguing in Publication Data

Data available

ISBN: 978-0-19-838915-6
10 9 8 7 6 5 4 3 2 1

Printed in Great Britain by Bell & Bain Ltd, Glasgow

Paper used in the production of this book is a natural, recyclable product made from wood grown in sustainable forests. The manufacturing process conforms to the environmental regulations of the country of origin

Acknowledgments

We are grateful to the following to reproduce the following copyright material.

Garrett Nagle

Cover photo: Michael Layefsky/Flickr/Getty Images

We have tried to trace and contact all copyright holders before publication. If notified the publishers will be pleased to rectify any errors or omissions at the earliest opportunity.

Dedication: to Angela, Rosie, Patrick, Bethany, Henry and Chris

Contents

Introduction to the International Baccalaureate Diploma Course

Course structure

Students take six academic subjects; three at higher level (HL) and three at standard level (SL). These six include two languages, one experimental science, mathematics, humanities and one further subject of the student's choice.

Aims

The aims of the geography syllabus at HL and SL are to enable students to:

- develop an understanding of the interrelationships between people, places, spaces and the environment
- develop a concern for human welfare and the quality of the environment and an understanding of the need for planning and sustainable management
- appreciate the relevance of geography in analysing contemporary issues and challenges and develop a global perspective of diversity and change.

Geographic skills SL/HL

Geographic skills are learnt throughout the course as an integral part of the syllabus. Many skills are universal and can be applied to all topics and demonstrated in all components of the exam, while others are more specific. The skills are listed below with the relevant examination components.

Skills	Exam component
Images Interpret and analyse maps using latitude, longitude, direction, scale and grid references. Interpret and (where appropriate) draw and annotate isoline, chloropleth, flow, dot, topographical and topological maps; interpret satellite and aerial photos and cross-profiles.	Paper 1 Section A* Paper 2, IA
Graphs Interpret and construct: line, bar, scatter, triangular, logarithmic and bi-polar graphs, pie and flow charts, radial diagrams, population pyramids and Lorenz curves.	Paper 1 Section A* Paper 2*, IA
Statistical tests and indices Interpret and calculate: total, mode, mean, median, frequency, range, density, percentage, ratio, Spearman's rank correlation, chi squared test, nearest neighbour index, location quotient, diversity index, HDI, dependency ratio and measures of spatial interaction (gravity models).	Paper 1 Section A* Paper 2*, IA
Research methods Observe and record information by: interviewing, drawing a field sketch and taking photgraphs, measuring, judging, recording, classifying, describing trends, patterns and relationships in data, predicting, identifying anomalies, making decisions, concluding and evaluating research methods.	Internal assessment
Writing skills Evaluate geographic information in terms of reliability, bias, relevance and accuracy. Synthesize information, respond appropriately to command terms and present a coherent argument.	All components

** Calculation is required only in the internal assessment (IA) component*

Terminology used in the syllabus

Contemporary within your lifetime
Recent an event that has occurred since the year 2000
Geographic demographic, environmental, social, cultural, economic and political factors
LEDC/MEDC are not used in this syllabus; alternative terms such as low/high income, poor/rich may be used
Example usually a named place e.g. Thailand would be an example of a country where tourism is important
Case study a more detailed description of a named place e.g. characteristics and location of Thailand's tourist industry
Global on a world scale e.g. global climate change affects the whole world
Region major world region e.g. South East Asia
National refers to one country
Local within a national boundary e.g. a town and its surrounding area

Population trends

GLOBAL POPULATION CHANGE 1930–2020

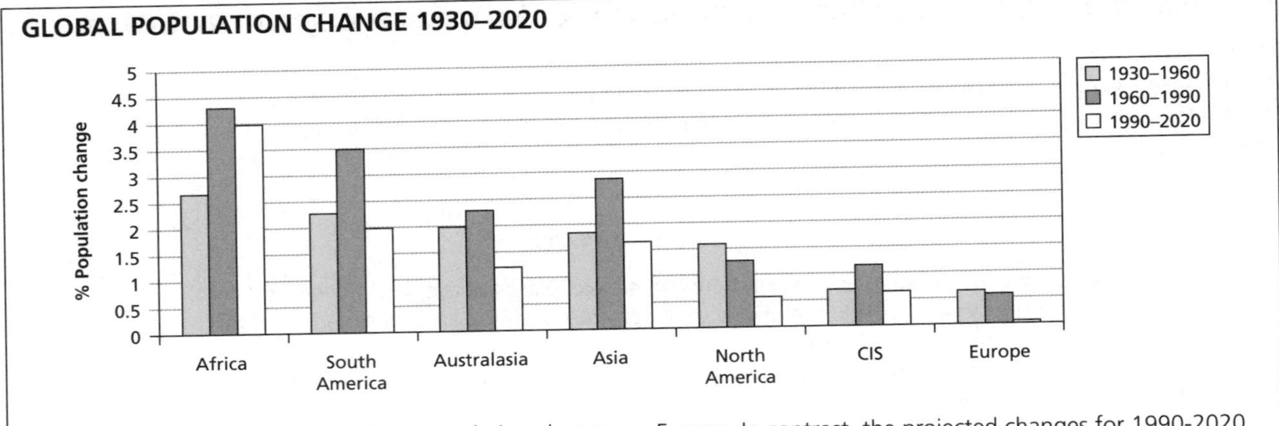

The graph shows that in most regions population change increased between 1930 and 1960, and again between 1960 and 1990. The exceptions were North America and Europe. In contrast, the projected changes for 1990-2020 show that population growth rates will fall in all regions, notably South America, Asia and Australasia.

EXPONENTIAL GROWTH

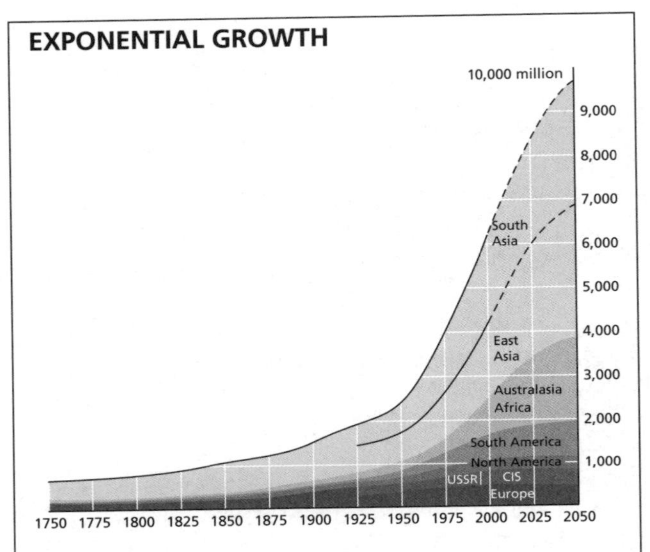

Exponential growth of the world's population, 1750–2050

The world's population is growing very rapidly. Most of this growth is quite recent. Global population doubled between 1650 and 1850, 1850 and 1920, and 1920 and 1970. It is thus taking less time for the population to double.

Up to 95% of population growth is taking place in less economically developed countries (LEDCs). An increasing or accelerating rate of growth is known as **exponential growth**. However, the world's population is expected to stabilize at about 12 billion by around 2050–80.

Global population growth creates:

- great pressures on governments to provide for their people
- increased pressure on the environment
- increased risk of famine and malnutrition
- greater differences between the richer countries and the poorer countries.

DEMOGRAPHIC CHANGE AND GLOBAL TRENDS

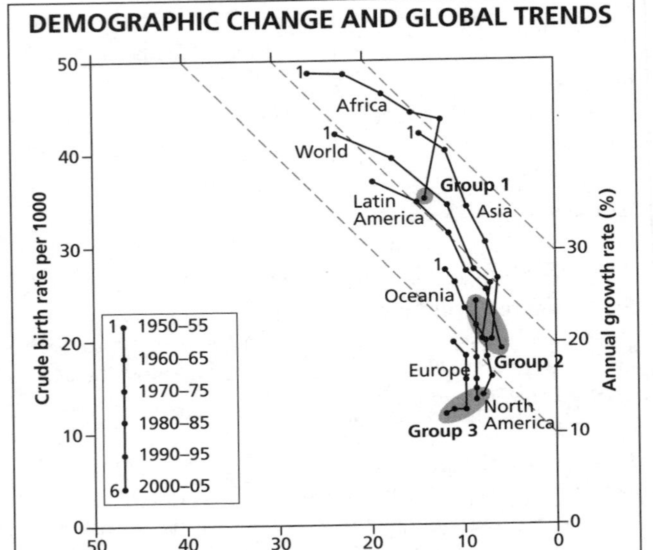

Demographic paths of the world's major regions

The **annual growth rate** is found by subtracting the crude death rate (‰ – per thousand) from the crude birth rate (‰) and is then expressed as a percentage (%). Percentages are used for growth rates rather than per thousand, partly due to familiarity of the term and because they are easier to use in calculations. Remember that 20‰ is the same as 2%.

Highest growth rates are found in Africa, while lowest growth rates are in North America and Europe.

EXTENSION

Visit

www.imf.org/external/pubs/ft/fandd/2006/09/picture.htm

This is a useful site for a number of features in global demographic (population) trends, such as population growth, changes in birth rates and death rates, migration ratios, and contrasts between rich and poor countries.

Birth rates (1)

WORLD BIRTH RATES

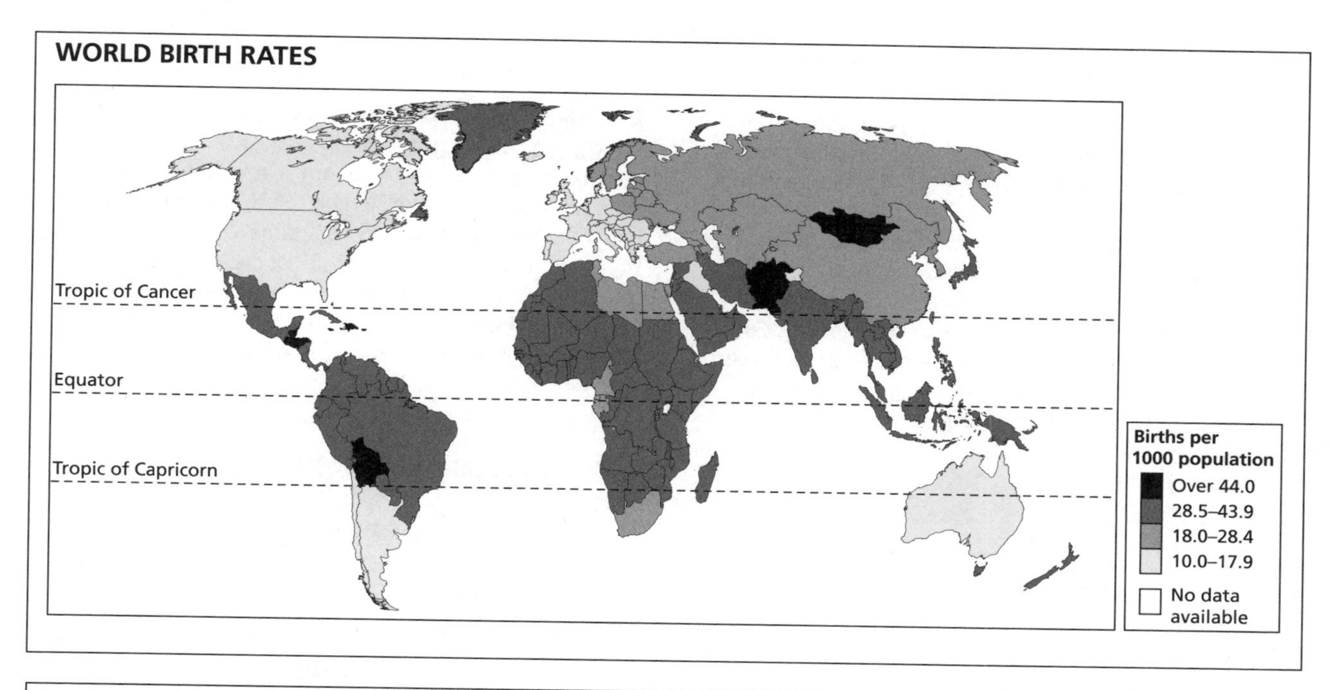

MEASUREMENTS OF FERTILITY

The **crude birth rate** (CBR) = $\frac{\text{total no. of births}}{\text{total population}} \times 1000$ per year

Mauritius, 2001 = $\frac{19{,}600}{1{,}189{,}000}$ = 16.5‰

The CBR is easy to calculate and the data are readily available. However, it does not take into account the age and sex structure of the population. By contrast, the **standardized birth rate** (SBR) gives a birth rate for a region on the basis that the region's age composition is the same as that of the whole country.

The **total fertility rate** (TFR) is the average number of births per thousand women of childbearing age. In Mauritius in 2001 it was 2.01. It is the completed family size if fertility rates remain constant.

The **general fertility rate** is the number of births per thousand women aged 15–49 years (sometimes 15–44 years). This can be shown in the following formula:

General fertility rate = $\frac{\text{no. of births}}{\text{women in 15–49 year age range}} \times 1000$ per year

The **age-specific birth rate** (ASBR) = $\frac{\text{no. of births}}{\text{women of any specified year group}} \times 1000$ per year

	15–19 yrs	20–24 yrs	25–29 yrs	30–34 yrs	35–39 yrs	40–44 yrs	45–49 yrs	TFR
MEDCs	32	96	111	71	26	5	0	1.7
LEDCs	140	275	273	218	149	79	27	5.8

Variations in birth rate by age of woman

In general, the highest fertility rates are found among the poorest countries, and very few LEDCs have made the transition from high birth rates to low birth rates. Most MEDCs, by contrast, have brought the birth rate down. In MEDCs, fertility rates have fallen as well – the decline in population growth is not therefore due to changing population structure.

CHANGES IN FERTILITY

Changes in fertility are a combination of both **sociocultural** and **economic** factors. While there may be strong correlations between these sets of factors and changes in fertility, it is impossible to prove the linkages or to prove that one set of factors is more important than the other.

Birth rates (2)

SOCIOCULTURAL FACTORS AND FERTILITY

Status of women

The status of women is assessed by the **gender-related development index** (GDI), which measures the inequality between the sexes in life expectancy, education and the standard of living.

In countries where the status of women is low and few women are educated or involved in paid employment, birth rates are high.

In countries such as Singapore, where the status of women has improved, the birth rate has fallen. Between 1960 and 2000 there were great social and economic changes there, resulting in full employment, including female employment. As a result, the total fertility rate fell from over 3.0 to 1.5.

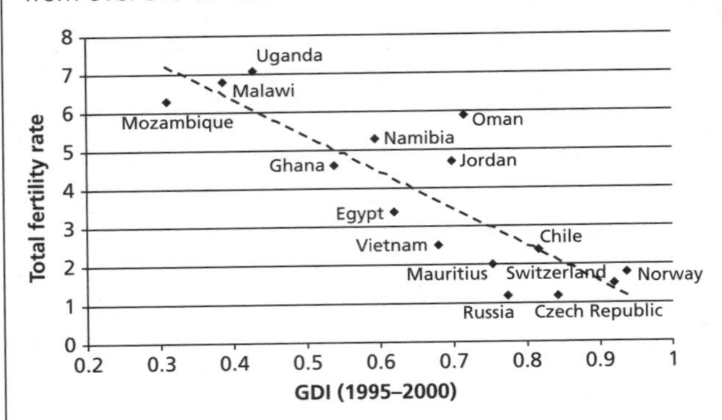

Level of education and material ambition

In general, the higher the level of parental education, the fewer the children. Middle-income families with high aspirations but limited means tend to have the smallest families. They wish to improve their standard of living, and will limit their family size to achieve this. Poor people with limited resources or ambition often have large families. Affluent people can afford large families.

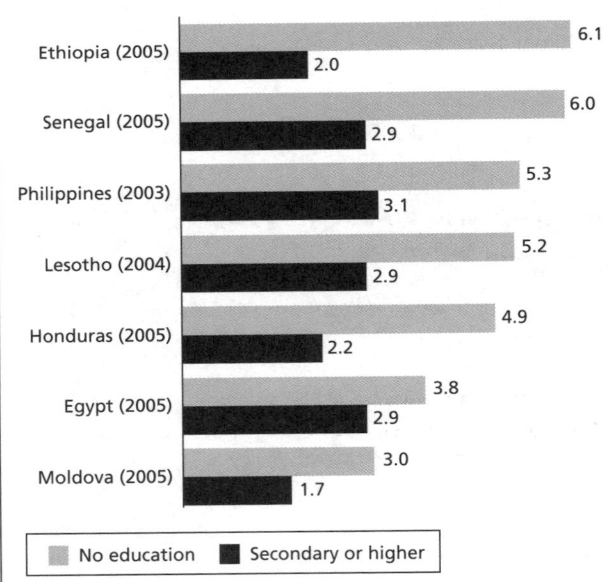

Lifetime births per woman by highest level of education

EXTENSION

Reading scatter graphs

When answering questions about scatter graphs, look for a number of points including trend, maximum value, minimum value and exceptions/anomalies.

Here the trend is negative – as GDI increases the fertility rate decreases. For example, Uganda has the highest TFR (7) and a low GDI. In contrast the Czech Republic has a low TRF (1) but a high GDI. An exception is Oman with a high TFR and GDI.

Level of education	CBR	TFR
University	42.18	1.15
Senior middle school	63.88	1.23
Junior middle school	67.43	1.44
Primary school	86.25	2.02
Illiterate	94.50	2.44

Women's educational level and births: evidence from China

Type of residence

People in rural areas tend to have more children than those in urban areas. Reasons for this include:

- more rigid social pressures on women
- greater freedom and less state control (e.g. China's one-child policy is enforced less rigorously in rural areas)
- females in rural areas have fewer educational and economic opportunities.
- In some urban areas, such as shanty towns, there are high levels of fertility because of their youthful population structure.

Religion

The role of religion in relation to fertility rates is commonly confused. The lowest birth rates in Europe include those of Italy and Spain, both Catholic countries. In contrast, some poor Catholic countries, such as Mexico and Brazil, have high birth rates. In general, most religions are pro-natalist (they favour large families), and are opposed to birth control, sterilization and contraception. In MEDCs, however, most people do not follow the dictates of religious beliefs very strongly.

Health of the mother

Although more pregnancies are successful for women who are well nourished and healthy, women who are not healthy may become pregnant more frequently. This is because they may experience a higher infant mortality and more unsuccessful pregnancies. Hence they become pregnant again in order to compensate for the child they have lost.

Birth rates (3)

ECONOMIC FACTORS AND FERTILITY

Economic prosperity

The correlation between economic prosperity and the birth rate is not total, but there are links. Economic prosperity favours an increase in the birth rate, while increasing costs lead to a decline in the birth rate. Recession and unemployment are also linked with a decline in the birth rate. This is related to the cost of bringing up children. Surveys have shown that the cost of bringing up a child in the UK can be over $300,000, partly through lost parental earnings. Whether the cost is real or imagined (perceived) does not matter. If parents believe they cannot afford to bring up a family, or that by having more children their standard of living will be reduced, they are less likely to have children.

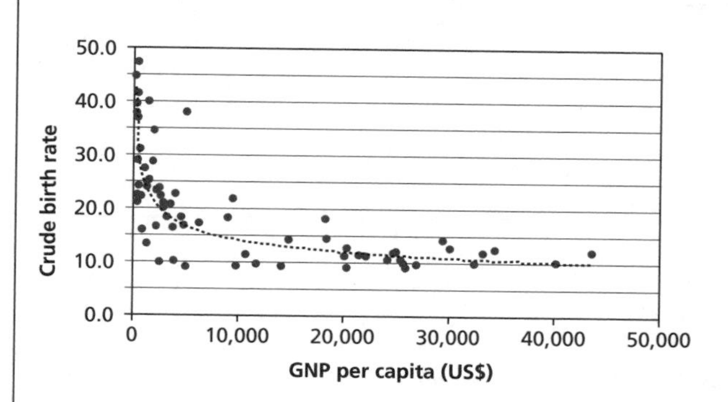

At the global scale, a strong link exists between fertility and the level of economic development. The UN and many civil societies, including non-government organizations (NGOs), believe that a reduction in the high birth rates in the LEDCs can be achieved only by improving the standard of living in those countries.

There is also evidence to suggest that the more equitable the distribution of wealth within a country, the lower the fertility rate (see below).

	Share of income			
	GNP per capita	Lowest 10%	Top 10%	Fertility rate
Nicaragua	US$2,300	0.7	49	4.3
Armenia	US$2,200	2.3	35	1.4

The need for children

High infant mortality rates (pages 22 and 119) increase the pressure on women to have more children. Such births, to offset the high mortality losses, are termed replacement births or compensatory births.

In some agricultural societies, parents have larger families to provide labour for the farm and as security for the parents in old age. This is much less important now as fewer families are engaged in farming, and many farmers work as labourers rather than own their own farms.

EXTENSION

Choose two or three countries at different levels of economic and social development and research data relevant to their fertility rates. Track changes over time. For example:

	GNP per capita (US$)	CBR (‰)	TFR	GDI
Canada	20,000	11	1.6	0.9
Poland	3900	10	1.5	0.8
Tanzania	200	40	5.5	0.4

Visit

The CIA website **www.cia.gov/library/publications/the-world-factbook/index.html** has excellent up-to-date data.

EXTENSION

Using tables

To give your answers more "weight" try using data to support them. For example, if you were asked to outline the relationship between GNP per capita and CBR, your answer might include the following:

Canada has a higher level of GNP (US$20 000) than Poland (US$3900) and Tanzania (US$200). Its crude birth rate (11‰) is higher than that of Poland but much lower than that of Tanzania.

Do not be concerned that the data are inconsistent and at variance to the expected pattern (we would expect the Canadian population to have a lower birth rate than Poland because they are wealthier, but in this case the difference is small). In geography there are many exceptions to general trends.

Mortality (1)

WORLD DEATH RATES

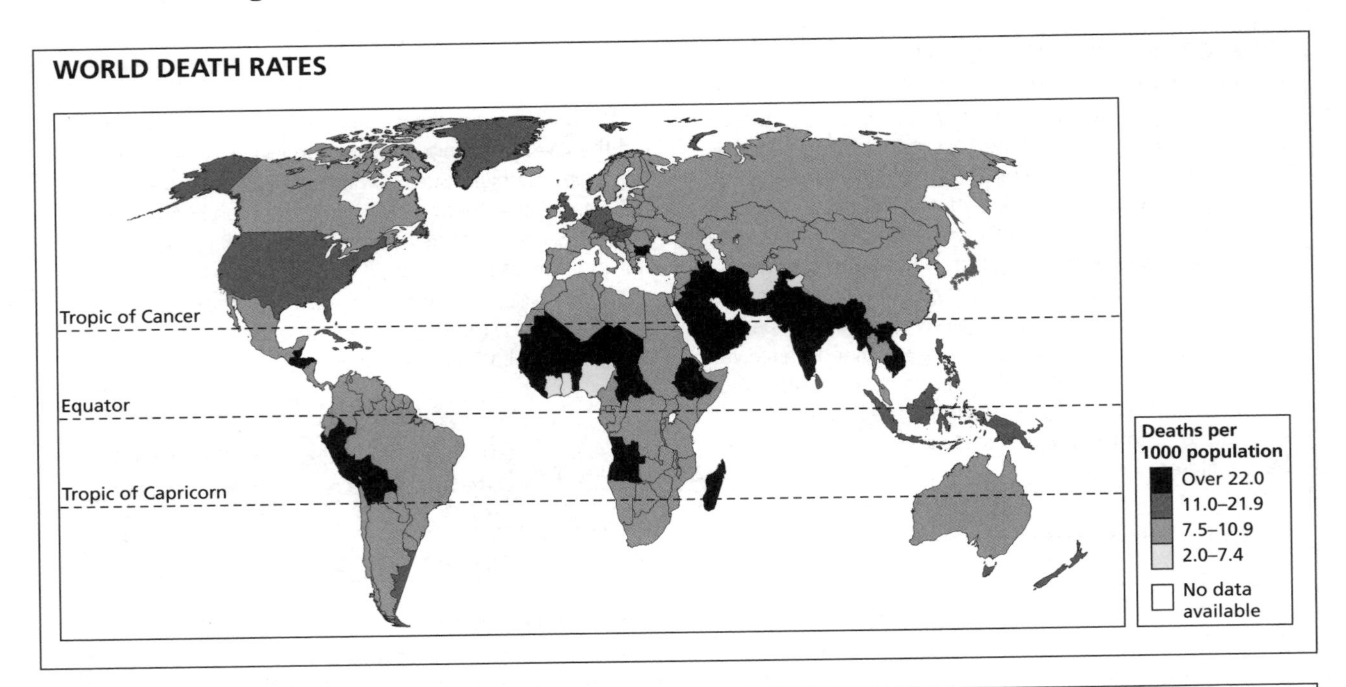

MEASUREMENTS OF MORTALITY

The **crude death rate** (CDR) = $\frac{\text{total no. of deaths}}{\text{total population}} \times 1000$ per year

The CDR is a poor indicator of mortality trends – populations with a large number of aged, as in most MEDCs, will have a higher CDR than countries with more youthful populations. Denmark, for example, has a CDR of 11‰; in Pakistan it is 7.8‰ (see page 7). Consequently, to compare mortality rates we use the **standardized mortality rate** (SMR) or **age-specific mortality rates** (ASMRs) such as the **infant mortality rate** (IMR).

The **infant mortality rate** (IMR) = $\frac{\text{total no. of deaths of children} < 1 \text{ year old}}{\text{total no. of live births per year}} \times 1000$

The **child mortality rate** (CMR) = $\frac{\text{total no. of deaths of children aged 1–5 years}}{\text{total number of children aged 1–5 years}} \times 1000$

Life expectancy (E_o) is the average number of years that a person can be expected to live, given that demographic factors remain unchanged.

PATTERNS OF MORTALITY

At the global scale, the pattern of mortality in MEDCs differs from that in LEDCs. In the former, as a consequence of better nutrition, healthcare and environmental conditions, the death rate falls steadily to a level of about 9‰, with very high life expectancies (75+ years). In many of the very poorer countries, high death rates and low life expectancies are still common, although both have shown steady improvement over the past few decades due to improvements in food supply, water, sanitation and housing. This trend, unfortunately, has been reversed as a consequence of AIDS.

CAUSES OF DEATH

As a country develops, the major forms of illness and death change. LEDCs are characterized by a high proportion of infectious diseases, many of which may be waterborne, for example cholera and gastroenteritis, or vector-borne, for example river blindness and malaria, diarrhoea and vomiting. These may prove fatal. By contrast, in MEDCs, fatal diseases are more likely to be degenerative conditions such as cancer, strokes or heart disease.

The change in disease pattern from infectious to degenerative is known as the **epidemiological transition model**. (Epidemiology is the study of diseases.) Such a change generally took about a century in the MEDCs, but is taking place faster in the LEDCs.

Mortality (2)

VARIATIONS IN MORTALITY RATES

Variations occur both at the global scale and on a more local scale:

- **Age structure:** Some populations, such as those in retirement towns and especially in the older industrialized countries, have very high life expectancies and this in turn results in a rise in the CDR. Countries with a large proportion of young people will have much lower death rates (Mexico, with 34% of its population under the age of 15 years, has a CDR of 5‰).
- **Social class:** The poorer people within any population have higher mortality rates than the more affluent. In some countries, such as South Africa, this will also be reflected in racial groups (see right).
- **Occupation:** Certain occupations are hazardous – the military, farming, oil production and mining, for example. Some diseases are linked to specific occupations – such as mining and respiratory disease.
- **Place of residence:** In urban areas, mortality rates are higher in areas of relative poverty and deprivation, such as inner cities and shanty towns. This is due to overcrowding, pollution, high population densities and stress. In many rural areas, where there is widespread poverty and limited farm productivity, mortality rates are high. For example, in the rural north-east of Brazil, life expectancy is 27 years shorter than in the richer south-east region.

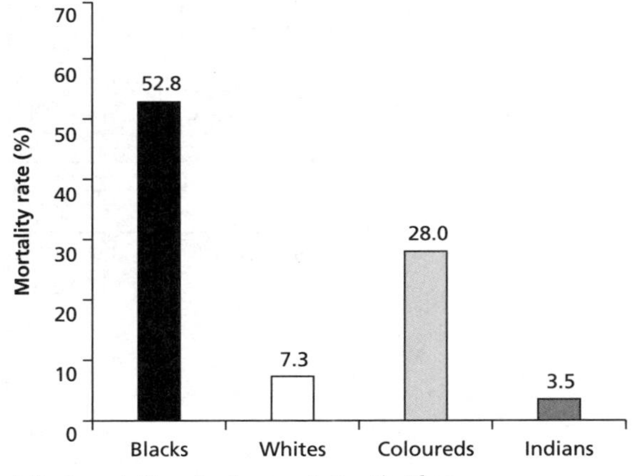

Infant mortality rates by race in South Africa

CHILD MORTALITY AND INFANT MORTALITY

While the CMR shows small fluctuations over time, the IMR can show greater fluctuations and is one of the most sensitive indicators of the level of development. This is due to the following:

- High IMRs are found only in the poorest countries.
- The causes of infant deaths are often preventable.
- IMRs are low where there is safe water supply and adequate sanitation, housing, healthcare and nutrition.
- The CMR is declining. It dropped by about a quarter between 1990 and 2006. In Latin America, central Europe, the former Soviet Union and east Asia, it fell by about a half. Progress in sub-Saharan Africa has been slower.

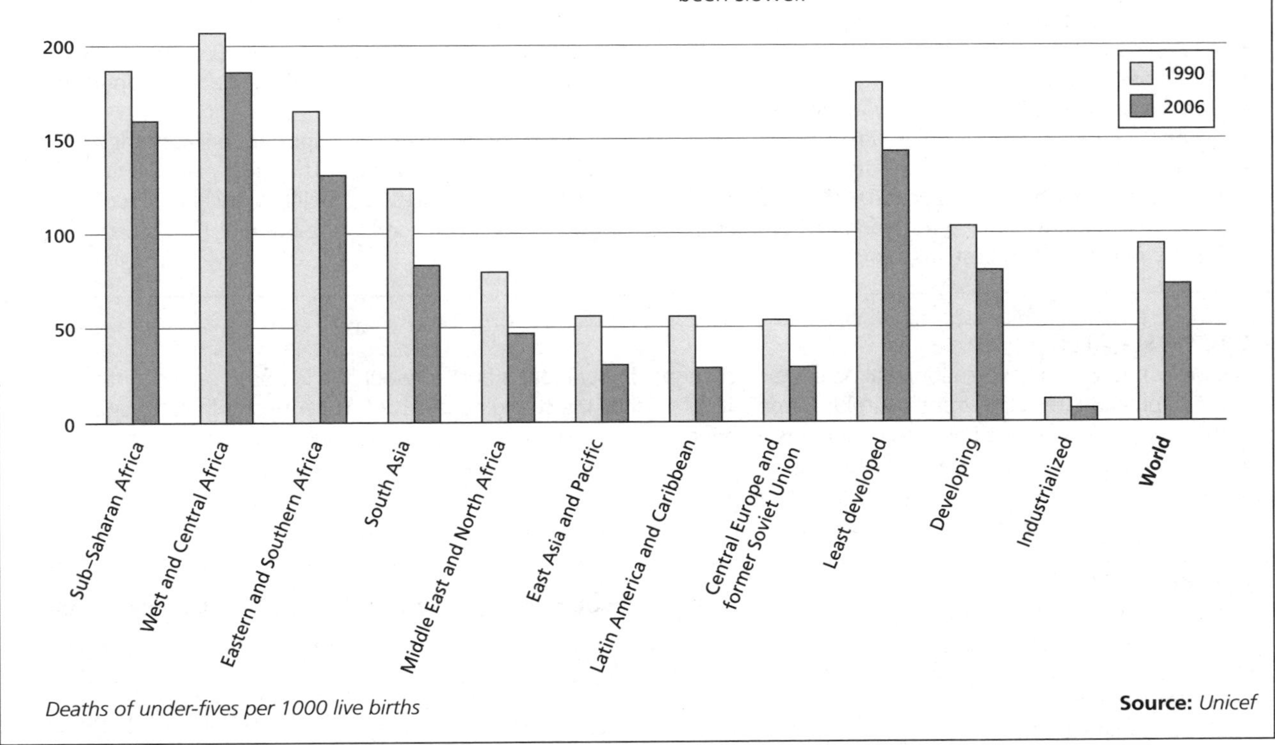

Deaths of under-fives per 1000 live births

Source: *Unicef*

Population pyramids (1)

Population structure or composition refers to any *measurable* characteristic of the population. This includes age, sex, ethnicity, language, religion and occupation.

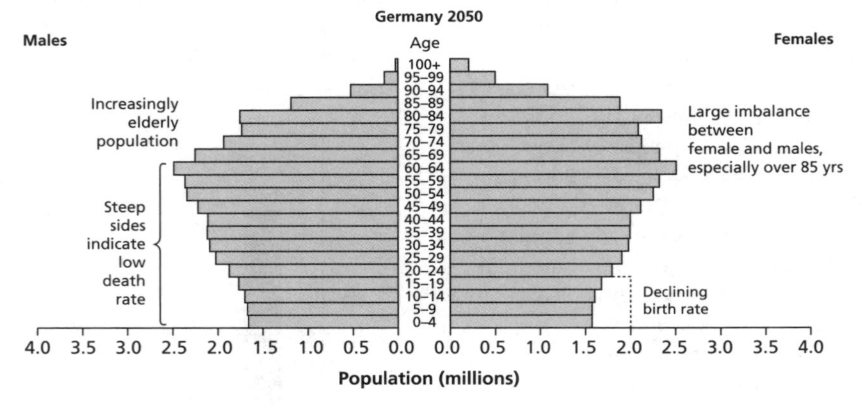

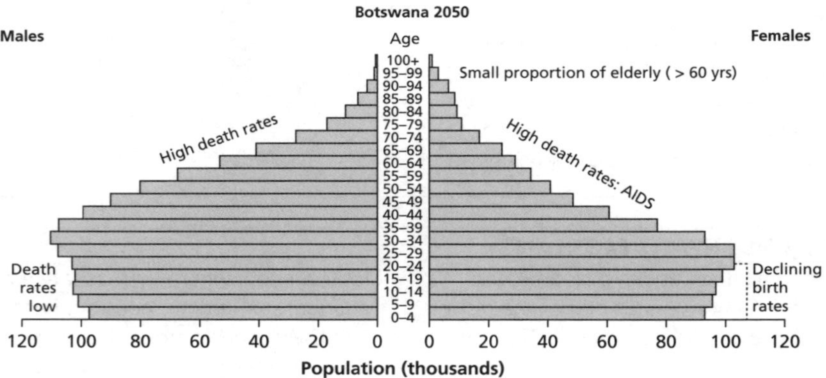

Source: *US Census Bureau*

Population pyramids tell us a great deal of information about the age and sex structure of a population:

- A wide base suggests a high birth rate.
- A narrowing base suggests a falling birth rate.
- Straight or near-vertical sides show a low death rate.
- A concave slope suggests a high death rate.
- Bulges in the slope indicate high rates of in-migration. (For instance, excess males aged 20–35 years will be economic migrants looking for work; excess elderly, usually female, will indicate retirement resorts.)
- Deficits in the slope show out-migration or age-specific or sex-specific deaths (epidemics, war).

Population pyramids can also be used to show the racial composition of a population or the employed population group.

Population pyramids are important because they tell us about population growth. They help planners to find out how many services and facilities, such as schools and hospitals, will be needed in the future.

GROWTH RATES

The growth rate is the average annual percentage change in the population, resulting from a surplus (or deficit) of births over deaths and the balance of migrants entering and leaving a country. The rate may be positive or negative. The growth rate is a factor in determining how great a burden would be imposed on a country by the changing needs of its people for infrastructure (e.g. schools, hospitals, housing, roads), resources (e.g. food, water, electricity) and jobs.

DOUBLING TIMES

The doubling time refers to the length of time it takes for a population to double in size, assuming its natural growth rate remains constant. Approximate values for it can be calculated using the formula:

$$\text{Doubling time (years)} = \frac{70}{\text{growth rate in percentage}}$$

Country	Growth rate (%)	Doubling time
Denmark	0.1	700 years
Brazil	0.9	78 years
Indonesia	1.6	44 years
Uganda	3.0	23 years

Doubling times for selected countries

Population pyramids (2)

POPULATION MOMENTUM

Population momentum is the tendency for a population to grow despite a fall in the birth rate or fertility levels. It occurs because of a relatively high concentration of people in the pre-childbearing and childbearing years. As these young people grow older and move through reproductive ages, the more the number of births will exceed the number of deaths in the older populations, and so the population will continue to grow.

Population projections are predictions about future population based on trends in fertility, mortality and migration.

THREE POPULATION PYRAMIDS

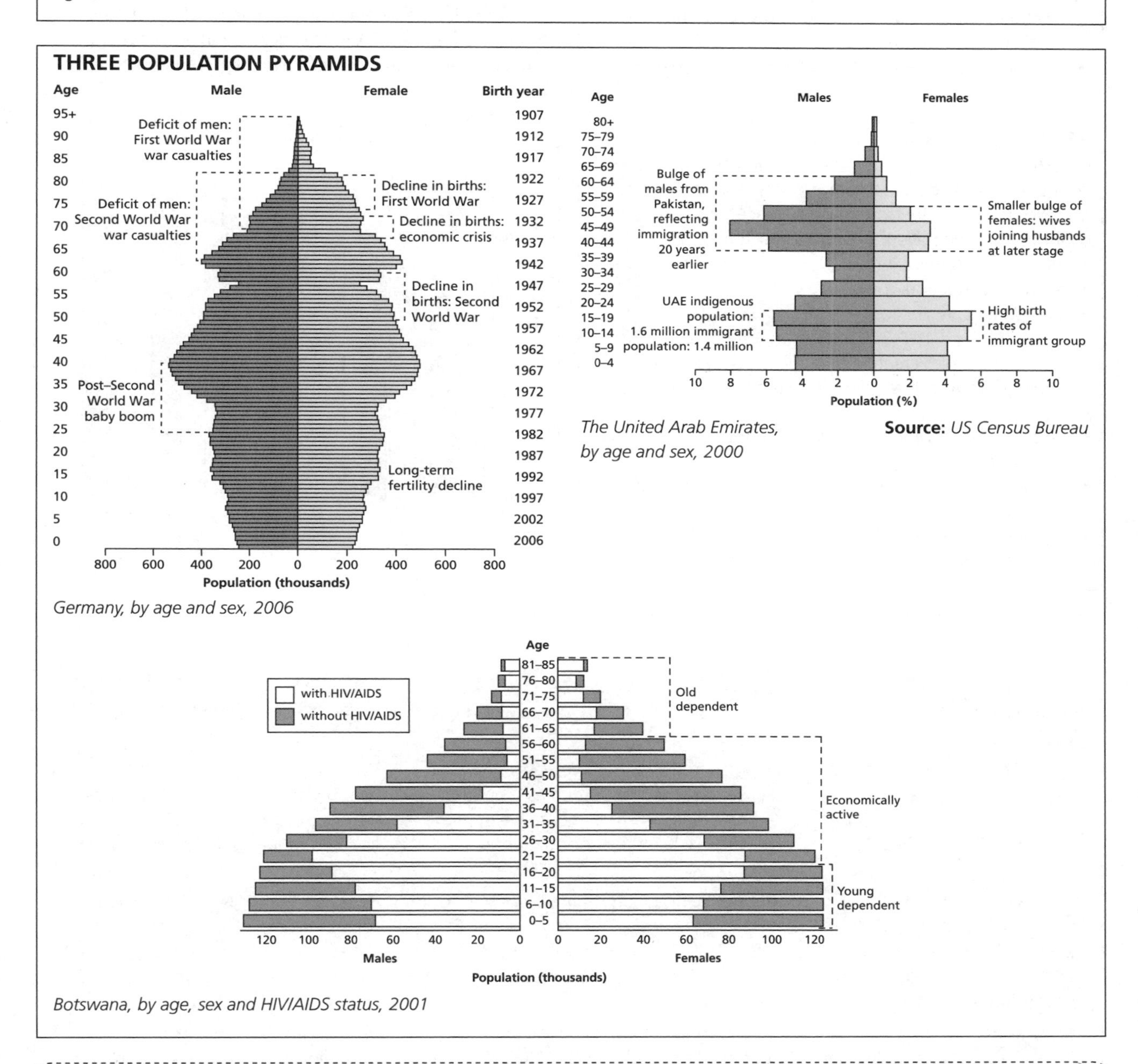

Germany, by age and sex, 2006

The United Arab Emirates, by age and sex, 2000

Source: *US Census Bureau*

Botswana, by age, sex and HIV/AIDS status, 2001

EXTENSION

Visit the US Census Bureau International database at **www.census.gov/ipc/www/idb/pyramids.html** and submit a query for the population pyramids of countries that you are interested in. Try to annotate the pyramids to describe and suggest reasons for the changes in the population structure.

Gender and change

GENDER AND POPULATION GROWTH

In many countries high rates of population growth are associated with a low status of women in society. Some of the reasons for this are listed on page 18.

The UN Decade for Women, from 1975 to 1985, recommended three important points for action:

- There should be legal equality for women.
- Further development needs to improve on the substandard role that women play.
- Women should receive an equal share of power.

GENDER AND SOCIAL ROLE

In 1970 Esther Boserup identified women as having been left behind in the development process. The social roles that women play vary from place to place, but in most countries women have three important functions:

- biological reproduction
- social reproduction
- economic reproduction.

These three roles create a great deal of physical and psychological stress. It is believed that in sub-Saharan Africa:

- up to one-third of women are pregnant or breastfeeding at any one time
- women comprise over half the workforce, sometimes over 70%
- women grow over 80% of the food eaten and contribute half of the region's cash crops.

WOMEN AND DEVELOPMENT

A number of approaches to the study of women and development have emphasized welfare, equality, anti-poverty, efficiency and empowerment. Strategic or political change is needed to attain equality and empowerment. In many countries this is highly unlikely.

Progress for sexual equality has been painfully slow. For example, the illiteracy rate is much higher for girls than for boys, over 70% of African countries have no female cabinet minister and, generally, women are becoming poorer.

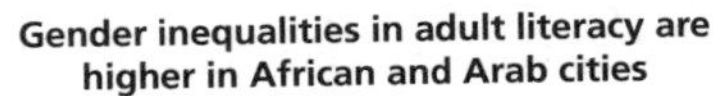

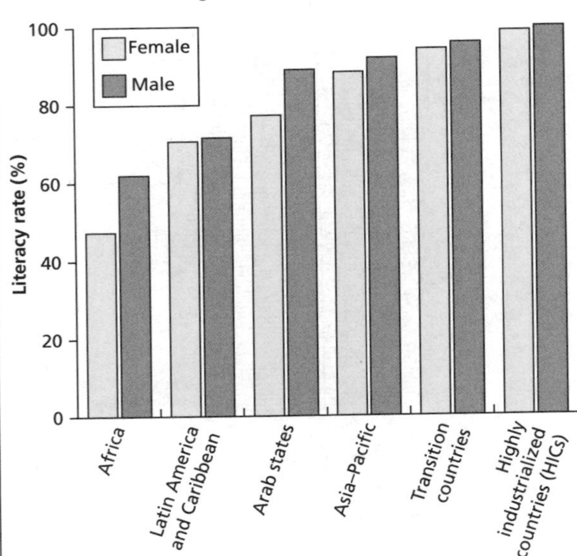

Inequalities in education

Issues	Welfare	Equality	Anti-poverty	Efficiency	Empowerment
Period most popular	1950–70	1975–85	1970s onward	post-1980s	1975 onward
Purpose	Women are given the resources to become better mothers	Women are seen as active participants in the development process	Women's poverty is seen as a problem of underdevelopment not of subordination	Women's economic participation is linked to equality	Women's subordination is seen as part of colonial oppression
Needs of women met and roles recognized	Food aid, malnutrition addressed and family planning	Reducing inequality with men by allowing political and economic autonomy	Allows women to earn an income in small-scale income-generating projects	Relies on the three roles of women to replace declining social services	Bottom-up role is recognized as women are empowered
Comment	Women are seen in a traditional reproductive role	Criticized as western feminism	Popular with small-scale NGOs	Women seen as potential workforce	Largely unsupported at present

Different policy approaches to women in Africa

THE REASONS FOR SLOW PROGRESS

- Conditions are deteriorating in a large part of Africa. As a result of structural adjustment programmes (SAPs) countries spend less money on health and social welfare – cuts that are disproportionately borne by women.
- There is a lack of commitment to women by many countries and by donors.
- Women have to work as well as be the head of the household, but they have little legal status.

Gender inequalities

THE GOAL OF GENDER EQUALITY

Gender equality has gained wide acceptance as an important goal for many countries around the world. Participants at the 1994 International Conference on Population and Development in Cairo agreed on the principle "that advancing gender equality and equity and the empowerment of women, and the elimination of all kinds of violence against women, and ensuring women's ability to control their own fertility, are cornerstones of population and development-related programmes."

When women have frequent and numerous births, their life choices are often restricted. When women have fewer children, they face fewer years of childcare and they are freer to work.

LIFE EXPECTANCY AT BIRTH

One area where the statistics for women are better than for men is life expectancy.

	Total	Males	Females
World	68	66	70
MEDCs	77	73	80
LEDCs	64	62	65
Africa	53	52	54
North America	78	75	81
Latin America and the Caribbean	73	70	76
Asia	68	67	70
Europe	75	71	79
Oceania	75	73	78

Global life expectancy (years), 2008

WOMEN AND UNIONS

Working women are increasingly becoming unionized. In India, for example, SEWA (Self-Employed Women's Association) operates as a trade union and as an economic empowerment group. Labour unions have historically been a male preserve, but women are now making up an increasing share of membership. The involvement of women in paid employment has also led to the politicization of women and gender issues.

EXTENSION

Visit

www.prb.org/pdf07/07WPDS_Eng.pdf for the 2007 World Population Data Sheet and see how life expectancy varies for the countries of your choice.

www.prb.org/pdf07/62.3Highlights.pdf for the World Population Highlights from the 2007 World Population Data Sheet.

WOMEN'S WORK

In order to remain competitive in the global marketplace, businesses in many countries have capitalized on women as a source of labour willing to work in poor conditions for low wages. More women join the workforce in unskilled, labour-intensive and poorly paid jobs. This situation is made worse by the burden of household, childcare and domestic responsibilities. In eastern Europe the status of working women worsened drastically with the economic transition.

Some 25% of the world's households have women as their heads; in urban areas, especially in Latin America and Africa, the numbers sometimes exceed 50%. Households with a woman head typically represent a high proportion of those in informal settlements worldwide and they are among the poorest.

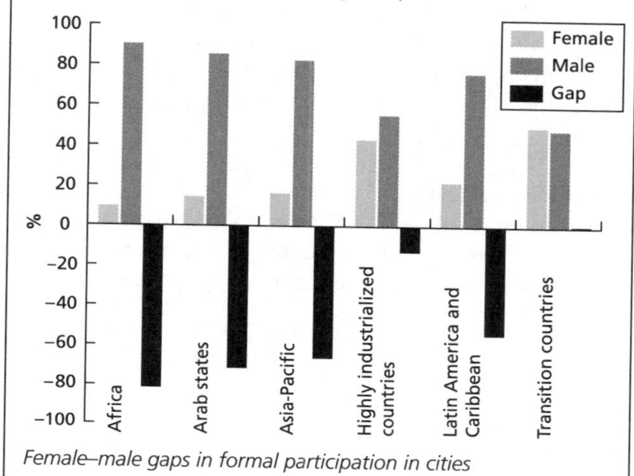

Female–male gaps in formal participation in cities

FEMINIZATION OF MIGRATION

Women account for almost half of immigrants around the world. Women now are increasingly likely to move for economic opportunity, rather than to join husbands or other family members as they did in the past.

Sending countries also differ in the percentage of women and men who emigrate, in part because of differential demand for labour in destination countries. For example, 70% of all Filipino labour migrants are women.

TENURE

Tenure is defined as the way in which the rights, restrictions and responsibilities that people have with respect to land (and property) are held. Comparatively few African countries have legislation in place to assure women's access to land and property. Those that do include Burkina Faso, Malawi, Mozambique, Niger, Rwanda, South Africa, Tanzania, Uganda and Zimbabwe.

Formal law, traditional legal systems and societal norms, including customary and religious laws, often deny women the right to acquire and inherit property, particularly in countries where shariah law applies.

Responses to high and low fertility

POLITICAL FACTORS AND FAMILY PLANNING

Most governments in LEDCs have introduced programmes aimed at reducing birth rates. Their effectiveness is dependent on:

- focusing on family planning in general and not just on birth control
- investing sufficient finance in the schemes
- working in consultation with the local population.

Where birth controls have been imposed by government, they are less successful (except in the case of China).

In MEDCs, financial and social support for children is often available to encourage a pro-natalist approach. However, in countries where there are fears of negative population growth (as in Singapore), more active and direct measures are taken by governments to increase birth rates.

DEPENDENCY RATIOS

The **dependency ratio** measures the working population and the dependent population. It is worked out by a formula:

$$\frac{\text{Population aged} <15 + \text{population aged} >60 \text{ (the dependents)}}{\text{Population aged 16–59 (the economically active)}}$$

It is very crude. For example, many people stay on at school after the age of 15 and many people work after the age of 60. But it is a useful measure to compare countries or to track changes over time.

- In the developed world there is a high proportion of elderly.
- In the developing world there is a high proportion of youth.

AGEING RATIOS

The future trends of fertility, mortality and migration shaping the pattern of population ageing in Europe are uncertain within certain ranges. Methods of statistical probability have been developed to describe these uncertainty ranges in a quantitative way.

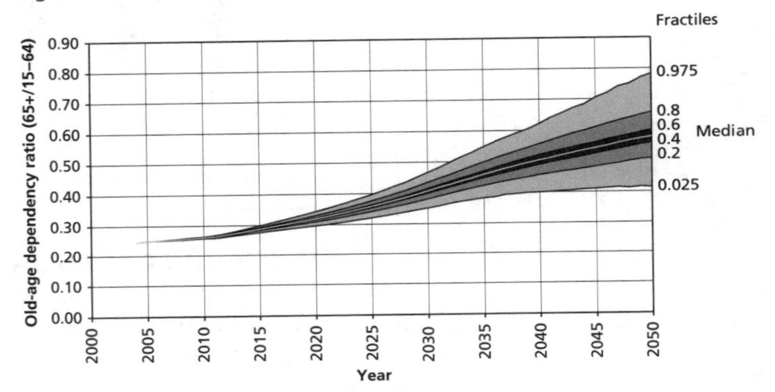

Old-age dependency ratio, EU

The graph shows the future trend in the old-age dependency ratio for all 27 EU member countries. Currently there are four people in the age group 15–64 (considered as the potential working age) for each person aged 65 or older. There is an 80% chance that the ratio will more than double by 2050, which means there will be fewer than two people of working age per person above age 65. At the high end there is about a 20% chance that there will be three people of working age for any two persons above age 65. Since not everyone between ages 15 and 64 will be working – due to education, unemployment, early retirement or other reasons – the actual ratio of contributors to beneficiaries of the pension system may be even less favourable.

There is significantly more demographic uncertainty as to the future trend in the proportion of the population above age 80. Only 4% of the population are currently of this advanced age. Over the next 20 years this proportion might well increase to about 6–7%, but then the increase accelerates due to the strong baby boom cohorts gradually entering this age group. By 2050, estimates range from a low of 7% to a high of 20% of the population above the age of 80.

EXTENSION

Visit
www.oeaw.ac.at/vid/download/edrp_1_06.pdf for the *European Demographic Research Report* 2006 No. 1.

Impacts of youthful and ageing populations

YOUTHFUL POPULATION STRUCTURE

In many LEDCs rapid youthful age structures are creating demand for many services and facilities. Much depends on whether a country has the resources to deal with this demand. In LEDCs this is often not the case, and there are at present problems in the provision of schooling, healthcare for children, childcare facilities, as well as leisure and recreational facilities. Although a large youthful population means there will be a large labour force in the near future, it also means that jobs will have to be created, or else unemployment will be high.

Even within the same country, there are variations in the age structure of a population. In Korea, for example, the migration of young workers to large, rapidly growing cities is altering the age structure of the cities. In contrast, in many of the smaller towns and villages there is a large elderly population.

ADVANTAGES AND DISADVANTAGES OF A YOUTHFUL POPULATION

Potential advantages	Potential disadvantages
Large potential workforce	Cost of supporting schools and clinics
Lower medical costs	Need to provide sufficient food, housing and water to a growing population, e.g. Rocinca, Rio de Janeiro
Attractive to new investment	High rates of unemployment
Source of new innovation and ideas	Large numbers living in poor quality housing, e.g. in shanty towns
Large potential market for selected goods	High rates of population growth
Development of services such as schools, crèches	High crime rates

AGEING POPULATION IN JAPAN

The number of elderly who are living alone increased from 0.8 million in 1975 to over 2.5 million in 2000.

By 2020 over 25% of the Japanese population will be over the age of 65. At present it is 15% of the population. There are a number of problems, including:

- inadequate nursing facilities
- depletion of the labour force
- deterioration of the economy
- migration of Japanese industry overseas
- cost of funding pensions and healthcare.

ADVANTAGES OF AN AGEING POPULATION

There are certain advantages of an ageing population.

- The elderly may have skills (including social skills) and training, and are sometimes preferred over younger workers.
- The elderly may look after their grandchildren and therefore allow both parents to work, for example in Japan and South Africa.
- In many MEDCs the elderly are viewed as an important market – the "grey economy". Many firms, ranging from holiday companies to healthcare providers, have developed to target this market.

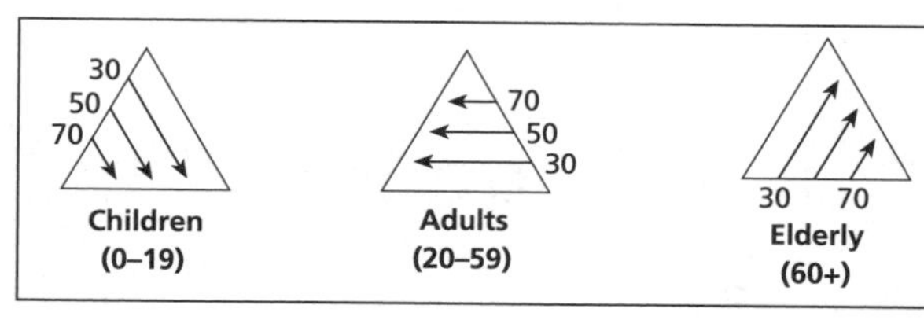

LEDCs	Less economically developed countries	UK	United Kingdom
MEDCs	More economically developed countries	Fr	France
		Sw	Sweden
		Jp	Japan
		Bo	Bolivia

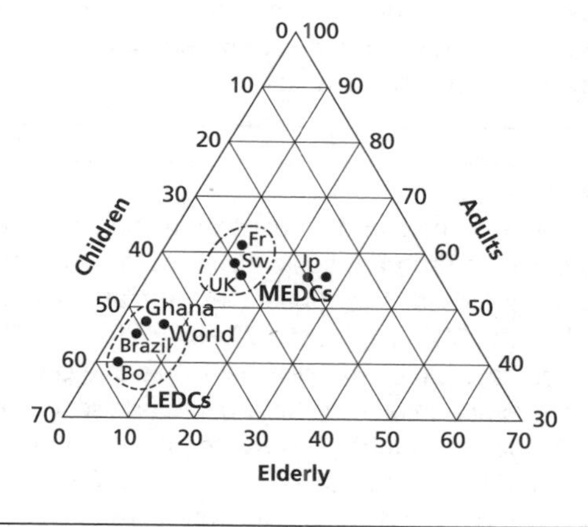

Age structures for selected countries

Managing population change

There are a number of ways in which governments attempt to control population numbers. There are contrasting strategies depending on whether the country wishes to increase its population size (**pro-natalist**) or whether it wants to limit it (**anti-natalist**).

FAMILY PLANNING IN DEVELOPING COUNTRIES

Family planning refers to attempts to limit family size. Family planning methods include contraceptives such as the pill and condoms, as well as drastic methods such as forced sterilization, abortion and infanticide.

POPULATION GROWTH AND THE STATUS OF WOMEN

High rates of population growth are often associated with a low status of women. Reasons for this include the following:

- A wife continues to bear children to prove her fertility, and to prevent the husband from marrying another wife.
- Wives in polygamous families compete with each other to produce the most children.
- Children provide labour for fetching firewood and water and for digging holes in the fields.
- Children are an investment as they provide old-age security for their parents.
- In large families there are likely to be not only rogues and robbers but also professionals such as doctors, lawyers, engineers, etc.
- Women have no say in determining the size of the family.

SINGAPORE

Between the 1960s and 1970s the government of Singapore pursued an anti-natalist policy. However, as the economy prospered and the population growth rate fell, it adopted a pro-natalist policy.

Despite incentives such as the "love cruises" arranged to help couples meet, the Singapore government found it difficult to raise population growth. Although the number of marriages increased, the birth rate did not rise.

The government has now realized that by increasing the status of women, and having more working women, women themselves do not want to have as many children as previous generations. They prefer to enjoy for themselves the fruits of their newly earned occupational status and material possessions.

CHINA

China operates the world's most severe and controversial family planning programme. In 1979 the "one-child" policy was imposed. The impact was dramatic. The birth rate fell from 33‰ in 1970 to 17‰ in 1979.

In urban areas most families have only one child, and the growing middle classes no longer discriminate so much against daughters. However, the countryside remains traditionally focused on male heirs. But the policy is being relaxed. In most provincial rural areas, couples can have two children without penalties.

The one-child policy is not an all-encompassing rule. It has always been restricted to ethnic Han Chinese living in urban areas; citizens living in rural areas and minorities living anywhere in China are not subject to the law. A special provision allows millions of couples to have two children legally: if a couple is composed of two people without siblings, then they may have two children of their own. Notwithstanding the above, the rule has been estimated to have reduced population growth in a country of 1.3 billion by as much as 300 million people over its first 20 years.

The policy has caused a disdain for female infants; abortion, neglect, abandonment and even infanticide have been known to occur to female infants. The result of such draconian family planning has resulted in the disparate ratio of 114 males for every 100 females in the 0–4 years age group. Selective abortion is a major cause, but many baby girls are probably not registered.

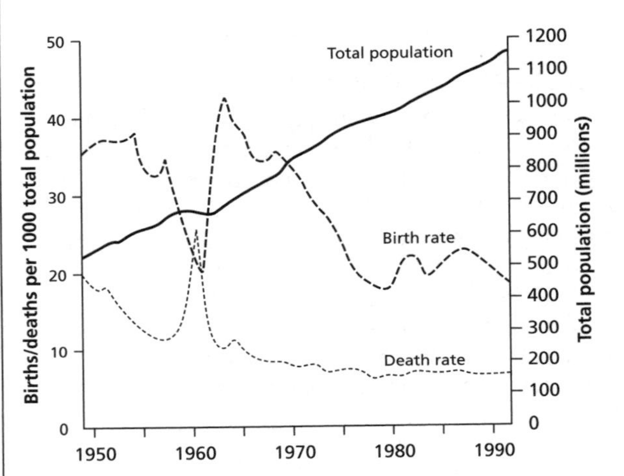

Changes in China's population, birth and death rates, 1949–1990s

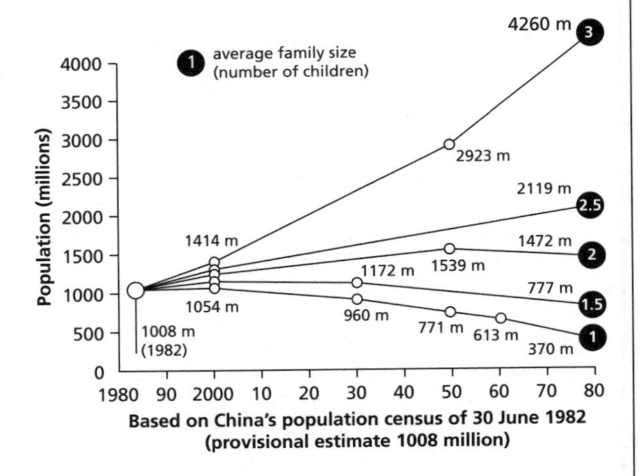

Five possible options for China's future population

Migration

TYPES OF MOVEMENT

Migration is the movement of people, involving a permanent (more than one year) change of residence. It can be internal or external (international), and voluntary or forced. It does not include temporary circulations such as commuting or transmigration.

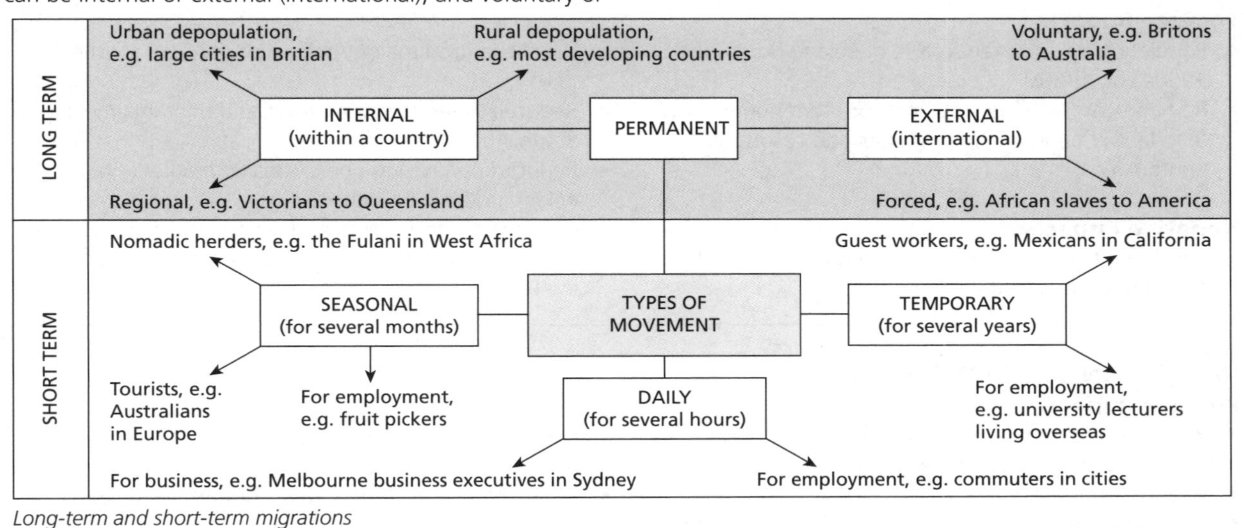

Long-term and short-term migrations

PATTERNS OF MIGRATION ACCORDING TO RAVENSTEIN

Findings	Explanation
Most migrants proceed over a short distance	Due to limited technology and transport, and poor communications, people know more about local opportunities
Migration occurs in a series of steps or stages	Typically from rural to small town, to large town to city, i.e. once in an urban area, people become "locked in" to the urban hierarchy
As well as movement to large cities, there is movement away from them (dispersal)	The rich move away from the urban areas and commute from nearby villages and small towns
Urban dwellers migrate less than rural dwellers	There are fewer opportunities in rural areas
Women are more migratory than men over short distances	Especially for marriage and in societies where the status of women is low
Migration increases with advances in technology	E.g. transport, communications and the spread of information

MIGRATION ACCORDING TO LEE (1966)

Lee described migration in terms of **push** and **pull** factors.

- Push factors are the negative features that cause a person to move away from a place (e.g. unemployment, low wages, natural hazards).
- Pull factors are the attractions (whether real or imagined) that exist at another place (e.g. better wages, more jobs, good schools).

The term "perceived" means what the migrant imagines exists, rather than what actually exists. The perceived and the real may be quite close, or they can be very different.

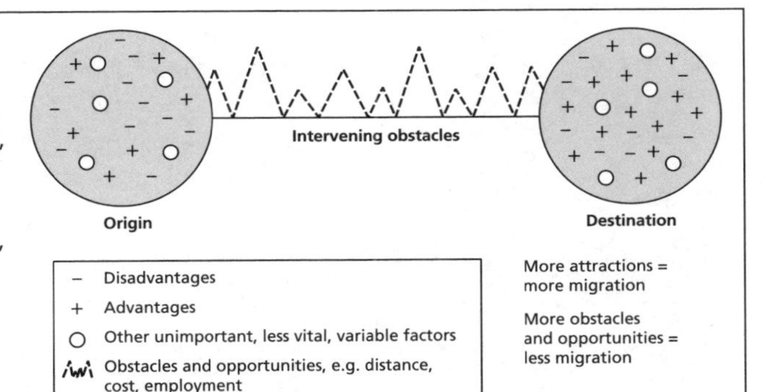

LIMITATIONS OF MODELS

All of these models are simplifications, and they contain hidden assumptions. These assumptions may be very unrealistic. For example:

- Are all people free to migrate?
- Do all people have the skills, education and qualifications that allow them to move?
- Are there barriers to migration – such as race, class, income, language, gender?
- Is distance a barrier to migration?

Impacts of international migration (1)

International migrations can have a range of positive and negative impacts on both the source area and the destination.

IMPACTS ON SOURCE AREA

Positive impacts

- Population pressure reduced (e.g. Ireland during the 1950s and 1960s)
- Remittances (see pages 28 and 165) sent home (e.g. labour migrants from Malawi and Lesotho in South Africa)

Negative impacts

- Removal of younger, more educated people (e.g. Indian software experts to the USA)
- Decline in local market/pulling power (e.g. southern Italy)
- Reduced workforce (e.g. Swaziland migrants moving to South Africa)
- Reduced purchasing power/smaller market (e.g. rural Ireland in the 1950s and 1960s)
- Closure of local services such as schools, hospitals (e.g. following the expulsion of Ugandan Asians to the UK in 1973)

IMPACTS ON DESTINATION

Positive impacts

- Population growth (e.g. Turks to West Germany in the 1970s and Portuguese to Switzerland)
- Larger workforce (e.g. the USA)
- Increased demand for housing (e.g. Silicon Valley in California)
- Increased demand for services (e.g. the M4 corridor in the UK)
- New industry and investment attracted to the area
- New skilled, young workforce (e.g. Italians in Bedford in the 1950s)
- Multicultural enrichment (e.g. Toronto, Canada)

Negative impacts

- Racism and segregation (e.g. Los Angeles, USA)
- Cultural disharmony (e.g. Bradford and Oldham, UK)
- Overcrowding and ghettoization (e.g. blacks in New York
- Spread of diseases (e.g. flu to Amazonian tribes or those of Easter Island)

BENEFITS AND COSTS

	Benefits		Costs	
	Individual	**For the country**	**Individual**	**For the country**
Emigrant countries	Increased earning and employment opportunities	Increased human capital with return migrants*	Transport costs	Loss of social investment in education
	Training (human capital)*	Foreign exchange for investment via migrant remittances	Adjustment costs abroad	Loss of "cream" of domestic labour force
	Exposure to new culture, etc.*	Increased output per head due to flow of unemployed and underemployed labour	Separation from relatives and friends	Social tensions due to raised expectations of return migrants*
		Reduced pressure on public capital stock		Remittances generate inflation by easing pressure on financing public sector deficits*
Immigrant countries	Cultural exposure, etc.(*)	Permits growth with lower inflation	Greater labour market competition in certain sectors	Dependence on foreign labour in particular occupations*
		Increased labour force mobility and lower unit labour costs		Increased demands on the public capital stock
		Rise in output per head for indigenous workers		Social tension with concentration of migrants in urban areas*

Source: *The Economist,* 15 November 1988

** indicates uncertain effects*

Impacts of international migration (2)

MIGRANT WORKERS

Migrant workers are those who migrate to find work. Such a movement can be:

- permanent or temporary
- long- or short-distance
- internal or across an international boundary.

Migrant labour has been vital for economic development in many countries, and it remains important today for many MEDCs, such as the USA, Australia and the UK.

FREEDOM OF MOVEMENT

Unlike other forms of migration, the main motive for migrant labourers is the search for better working conditions. As such, workers move freely or voluntarily to other countries. In fact, some countries openly advertise for migrant workers.

There are many well-established patterns of migrant labour, such as the migration of the Irish to mainland Britain. In this case, much of the migration was permanent and many of the descendants of the original migrants are fully integrated into British society. Within the European Union there is freedom of movement, so in theory nationals of any EU country can migrate to another.

SPATIAL AND TEMPORAL VARIATIONS

Migrant labour is important for capitalist development. As development is uneven spatially and temporally, labour must be mobile in order to match demand with supply.

Migrant labour has been very important in western Europe and the USA. Britain has relied on Ireland, eastern Europe, the New Commonwealth and Pakistan for its migrant labour, while Germany has depended to a large extent on Greece and Turkey. The USA, as well as using cheap labour from Mexico, has relied heavily on sources of labour from the Caribbean. Increasingly, skilled ICT labour from Asian countries, notably India, has been fuelling growth in the computer industry in California's Silicon Valley. A report in 2002 suggested that without labour migration the US economy would be in a far less healthy position.

TRENDS

The main trends with migrant labour are:

- the globalization of migrant labour
- the acceleration of migration
- the differentiation of migration into different types
- the feminization of migration (e.g. the migration of nurses from the Philippines to the UK).

ADVANTAGES AND DISADVANTAGES OF MIGRANT LABOUR

One of the main attractions of migrant labour, for the receiving country, is that it does not bear the costs of the social reproduction of labour. The costs of childrearing, education, housing and healthcare for the labourers are borne by another country. However, when the labourers are of age, their labour is used by the receiving country, thus depriving the source country of trained workers.

Source country		Destination	
Economic costs	**Economic benefits**	**Economic costs**	**Economic benefits**
Loss of young labour	Reduced un-/ underemployment	Costs of educating children	Undesirable posts often filled
Loss of skilled labour slows development	Returning migrants bring back new skills	Displacement of local labour	Skills gained at little cost (e.g. doctors to the USA)
Out-migration leads to a vicious circle of decline	Money sent home (remittances)	Money sent to the country of origin; pension outflow	Some retirement costs transferred to source country
Loss of skilled labour deters investment	Less pressure on resources such as land	Increased pressure on resources	Dependence on guest workers
Social costs	**Social benefits**	**Social costs**	**Social benefits**
Creates a culture of out-migration	Lower birth rates and reduced population pressure	Racism, discrimination and conflict	Creation of multicultural societies
Females left as head of household, mother and main provider	Remittances may improve welfare and education	Male-dominated states (e.g. oil-rich economies)	Cultural awareness and acceptance
Unbalanced population pyramid	Retiring population may build new homes	Loss of cultural identity, especially among second generation	Providers of local services
Returning on retirement places a burden on services	Some returnees may develop new activities such as recreation, leisure and tourism	Creation of ghettos and ghettoized schools	Growth of ethnic retailing and restaurants

Economic and social costs and benefits

Measurement of regional and global disparities (1)

THE HUMAN DEVELOPMENT INDEX (HDI)

Since 1990 the United Nations (UN) has urged the use of the HDI as a measure of development. It is a more reliable and comprehensive measure of human development and well-being than GNI per head. (GNI, gross national income, was previously known as GNP, gross national product. The two are virtually the same.) The HDI includes three basic components of human development:

- longevity (life expectancy)
- knowledge (adult literacy and average number of years' schooling)
- standard of living (purchasing power adjusted to local cost of living).

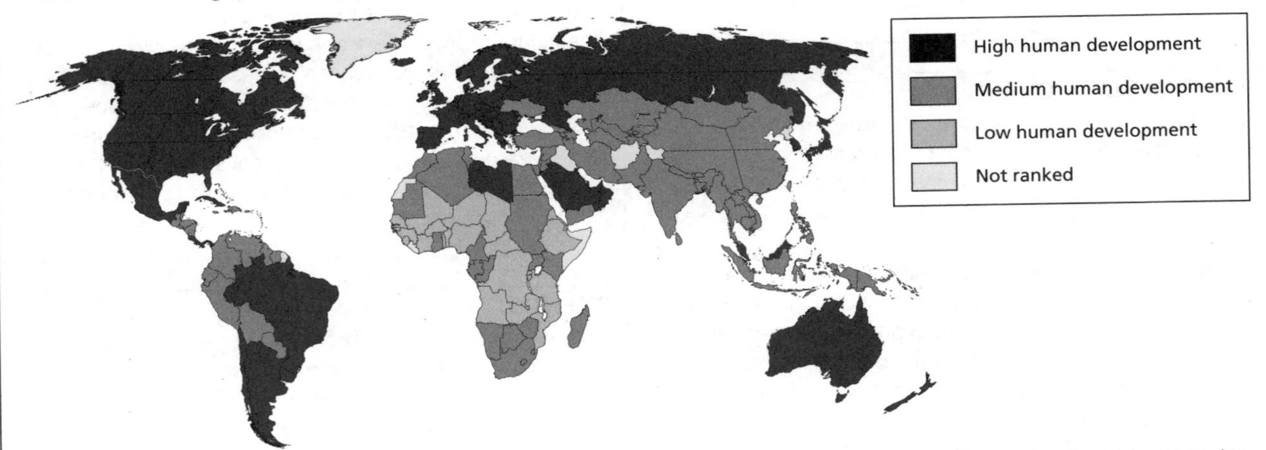

Global HDIs

Note: *HDI rankings for the 2007/2008 report are based on 2005 data.*

The UN 2007 table of HDIs shows Iceland at the top, closely followed by Norway and Australia. At the other end, Sierra Leone, Burkina Faso and Guinea–Bissau had the lowest HDI scores.

National averages can conceal a great deal of information. HDIs can be created to show regional and ethnic variations as shown below.

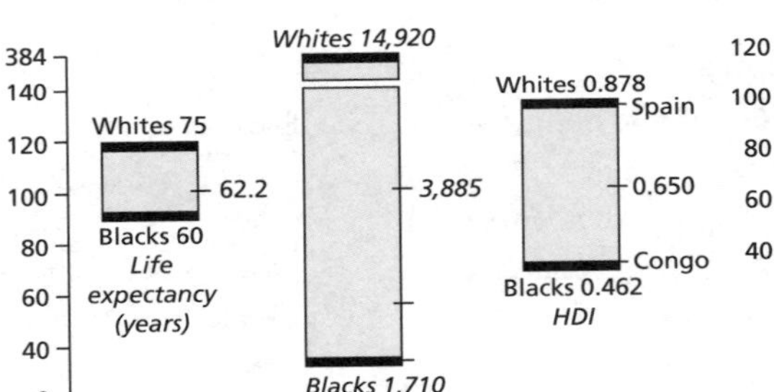

Some regional and ethnic disparities

THE INFANT MORTALITY RATE (IMR)

Another widely used indicator of development is the IMR. This refers to the number of children that die before their first birthday. It is expressed per thousand live births. It is widely used as an indicator of development for a number of reasons:

- High IMRs are found in the poorest LEDCs.
- The causes of death are often preventable.
- Where water supply, sanitation, housing, healthcare and nutrition are adequate, IMRs are low.

EXTENSION

Visit
http://hdr.undp.org/external/flash/hdi_map/
and see how the human development index has changed over time. The same link will provide you with data for each country's HDI. The data are organized into high HDI, medium HDI and low HDI.

Measurement of regional and global disparities (2)

GLOBAL INEQUALITIES

The gap between rich and poor people in the world has been increasing for the last two centuries. In 1820, the difference between the richest and the poorest country was about 3:1. By 1913 this had risen to 11:1, while by 1950 it had broadened to 35:1. In 1999 the wealthiest country was about 95 times richer than the poorest country. Indeed, Britain's income in 1820 was four times greater than that of Sierra Leone in 1999! Nevertheless, many poor countries have improved their GNI in recent decades.

- The assets of the world's three richest people are more than the combined GNI of all poor countries.
- The assets of the world's 200 richest people are more than the combined incomes of 41% of the world's people. By making an annual contribution of just 1% of their wealth, those 200 people could provide access to primary education for every child in the world.

1960	30:1
1970	32:1
1980	45:1
1989	59:1
1991	61:1

Ratio of income of richest 20% of the population to the poorest 20% of the population

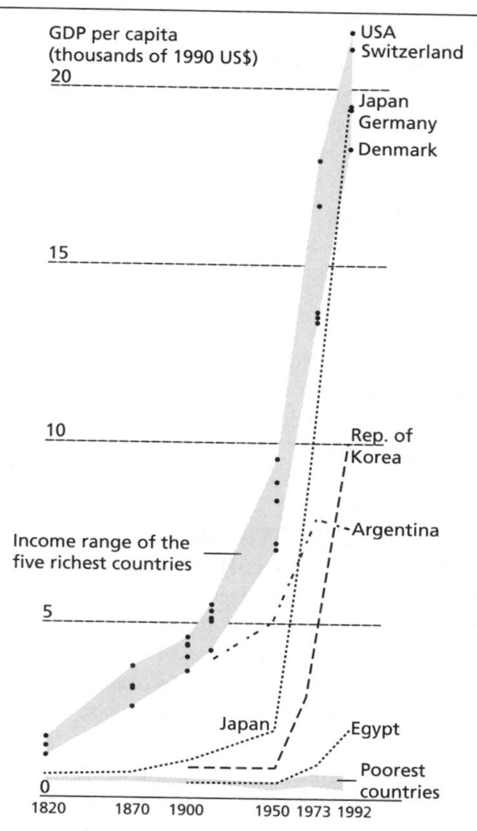

The development gap

The world's richest and poorest countries, 1820–2007 (GDP per capita, US$)

Richest					
1820		**1900**		**2007**	
UK	1,756	UK	4,593	Luxembourg	80,800
Netherlands	1,561	New Zealand	4,320	Qatar	75,900
Australia	1,528	Australia	4,299	Bermuda	69,900
Austria	1,295	USA	4,096	Jersey	57,000
Belgium	1,291	Belgium	3,652	Norway	55,600
Poorest					
1820		**1900**		**2007**	
Indonesia	614	Burma (Myanmar)	647	Somalia	600
India	531	India	625	Guinea–Bissau	600
Bangladesh	531	Bangladesh	581	Liberia	500
Pakistan	531	Egypt	509	Zimbabwe	500
China	523	Ghana	462	Congo	300

Source: *Updated from* **www.cia.gov/library/publications/the-world-factbook/rankorder/2004rank.html**

EXTENSION

Excellent up-to-date data are available from the *Human Development Report*: **http://hdr.undp.org/en/statistics/**

EXTENSION

https://www.cia.gov/library/publications/the-world-factbook/docs/rankorderguide.html is a list of IMRs ranked from highest to lowest. How does this compare with the list of countries arranged by HDI?

Origin of disparities

INEQUALITIES IN DEVELOPMENT

Though some parts of the world have experienced unprecedented growth and improvement in living standards in recent years, poverty remains entrenched and much of the world is trapped in an inequality predicament. Despite considerable economic growth in many regions, the world is more unequal than it was 10 years ago.

Within the group of countries that are commonly thought of as poor, there is considerable variation. Some countries are relatively well-off. NICs such as South Korea and Taiwan have quite high levels of GNI per capita. The development of the original Asian "tigers" is the result of a combination of state-led industrialization, spontaneous industrialization and industrialization led by transnational corporations (TNCs).

LAND OWNERSHIP (TENURE)

The case of black agriculture in South Africa

The decline of black subsistence agriculture has traditionally been put down to the shortage of land relative to the growing population and the increasing poverty of that population. Shortage of land led to overcrowding, overgrazing, use of poor land, soil erosion, denudation and, ultimately, declining yields.

In the period before the black population was forced into reserves and, later, homelands, tribal groups were not confined to small areas. The loss of their traditional lands led to the decline of the black rural economy. Increased poverty prevented black farmers from affording the inputs necessary to improve yields. As the reserves were unable to feed the needs of the black population, many black people resorted to the only thing possible – they became migrant labourers and entered the cash economy. Thus, migrant labour was a result and a cause of low productivity in black agriculture. The failure of many migrants to send much of their wages back to homeland areas further weakened the agricultural base.

EMPLOYMENT

The UN *Report on the World Social Situation 2005: The Inequality Predicament* focuses on the gulf between the formal and informal economies, the widening gap between skilled and unskilled workers, and the growing disparities in health, education and opportunities for social, economic and political participation.

The report notes that a focus on growth and income generation neither sufficiently captures nor addresses the intergenerational transmission of poverty; it can lead to the accumulation of wealth by a few and deepen the poverty of many.

The report further notes the following:

- Inequalities between and within countries have accompanied globalization. These inequalities have had negative consequences in many areas, including employment, job security and wages.
- Unemployment remains high in many contexts and youth unemployment rates are particularly high. Youths are two to three times more likely than adults to be unemployed and currently make up as much as 47% of the total 186 million people out of work worldwide. Most labour markets are unable to absorb all of the young people seeking work.
- Millions are working but remain poor; nearly a quarter of the world's workers do not earn enough to lift themselves and their families above the $1 per day poverty threshold. A large majority of the working poor are informal non-agricultural workers. Changing labour markets and increased global competition have led to an explosion of the informal economy and a deterioration in wages, benefits and working conditions, particularly in developing countries.
- In many countries wage inequalities, especially between skilled and unskilled workers, have widened since the mid-1980s, with falling real minimum wages and sharp rises in the highest incomes. China and India have seen considerable income growth, but differentials remain wide. In rich countries, the income gap has been especially pronounced in Canada, the UK and the USA.

PARENTAL EDUCATION AND INEQUALITY

The link between investment in education and poverty is one of the most important dimensions of policies towards poverty. Education may affect poverty in two ways. It may raise the incomes of those with education. In addition, by promoting growth in the economy, it may increase income levels for those with higher levels of education. Those with higher qualifications tend to have fewer children.

EXTENSION

Summarizing skills

When faced with a large amount of text – such as on this page – it is important to be able to break it down into manageable chunks. There are a number of ways of summarizing data – which one you use depends on which you prefer! For example you could:

- highlight notes with a hightlighter
- create spider diagrams or mind maps
- develop mnemonics – using the first letters of words to create a new word that you can remember easily
- create a shorthand language e.g. "Blk Ag in SA" and summarize notes in the margin
- create revision cards of the key terms/concepts/case studies.

Remember, the more you practice the better you will become. The briefer the notes, the easier it is to revise.

Millennium Development Goals

The eight Millennium Development Goals (MDG) were agreed at the UN MIllennuim Development Summit in September 2000. Nearly 190 countries have signed up to them.

Goal	Target
1 Eradicate extreme poverty and hunger	• Reduce by 50% the proportion of people living on less than $1 a day • Reduce by 50% the proportion of people suffering from hunger
2 Achieve universal primary education	• Ensure all children complete a full course of primary schooling
3 Promote gender equality and empower women	• Eliminate gender disparity in primary and secondary education by 2005 (all levels by 2025) • Ensure literacy parity between young men and women • Women's equal representation in national parliaments
4 Reduce child mortality	• Reduce by two-thirds the under-5 mortality rate • Universal child immunization against measles
5 Improve maternal health	• Reduce the maternal mortality ratio by 75%
6 Combat HIV/AIDS, malaria and other diseases	• Halt and begin to reverse the spread of HIV/AIDS • Halt and begin to reverse the incidence of malaria • Halt and begin to reverse the incidence of tuberculosis
7 Ensure environmental sustainability	• Reverse loss of forests • Halve proportion without improved drinking water in urban areas • Halve proportion without improved drinking water in rural areas • Halve proportion without sanitation in urban areas • Halve proportion without sanitation in rural areas • Improve the lives of at least 100 million slum dwellers by 2020
8 Develop global partnership for development	• Reduce youth unemployment

The UN Millennium Development Goals (MDGs)

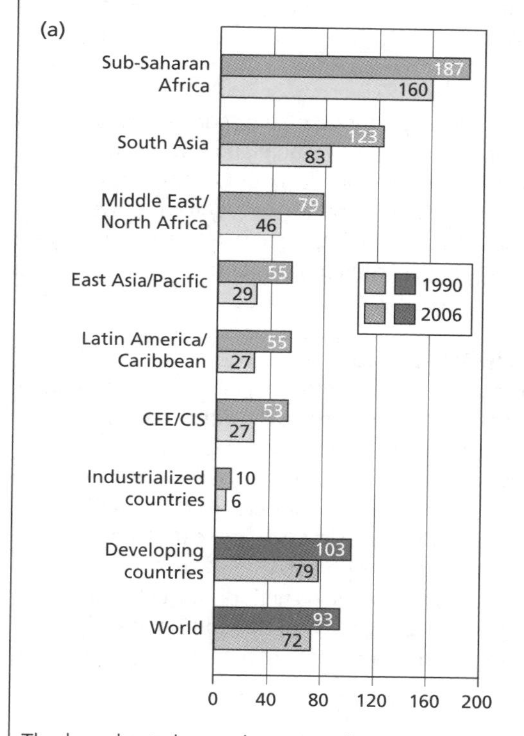

The bar chart shows that mortality rates are falling while the line graph shows that there is still some way to go in improving access to sanitation.

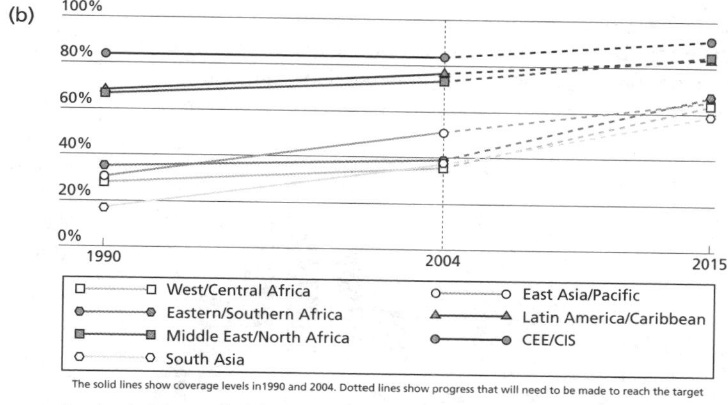

The solid lines show coverage levels in1990 and 2004. Dotted lines show progress that will need to be made to reach the target

A mixed picture of global progress: (a) mortality rates by region, (b) regional trends towards the MDG sanitation target

EXTENSION

Visit

www.mdgmonitor.org/

for the eight Millennium Development Goals (MDGs)

www.mdgmonitor.org/factsheets.cfm

to track the progress of the country of your choice

www.mdgmonitor.org/map.cfm?goal=&indicator=&cd

for interactive maps of the MDGs.

Global disparities and change

CHANGING GLOBAL INEQUALITIES

- **Purchasing power parity** (PPP): what a person can buy with their income at local prices

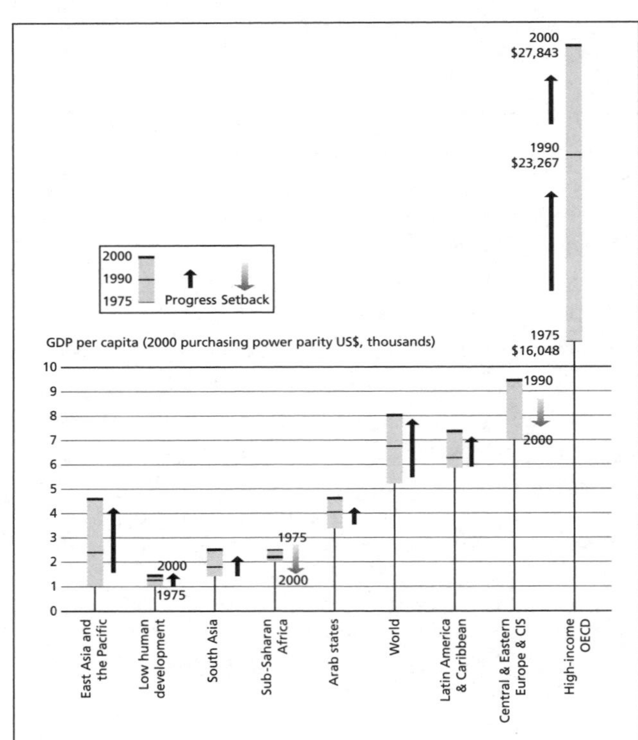

Global disparities in income: are regions closing the gap?

Until 200 years ago, Asia was the dominant world economic power. Today, rapid economic growth rates are helping the region regain its former position, although progress varies widely among and within countries in Asia.

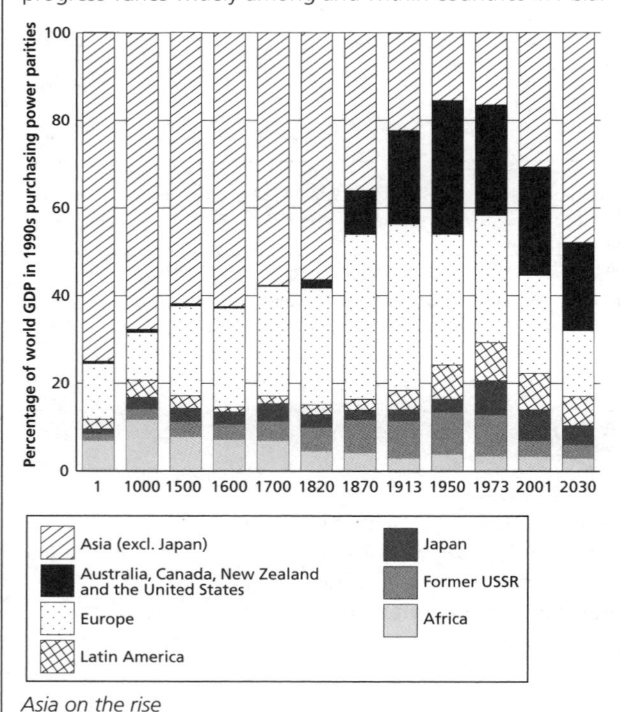

Asia on the rise

INCOME INEQUALITIES

The "Twin Peaks" of rich and poor

The greatest contributors to world income inequality are the large countries at either end of the spectrum, the "Twin Peaks":

- One pole represents the 2.4 billion people whose mean income is less than $1000 a year and includes people living in India, Indonesia and rural China. With 42% of the world's population, this group receives just 9% of the world PPP income.
- The other pole is the group of 500 million people whose annual income exceeds $11,500. This group includes the USA, Japan, Germany, France and the UK. Combined, these countries account for 13% of the world's population, yet use 45% of the world PPP income.

Changes in income

In the last 25 years, the main changes in income between different regions of the world include:

- the continued rapid economic growth in the already rich countries relative to most of the rest of the world
- the decline in real income of sub-Saharan Africa and eastern Europe
- the relatively modest gains in Latin America and the Arab states.

Some of the most important global disparities relate to the lack of decent work available and low incomes. According to the International Labour Organization (ILO), about 200 million people don't have any form of work. Many millions more, including some who are reasonably educated, face inadequate employment.

SOCIAL INEQUALITIES

Despite progress in some contexts, health and education inequalities have widened, especially within countries. Sub-Saharan Africa and parts of Asia are in the worst predicament. There are wide gaps in access to immunization, maternal and childcare, nutrition and education. Gender gaps in access to education have narrowed somewhat, but persist.

Indigenous peoples, persons with disabilities, older persons and youth are typically excluded from decision-making processes that affect their welfare.

ENVIRONMENTAL IMPACTS

Today's disparities are also closely linked to the human impact on the environment. It is the poor who frequently end up with poor quality land, water, fuel and other natural resources, which in turn limit their productivity.

Trends in life expectancy, education and income

LIFE EXPECTANCY

There are certain interesting trends in life expectancy:

- For most countries in the world, more babies are surviving infancy and childhood.
- During the first half of the 20th century, rich countries saw the average life expectancy of their population increase by over 20 years. The graph depicts the speed of population ageing. The values represent the number of years required or expected for the percentage of population aged 65 or over to rise from 7% to 14%. Spain, which had a comparatively low life expectancy in 1900, saw it double by 1995 and equal that of other rich nations.
- In the 1950s female life expectancy continued to rise, but gains in male life expectancy slowed significantly or levelled off. In most MEDCs, women outlive men by 5–9 years.
- The oldest old (aged 80+) are the fastest growing segment of many nations' populations. For the Scandinavian countries, France and Switzerland, the 80+ are approximately 4% of the total population.
- Increases in life expectancy are not uniform for all people living within a country. Indigenous populations living in rich countries have population pyramids that are more typical of developing countries. For example, American Indian, Inuit and Aleut populations have an age structure more like Morocco than the USA; and the Aborigines and Torres Strait Islanders of Australia have a population pattern that is roughly the same as that of Ethiopia.

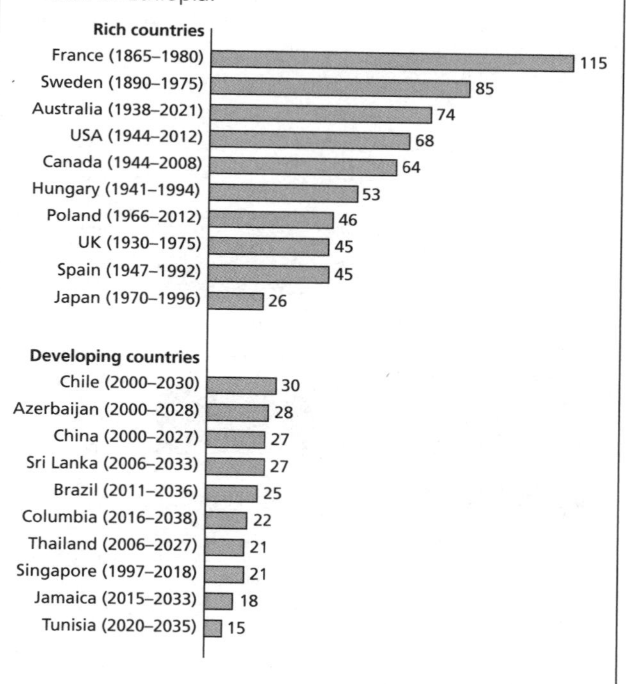

- From 1900 to 1995, females in LEDCs added more than 30 years to their life expectancy.
- In MEDCs, not only do more people survive to old age, but those who do can expect to live longer than their predecessors.

However, in some LEDCs, life expectancy is falling as a result of AIDS.

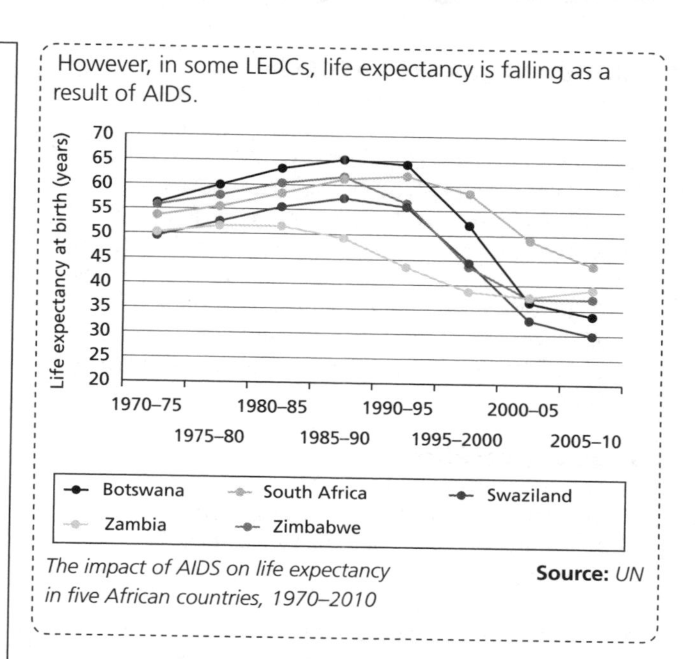

The impact of AIDS on life expectancy in five African countries, 1970–2010

Source: *UN*

EDUCATION AND INCOME

The inequality of education in India, in terms of both educational opportunities and education standards, implies a huge social loss from the underdevelopment and underutilization of human capital. Korea, since the 1960s, has channelled two-thirds of its education spending into compulsory basic education. In the 1990s, subsidies to primary students were two to three times those for college students.

Inequality of education in China

Before economic reforms in 1978, China had achieved a higher human development level than countries at similar income levels. There has been continued progress, but regional disparities have widened. Public expenditure for education is inadequate, at 2.4–2.8% of GDP, and there is an urban bias in provision. There is underinvestment in primary education, and oversubsidization of tertiary education.

Gini coefficients

Gini coefficient measures inequality – the higher the Gini coefficient, the greater the inequality.

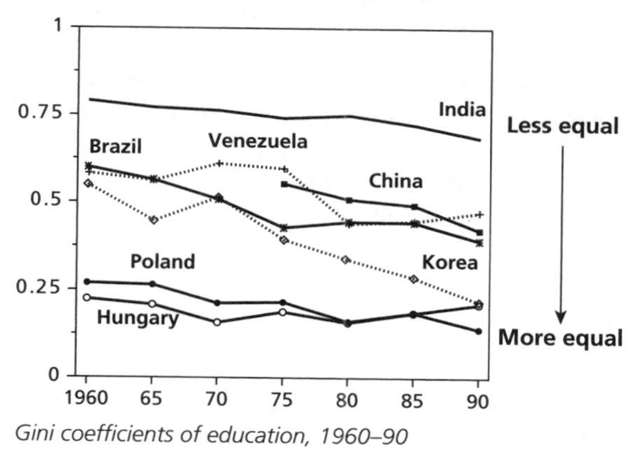

Gini coefficients of education, 1960–90

Reducing disparities (1)

TRADE AND MARKET ACCESS

There are many ways in which development disparities may be addressed. Unfair trading patterns are one of the causes of the development gap. MEDCs account for 75% of the world's exports and over 80% of manufactured exports. The pattern is complicated by flows of foreign direct investment (FDI), and the internal trade within transnational or multinational corporations (TNCs or MNCs). Most of the flow of profits is back to MEDCs, while an increasing share of FDI is to NICs. Reform of trade is necessary to protect LEDCs and small countries.

Regulatory bodies

The main regulatory bodies include:

- international regulators such as the International Monetary Fund (IMF) and the World Trade Organization (WTO)
- coordinating groups of countries such as the G8
- regional trading blocs such as the European Union (EU), North American Free Trade Association (NAFTA) and Association of South East Asian Nations (ASEAN)
- national governments.

However, much of the trade and money exchange that takes place is run by stock exchanges and the world's main banks. For example, Barclays Capital is the investment-banking sector of Barclays Bank. It deals with over £360 billion of investment through its 33 offices located worldwide. Its regional headquarters are located mostly in MEDCs in cities such as London, Paris, Frankfurt, New York and Tokyo. Hong Kong is the exception, although it is an important financial centre, like most of the other places on the list.

There is widespread criticism that many of the regulatory bodies have limited power, and that when faced with a powerful MEDC or TNC they capitulate.

FAIR OR ETHICAL TRADE

Fair or ethical trade can be defined as trade that attempts to be socially, economically and environmentally responsible. It is trade in which companies take responsibility for the wider impact of their business. Ethical trading is an attempt to address the failings of the global trading system.

Good examples of fair trading include Prudent Exports and Blue Skies, both pineapple-exporting companies in Ghana. Prudent Exports, which grows as well as exports pineapples, has introduced better working conditions for its farmers, including longer contracts and better wages. The company has its own farms, buys pineapples from smallholders and exports directly to European supermarkets. It has also responded to requests to cut back on the use of pesticides and chemical fertilizers. The result has been an increase in productivity and sales, supplying a leading British supermarket. Indeed, some retailers appear to be the driving force behind fair trade as they seek out good practice in their suppliers in terms of health and safety at work, employment of children, pay and conditions, and even the freedom of association of workers.

Nevertheless, there are conflicts of interest. For many western consumers, fair trade means banning pesticides or banning the use of child labour. Yet in many LEDCs it is normal for children to help out on farms, just as it was in the UK in the late 19th and early 20th centuries. Most LEDC farmers would prefer to send their children to school, but if the price they receive for their produce is low then they cannot afford the school fees. If western consumers want to stop child labour on farms, they may have to pay high prices for the food they buy.

REMITTANCES

Remittances are the transfer of money and/or goods by foreign workers to their home countries. Total global remittances from workers to their families reached $318 in 2007, up from $170 billion in 2002. Most of the money goes to LEDCs, more than double the value of foreign aid. The three countries receiving the most are India, China and Mexico, which together account for nearly one-third of remittances to the developing world. However, Mexico has been affected by the slowdown in the US economy. The largest recipient region was Latin America and the Caribbean, but since 2002 transfers to Europe and central Asia have increased the fastest.

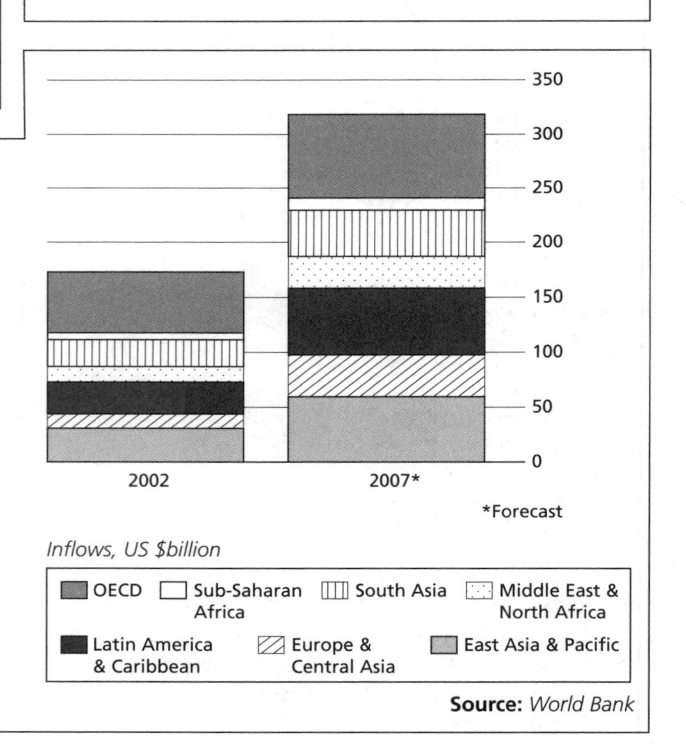

Source: *World Bank*

Reducing disparities (2)

TRADING BLOCS

A **trading bloc** is an arrangement among a group of nations to allow free trade between member countries but to impose tariffs (charges) on other countries who may wish to trade with them. The EU is an excellent example of a trading bloc. Many trading blocs were established after the Second World War as countries used political ties to further their economic development. There are a number of regionally based trading blocs.

Within a trading bloc, member countries have free access to each other's markets. Thus, within the EU, the UK has access to Spanish markets, German markets and so on. However, Spain, Germany and the other countries of the EU have access to the UK's market. Being a member of a trading bloc is important as it allows greater market access – in the case of the EU this amounts to over 470 million wealthy consumers.

Some critics believe that trading blocs are unfair as they deny access to non-members; countries from the developing world, for example, have more limited access to the rich markets of Europe. This makes it harder for them to trade and to develop. In order to limit the amount of protectionism the World Trade Organization has tried to promote free trade. This would allow equal access to all producers to all markets.

The creation of EPZs has been a popular policy for governments of LEDCs because they represent a relatively easy path to begin industrialization in a country. The MNC normally provides technology, capital, inputs and the export markets.

Although the establishment of an EPZ could be seen as beneficial in the short term for the LEDC, in the long term it offers a major problem as regards economic sustainability. MNCs are normally attracted by trade and tax incentives, low labour costs and labour flexibility to locate a branch plant in an EPZ. However, they tend to pull out when economic conditions deteriorate. Thus a reliance on simple export processing would at best perpetuate a reliance on low-skilled, labour-intensive assembly and at worst see the premature end of this type of manufacturing activity within the developing country.

EXPORT PROCESSING AND FREE TRADE ZONES

Export processing zones (EPZs) and free trade zones (FTZs) are important parts of the so-called new international division of labour, and represent what are seen as relatively easy paths to industrialization. By the end of the 20th century, over 90 countries had established EPZs as part of their economic strategies.

- **Export processing zones** have been defined as labour-intensive manufacturing centres that involve the import of raw materials and the export of factory products.
- **Free trade zones** can be classified as zones in which manufacturing does not have to take place in order for trading privileges to be gained and, hence, such zones have become more characterized by retailing.

Popularity of EPZs

The popularity of EPZs is due to three groups of factors that link the economies of LEDCs with those of the world economy in general and the advanced economies in particular:

1. Problems of indebtedness and serious foreign exchange shortfalls in LEDCs since the 1980s
2. The spread of new-liberal ideas in the 1990s that encouraged open economies, foreign investment and non-traditional exports
3. The search by MNCs for cost-saving locations, particularly in terms of wage costs, in order to shift manufacturing, assembly and component production from locations in the advanced economies

The feasibility of MNCs relocating manufacturing capacity to EPZs was also improved by standardization in production processes. It proved profitable for MNCs to shift standardized production to low labour-cost locations.

In EPZ locations there was normally an added bonus for the MNC, as LEDC governments offered them concessions including:

- trade – the elimination of customs duties on imports
- investment – liberalization of capital flows and occasionally access to special financial credits
- important investments in the provision of local infrastructure by the central and/or local government of the host country
- taxation – reduction or exemption from federal, state and local taxes
- labour relations – limitations on labour legislation that apply in the rest of the country, such as the presence of trade unions and adherence to minimum wage and working hours legislation.

Location of EPZs

Within LEDCs, EPZs have been established in a wide range of environments – from border areas (as in north Mexico), to relatively undeveloped regions, to locations adjacent to large cities. The most common location has been on the coast, as in the case of China. EPZs have been most concentrated in the Asia-Pacific region, where in the 1990s approximately 40% of EPZs were located but where two-thirds of employment in EPZs was generated. Latin America and the Caribbean is the next most significant region for EPZs.

The impact of aid and debt relief (1)

THE EFFECTIVENESS OF AID

When aid is effective	When aid is ineffective
It provides humanitarian relief	Aid might allow countries to postpone improving economic management and mobilization of domestic resources
It provides external resources for investment and finances projects that could not be undertaken with commercial capital	Aid can replace domestic saving, direct foreign investment and commercial capital as the main sources of investment and technological development
Project assistance helps expand much-needed infrastructure	The provision of aid might promote dependency rather than self-reliance
Aid contributes to personnel training and builds technical expertise	Some countries have allowed food aid to depress agricultural prices, resulting in greater poverty in rural areas and a dependency on food imports; it has also increased the risk of famine in the future
Aid can support better economic and social policies	Aid is sometimes turned on and off in response to the political and strategic agenda of the donor country, making funds unpredictable, which can result in interruptions in development programmes
	The provision of aid might result in the transfer of inappropriate technologies or the funding of environmentally unsound projects
	Emergency aid does not solve the long-term economic development problems of a country
	Too much aid is tied to the purchase of goods and services from the donor country, which might not be the best or the most economical
	A lot of aid does not reach those who need it, that is, the poorest people in the poorest countries

POOR COUNTRIES' DEBT

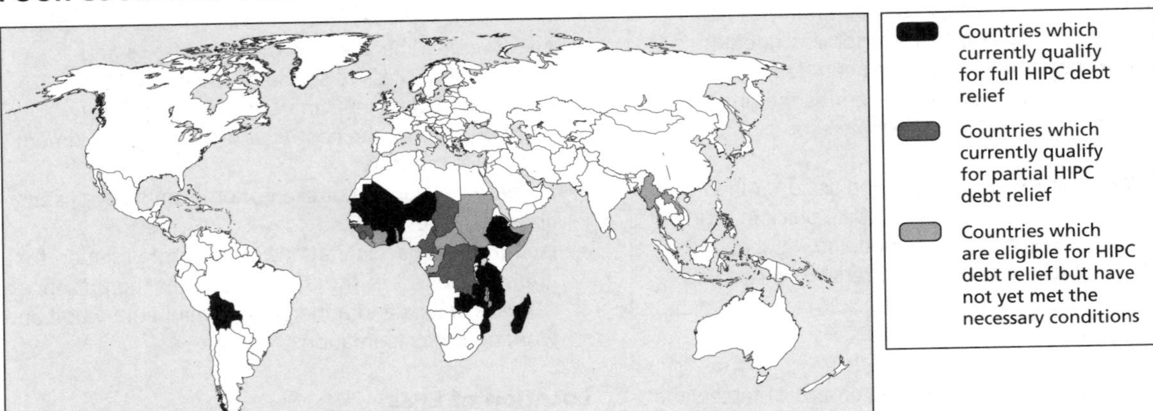

Heavily indebted poor countries (HIPCs)

Sub-Saharan Africa (SSA) includes most of the 42 countries classified as heavily indebted and 25 of the 32 countries rated as severely indebted. In 1962, SSA owed $3 billion (£1.8 billion). Twenty years later this debt had reached $142 billion. Today it is about $235 billion. The most heavily indebted countries are Nigeria ($35 billion), Côte d'Ivoire ($19 billion) and Sudan $18 billion).

Many developing countries borrowed heavily in the 1970s and early 1980s, encouraged to do so by western lenders, including export credit agencies. They soon ran into problems:

- low growth in industrialized economies
- high interest rates between 1975 and 1985
- a rise in oil prices
- falling commodity prices.

EXTENSION

Visit **http://www.imf.org/external/np/exr/facts/hipc.htm** for a fact sheet on debt relief under the Heavily Indebted Poor Countries (HIPC) initiative.

The impact of aid and debt relief (2)

WHAT HAS BEEN DONE TO DEAL WITH THE PROBLEM?

Since 1988, the Paris Club of government creditors has approved a series of debt relief initiatives.

- The World Bank has lent more through its concessional lending arm.
- The International Development Agency has given loans for up to 50 years without interest but with a 3–4% service charge.
- Lending has risen from $424 million in 1980 to $2.9 billion, plus a further $928 million through the African Development Bank.
- The IMF has also introduced a soft loan facility conditional on wide-ranging socio-economic reforms.

Structural adjustment programmes (SAPs)

SAPs were designed to cut government expenditure, reduce the amount of state intervention in the economy, and promote liberalization and international trade. SAPs were explicit about the need for international trade.

SAPs consist of four main elements:

1. Greater use of a country's resource base
2. Policy reforms to increase economic efficiency
3. Generation of foreign income through diversification of the economy and increased trade
4. Reducing the active role of the state

These were sometimes divided into two main groups:

- **stabilization measures**: short-term steps to limit any further deterioration of the economy (e.g. wage freezes; reduced subsidies on food, health and education)
- **adjustment measures**: longer-term policies to boost economic competitiveness (e.g. tax reductions, export promotion, downsizing of the civil service, privatization, economic liberalization).

THE ACHIEVEMENTS OF LEDCS

People in the West tend to forget about the achievements of the developing world. For example:

- average real incomes in the poor world have more than doubled in the past 40 years despite population growth
- under-5 death rates have been cut by 50% or more in every region over the past 40 years
- average life expectancy has risen by more than one-third in every region since 1950
- the percentage of people with access to safe water supply has risen from about 10% to 60% in rural areas of the developing world since 1975.

EXTENSION

Visit **http://imf.org/external/np/exr/facts/mozam/mozam.htm** for facts on Mozambique and debt service.

THE HEAVILY INDEBTED POOR COUNTRIES INITIATIVE

The HIPC initiative, launched in 1996 by the IMF and the World Bank and endorsed by 180 governments, has two main objectives:

- to relieve certain low-income countries of their unsustainable debt to donors
- to promote reform and sound policies for growth, human development and poverty reduction.

Debt relief occurs in two steps:

- At the decision point, the country gets debt service relief after demonstrating adherence to an IMF programme and progress in developing a national poverty strategy.
- At the completion point, the country gets debt stock relief upon approval by the World Bank and the IMF.

"Debt service" is the cash required over a given period for the repayment of interest and principal on a debt – monthly mortgage payments are a good example. "Stock relief" is the cancelling of specific debts; this will achieve a reduction in debt service over the life of a loan.

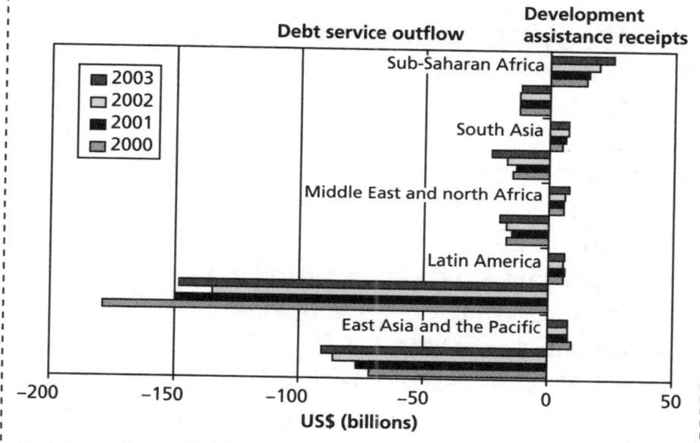

Debt service and development assistance, 2000–3

Source: *World Bank*

Of the 42 countries participating in the initiative, 34 are in sub-Saharan Africa. None had a PPP above $1500 in 2001, and all rank low on the HDI.

Expanding market access is essential to help countries diversify and expand trade. Trade policies in rich countries remain highly discriminatory against developing country exports.

MEDCs should set targets to:

- increase official development assistance
- remove tariffs and quotas on agricultural products, textiles and clothing exported by developing countries
- finance debt reduction for HIPCs having reached their completion points to ensure sustainability.

Atmosphere and change (1)

GLOBAL WARMING

- **Global warming** refers to the increase in temperatures around the world that has been noticed over the last 50 years or so, and in particular since the 1980s.
- The **greenhouse effect** is the process by which certain gases – water vapour, carbon dioxide, methane and chlorofluorocarbons (CFCs) – allow short-wave radiation from the sun to pass through and heat up the earth, but trap an increasing proportion of long-wave radiation from the earth. This radiation leads to a warming of the atmosphere.
- The **enhanced greenhouse effect** is the increasing amount of greenhouse gases in the atmosphere as a result of human activities, and their impact on atmospheric systems, including global warming.

One concern about global warming is the build-up of greenhouse gases (GHGs).

Carbon dioxide (CO_2) levels have risen from about 315 parts per million (ppm) in 1950 to 355 ppm and are expected to reach 600 ppm by 2050. The increase is due to human activities – burning fossil fuels (coal, oil and natural gas) and deforestation. Deforestation of the tropical rainforest is a double blow – not only does it increase atmospheric CO_2 levels, it removes the trees that convert CO_2 into oxygen.

Methane is the second largest contributor to global warming, and is increasing at a rate of 1% per annum. It is estimated that cattle convert up to 10% of the food they eat into methane, and emit 100 million tonnes of methane into the atmosphere each year. Natural wetland and paddy fields are another important source – paddy fields emit up to 150 million tonnes of methane annually. As global warming increases, bogs trapped in permafrost will melt and release vast quantities of methane.

Chlorofluorocarbons (CFCs) are synthetic chemicals that destroy ozone, as well as absorb long-wave radiation. CFCs are increasing at a rate of 6% per annum, and are up to 10,000 times more efficient at trapping heat than CO_2.

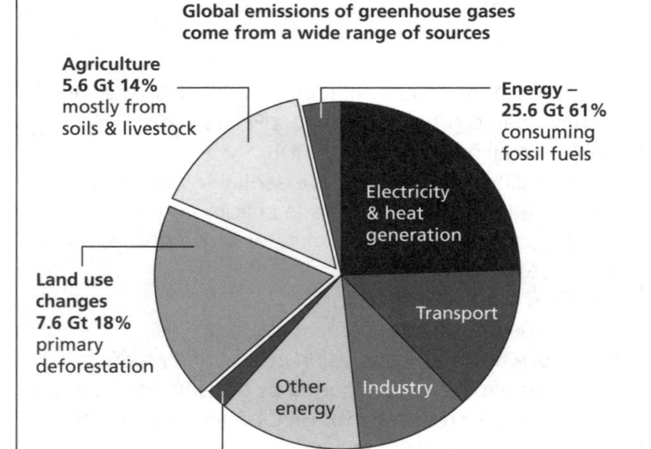

Main sources of CO_2 emissions

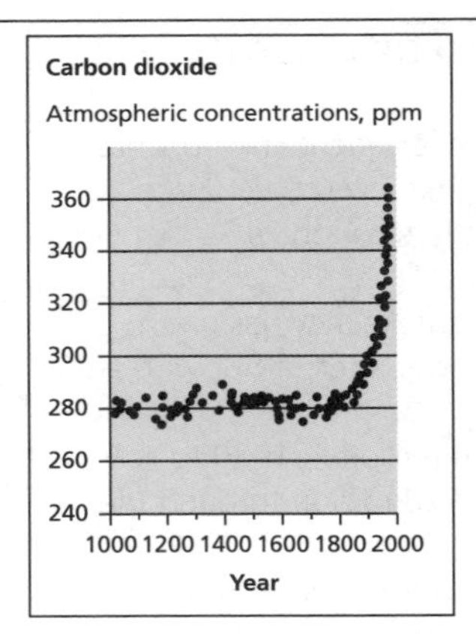

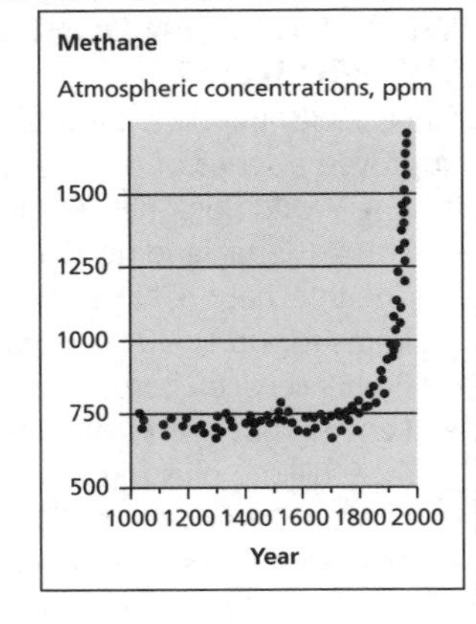

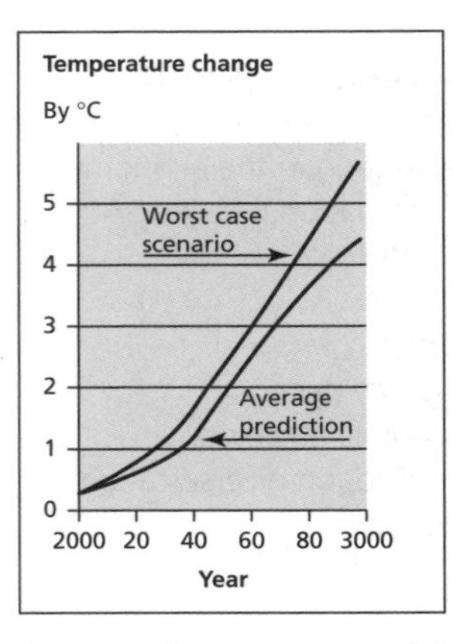

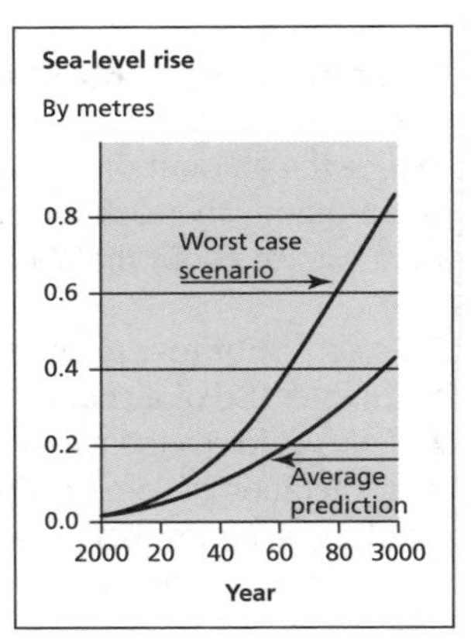

Causes and consequences of global warming

The effects of global warming

- A rise in sea levels, causing flooding in low-lying areas such as the Netherlands, Egypt and Bangladesh – up to 200 million people could be displaced
- An increase in storm activity (owing to more atmospheric energy)
- Changes in agricultural patterns (e.g. a decline in the USA's grain belt, but an increase in Canada's growing season)
- Reduced rainfall over the USA, southern Europe and the Commonwealth of Independent States (CIS)
- Extinction of up to 40% of species of wildlife

EXTENSION

Pie charts – absolute and relative scale

Pie charts are a great way of showing *relative* data. They are quite easy to draw and label and show clearly the biggest contributors – in this case energy.

However, pie charts are not very good at representing *absolute* data. Sometimes, as here, it is important to add the absolute size (25.6 Gt for energy) to give some idea of the scale of the data.

Two pie charts may be drawn at the same size but may have very different absolute scales.

Atmosphere and change (2)

THE IMPLICATIONS OF CLIMATE CHANGE

The effects of global warming on the natural, social and economic environment are mixed:

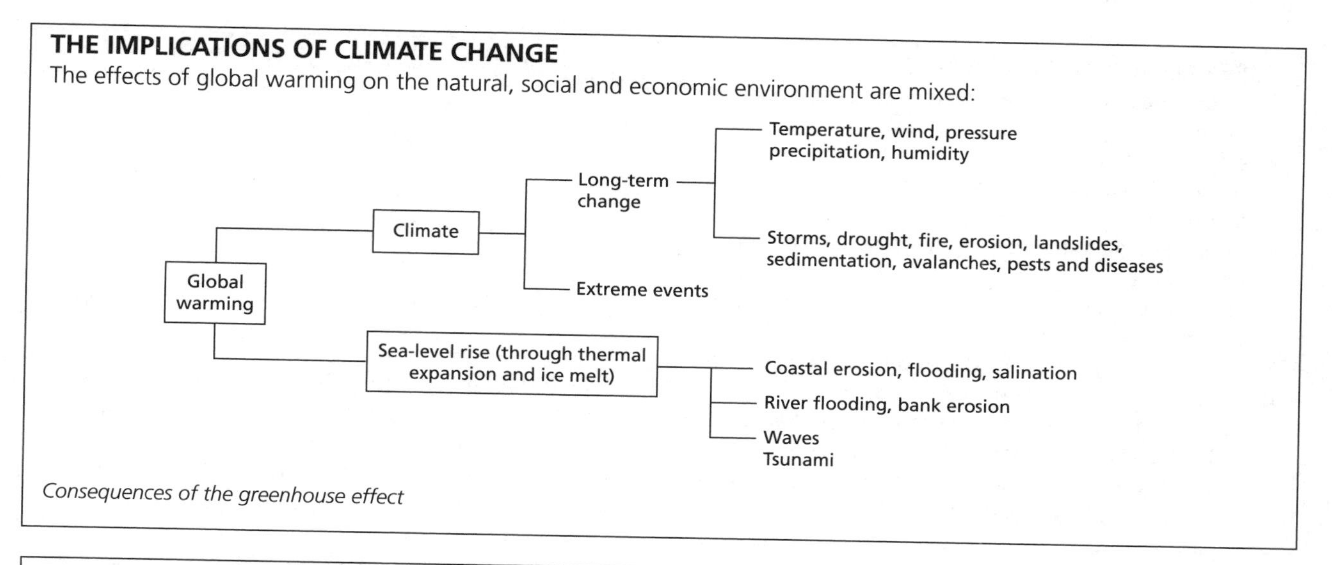

Consequences of the greenhouse effect

POLICIES TO COMBAT CLIMATE CHANGE

Emissions of the main anthropogenic (man-made) greenhouse gas, CO_2, are influenced by:

- the size of the human population
- the amount of energy used per person
- the level of emissions resulting from that use of energy.

A variety of technical options which could reduce emissions, especially from use of energy, are available. Reducing CO_2 emissions can be achieved through:

- improved energy efficiency
- fuel switching
- use of renewable energy sources
- nuclear power
- capture and storage of CO_2.

These options are most easily applicable to stationary plant. Another class of measure involves increasing the rate at which natural sinks take up CO_2 from the atmosphere, for example by increasing the number of forests.

Projected impacts of climate change

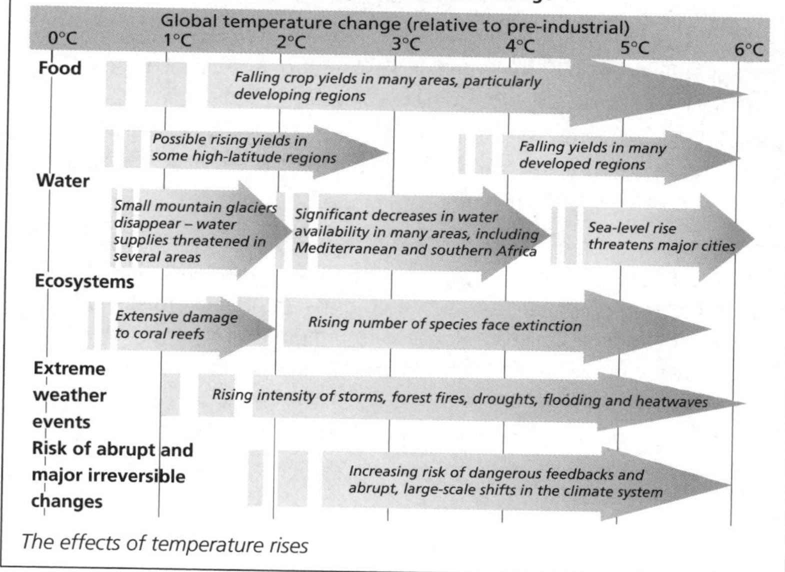

The effects of temperature rises

INTERNATIONAL POLICY TO PROTECT CLIMATE

See page 48 for an account of policy at an international scale.

EXTENSION

Uncertainty in geography

There is a great amount of uncertainty in geography. Try to avoid statements that are too forceful or dogmatic. For example, nobody knows what the impact of climate change will be. There are different scenarios based on possible termerature changes. Some people even suggest that certain areas might get colder, such as the northern UK if the Gulf Stream shuts down. We do not know what will happen – therefore it is wise to be aware that there is uncertainty and there may be very different results in the end.

Soil degradation (1)

TYPES OF SOIL DEGRADATION

Soil degradation is the decline in quantity and quality of soil. It includes:

- erosion by wind and water
- biological degradation (the loss of humus and plant/animal life)
- physical degradation (loss of structure, changes in permeability)
- chemical degradation (acidification, declining fertility, changes in pH, salinization and chemical toxicity).

There are many types of water erosion, including surface, gully, rill and tunnel erosion.

Water and wind erosion account for more than 80% of the 20 million km^2 of degraded land worldwide.

Acidification is the change in the chemical composition of the soil, which may trigger the circulation of toxic metals.

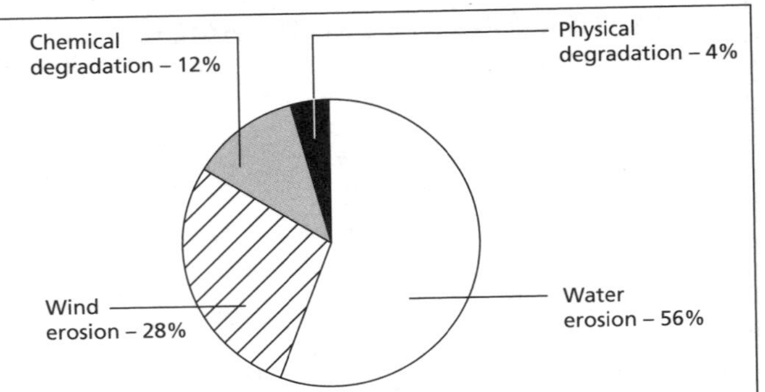

Salt-affected soils are typically found in marine-derived sediments, coastal locations and hot arid areas, where capillary action brings salts to the upper part of the soil. Soil salinity has been a major problem in Australia following the removal of vegetation in dryland farming.

THE UNIVERSAL SOIL LOSS EQUATION (USLE)

The universal soil loss equation **A = RKLSCP** is an attempt to predict the amount of erosion that will take place in an area on the basis of certain factors which increase susceptibility to erosion.

Factor	Description
Ecological conditions	
Erosivity of soil **R**	Rainfall totals, intensity and seasonal distribution. Maximum erosivity occurs when the rainfall occurs as high-intensity storms. If such rain is received when the land has just been ploughed or full crop cover is not yet established, erosion will be greater than when falling on a full canopy. Minimal erosion occurs when rains are gentle and fall onto frozen soil or land with natural vegetation or a full crop cover.
Erodibility **K**	The susceptibility of a soil to erosion. Depends on infiltration capacity and the structural stability of soil. Soils with high infiltration capacity and high structural stability, which allow the soil to resist the impact of rain splash, have lowest erodibility values.
Length-slope factor **LS**	Slope length and steepness influence the movement and speed of water down the slope, and thus its ability to transport particles. The steeper the slope, the greater the erosivity; the longer the slope, the more water is received on the surface.
Land-use types	
Crop management **C**	Most control can be exerted over the cover and management of the soil, and this factor relates to the type of crop and cultivation practices. Established grass and forest provide the best protection against erosion; of agricultural crops, those with the greatest foliage and thus greatest ground cover are optimal. Fallow land or crops that expose the soil for long periods after planting or harvesting offer little protection.
Soil conservation **P**	Soil conservation measures, such as contour ploughing, bunding, use of strips and terraces, can reduce erosion and slow runoff water.

Factors relating to the universal soil loss equation (USLE)

CAUSES OF DEGRADATION

Causes of soil or land degradation include:

- the reduction of the natural vegetative cover, which renders the topsoil more susceptible to erosion
- unsustainable land-use practices such as excessive irrigation, the inappropriate use of fertilizers and pesticides and overgrazing by livestock
- groundwater overabstraction, which may lead to dry soils, resulting in physical degradation
- atmospheric deposition of heavy metals and persistent organic pollutants, which make soils less suitable to sustain their original land cover and land use.

Climate change will probably intensify the problem. It is likely to affect hydrology and hence land use.

Soil degradation (2)

CAUSES OF DEGRADATION (CONTINUED)

Overgrazing and agricultural mismanagement affect more than 12 million km² worldwide. The situation is most severe in Africa and Asia, where 20% of the world's pasture and rangelands have been damaged. Huge areas of forest are cleared for logging, fuelwood, farming or other human uses.

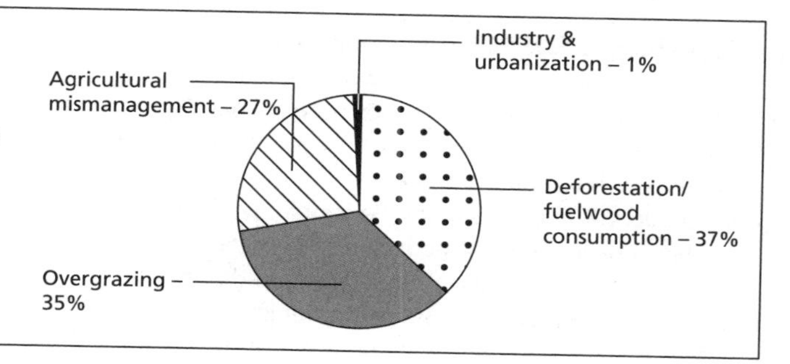

THE EFFECTS OF LOSS OF COVER

The removal of vegetation and topsoil has resulted in:

- increased surface runoff and stream discharge
- reduction of water infiltration and groundwater recharge
- development of erosional gullies and sand dunes
- change in the surface microclimate that enhances aridity
- drying up of wells and springs
- reduction of seed germination of native plants.

MANAGING SOIL DEGRADATION

Abatement strategies, such as afforestation, for combating accelerated soil erosion are lacking in many areas. To reduce the risk of soil erosion, farmers are encouraged towards more extensive management practices such as organic farming, afforestation, pasture extension and benign crop production. Nevertheless, there is a need for policymakers and the public to intensify efforts to combat the pressures and risks to the soil resource.

Methods to reduce or prevent erosion can be mechanical, for example physical barriers such as embankments and windbreaks, or they may focus on vegetation cover and soil husbandry. Overland flow can be reduced by increasing infiltration.

Mechanical methods

Mechanical methods include bunding, terracing and contour ploughing, and shelter belts such as trees or hedgerows. The key is to prevent or slow the movement of rainwater downslope. Contour ploughing takes advantage of the ridges formed at right angles to the slope to act to prevent or slow the downward accretion of soil and water. On steep slopes and in areas with heavy rainfall, such as the monsoon in South-East Asia, contour ploughing is insufficient and terracing is undertaken.

The slope is broken up into a series of flat steps, with bunds (raised levées) at the edge. The use of terracing allows areas to be cultivated that would not otherwise be suitable. In areas where wind erosion is a problem, shelter belts of trees or hedgerows are used. The trees act as a barrier to the wind and disturb its flow. Wind speeds are reduced, which therefore reduce the wind's ability to disturb the topsoil and erode particles.

Cropping techniques

Preventing erosion by different cropping techniques largely focuses on:

- maintaining a crop cover for as long as possible
- keeping in place the stubble and root structure of the crop after harvesting
- planting a grass crop – grass roots bind the soil, minimizing the action of the wind and rain on a bare soil surface.

Increased organic content allows the soil to hold more water, thus preventing aerial erosion and stabilizing the soil structure. In addition, care is taken over the use of heavy machinery on wet soils and ploughing on soil sensitive to erosion, to prevent damage to the soil structure.

Managing salt- and chemical-affected soils

There are three main approaches in the management of salt-affected soils:

- flushing the soil and leaching the salt away
- application of chemicals, such as gypsum (calcium sulphate) to replace the sodium ions on the clay and colloids with calcium ones
- a reduction in evaporation losses to reduce the upward movement of water in the soil.

Equally specialist methods are needed to decontaminate land made toxic by chemical degradation.

LAND DEGRADATION IN BARBADOS

The most significant area of land degradation in Barbados is within the Scotland District. Changing land-use practices and the application of inappropriate agricultural techniques (growing sugar cane on very steep slopes, for example) have resulted in significant and visible loss of soils.

Controlling land degradation

One of the most effective ways in which land degradation can be controlled is through increasing the vegetative cover within the affected area. Farmers in the region are taught methods which include keeping the soil covered, incorporating organic matter to assist with percolation and reducing the use of fertilizers.

Water usage and change (1)

CHANGING SUPPLY AND DEMAND

During the past century, while world population has tripled, the use of water has increased sixfold. Some rivers that formerly reached the sea no longer do so – all of the water is diverted before it reaches the river's mouth. The Colorado in the USA is a good example. Half the world's wetlands have disappeared in the same period, and today 20% of freshwater species are endangered or extinct. Many important aquifers are being depleted, and water tables in many parts of the world are dropping at an alarming rate. Worse still, world water use is projected to increase by about 50% in the next 30 years.

It is estimated that, by 2025, 4 billion people – half the world's population at that time – will live under conditions of severe water stress, with conditions particularly severe in Africa, the Middle East and south Asia. Many observers predict that disputes over scarce water resources will fuel an increase in armed conflicts. Water that is safe to drink remains as central to survival – and to improving the lives of the poor – as it has always been. Currently, an estimated 1.1 billion people lack access to safe water, 2.6 billion are without adequate sanitation, and more than 4 billion do not have their waste water treated to any degree. These numbers are likely to grow worse in the coming decades.

WATER SUPPLY

Water supply depends on several factors in the water cycle, including the rates of rainfall, evaporation, the use of water by plants (transpiration), and river and groundwater flows. It is estimated that less than 1% of all fresh water is available for people to use (the remainder is locked up in ice sheets and glaciers). Globally, around 12,500 km^3 of water are considered available for human use on an annual basis. This amounts to about 6600 m^3 per person per year.

If current trends continue, only 4800 m^3 will be available in 2025. This is an optimistic calculation because it is based on estimates of all the water flowing in rivers after evaporation and infiltration into the ground. It does not take into account the minimum required to maintain river ecosystems, for example. Nor does it reflect the difficulty in accessing all of this water or its extremely unequal distribution.

The world's available freshwater supply is not distributed evenly around the globe, either seasonally or from year to year. About three-quarters of annual rainfall occurs in areas containing less than one-third of the world's population, whereas two-thirds of the world's population live in the areas receiving only one-quarter of the world's annual rainfall. For instance, about 20% of the global average runoff each year is accounted for by the Amazon Basin, a vast region with fewer than 10 million people. India gets 90% of its rainfall during the summer monsoon season – at other times rainfall over much of the country is very low.

Water stress

When per capita water supply is less than 1700 m^3 per year, an area suffers from "water stress" and is subject to frequent water shortages. In many of these areas today, water supply is actually less than 1000 m^3 per capita, which causes serious problems for food production and economic development. Some 2.3 billion people live in water-stressed areas. If current trends continue, water stress will affect 3.5 billion – or 48% of the world's projected population – in 2025.

WATER USE

Currently, the quantity of water used for all purposes exceeds 3700 km^3 per year. Agriculture is the largest user, consuming almost two-thirds of all water drawn from rivers, lakes and groundwater. Since 1960, water use for crop irrigation has risen by 60–70%. Industry uses about 20% of available water, and the municipal sector uses about 10%. Population growth, urbanization and industrialization have increased the use of water in these sectors. As world population and industrial output have increased, the use of water has accelerated, and this is projected to continue. By 2025 global availability of fresh water may drop to an estimated 5100 m^3 per person per year, a decrease of 25% on the 2000 figure.

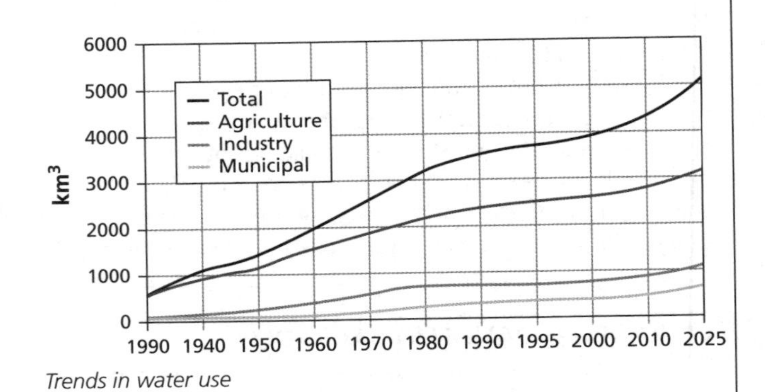

Trends in water use

Water usage and change (2)

WATER SCARCITY

Where water supplies are inadequate, two types of water scarcity affect LEDCs in particular:

- **Physical water scarcity** occurs where water consumption exceeds 60% of the usable supply. To help meet water needs, some countries such as Saudi Arabia and Kuwait import much of their food and invest in desalinization plants.
- **Economic water scarcity** occurs where a country physically has sufficient water resources to meet its needs, but additional storage and transport facilities are required – this will mean embarking on large and expensive water development projects, as in many in sub-Saharan countries.

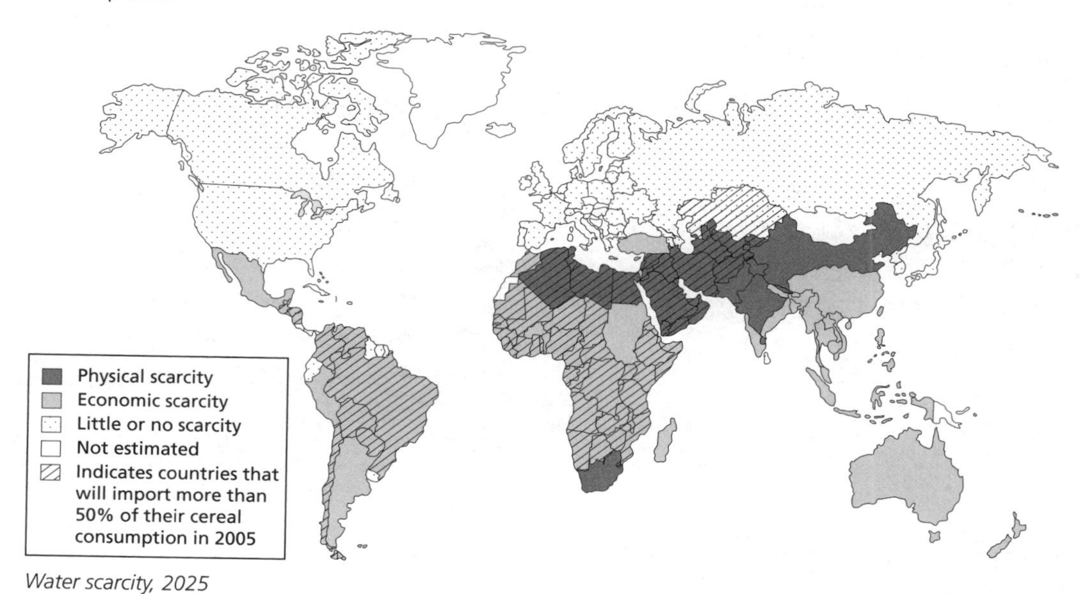

Water scarcity, 2025

In addition, in LEDCs access to adequate water supplies is most affected by the exhaustion of traditional sources, such as wells and seasonal rivers.

In many poor countries farmers use, on average, twice as much as water per hectare as in industrialized countries, yet their yields can be three times lower – a sixfold difference in the efficiency of irrigation.

WATER QUALITY

Water also needs to be of an adequate quality for consumption. However, the World Health Organization (WHO) estimates that around 4 million deaths each year can be attributed to water-related disease, particularly cholera, hepatitis, malaria and other parasitic diseases.

The real problem of drinking water and sanitation in developing countries is that too many people lack access to safe and affordable water supplies and sanitation.

GLOBAL WATER SUPPLY AND SANITATION

Urban areas are better served than rural areas, and countries in Asia, Latin America and the Caribbean are better off than African countries. Many piped water systems, however, do not meet water quality criteria, leading more people to rely on bottled water bought in markets for personal use (as in major cities in Colombia, India, Mexico, Thailand, Venezuela and Yemen).

In some cases, the poor pay more for their water than the rich. For example, in Port-au-Prince, Haiti, surveys have shown that households connected to the water system typically paid around $1.00 per cubic metre, while unconnected customers forced to purchase water from mobile vendors paid from $5.50 to a staggering $16.50 per cubic metre.

Sanitation and population growth

Fewer people have adequate sanitation than safe water, and the global provision of sanitation is not keeping up with population growth. Between 1990 and 2000 the number of people without adequate sanitation rose from 2.6 billion to 3.3 billion. Least access to sanitation occurs in Asia (48%), especially in rural areas.

There are still pressure points, especially in areas of rapid population growth. With squatter settlements in many of the world's poorest cities expanding rapidly, and local authorities unable to or legally prevented from providing sanitation, the situation is likely to deteriorate rapidly.

The world's riches: biodiversity and change (1)

BIODIVERSITY

Biodiversity means biological diversity. It is the variety of all forms of life on earth – plants, animals and micro-organisms. It refers to species (species diversity), variations within species (genetic diversity), and interdependence within species (ecosystem diversity) and habitat diversity.

It is estimated that there are up to 30 million species on earth. However, only 1.4 million species have yet been identified. The tropics are the richest area for biodiversity. Tropical forests contain over 50% of the world's species in just 7% of the world's land. They account for 80% of the world's insects and 90% of primates.

THE VALUE OF TROPICAL RAINFORESTS

Industrial uses	Ecological uses	Subsistence uses
Charcoal	Watershed protection	Fuelwood and charcoal
Saw logs	Flood and landslide protection	Fodder for agriculture
Gums, resins and oils	Soil erosion control	Building poles
Pulpwood	Climate regulation e.g. CO_2 and O_2 levels	Pit-sawing and saw-milling
Plywood and veneer		Weaving materials and dyes
Industrial chemicals		Rearing silkworms and bee-keeping
Medicines		Special woods and ashes
Genes for crops		Fruits and nuts
Tourism		

DEFORESTATION OF THE TROPICAL RAINFOREST

Tropical forests are being destroyed at a rate of over 11 million hectares a year (or 21 ha/minute). Increasingly, tropical rainforests are very scattered and fragmented. The Amazon rainforest is the main exception, although it is imploding.

Causes of deforestation in Brazil

There are five main causes of deforestation in Brazil:

- agricultural colonization by landless migrants and speculative developers along highways and agricultural growth areas
- conversion of the forest to cattle pastures, especially in eastern and south-eastern Para and northern Mato Grosso
- mining, for example the Greater Carajas Project in south-eastern Amazonia, which includes a 900 km railway and extensive deforestation to provide charcoal to smelt the iron ore; another threat from mining comes from the small-scale informal gold mines, *garimpeiros*, causing localized deforestation and contaminated water supplies
- large-scale hydroelectric power schemes such as the Tucurui Dam on the Tocantins River
- forestry taking place in Para, Amazonas and northern Mato Grosso.

Other causes include:

- drought (increases risk)
- climate change (can cause drought)
- timber exploitation (fires are used to overcome laws about clearing timber for sale, or to create a source for damaged and thus cheap timber)
- selective logging (can create artificially dry forests by opening up the canopy)
- lightning (the main natural cause)
- land clearing ("slash-and-burn" agriculture during dry and windy conditions can cause major fires).

Trends

Deforestation in Brazil shows five main trends:

1. It is a recent phenomenon.
2. It has partly been promoted by government policies.
3. There is a wide range of causes of deforestation.
4. Deforestation includes new areas of deforestation as well as the extension of previously deforested areas.
5. Land speculation and the granting of land titles to those who "occupy" parts of the rainforest is a major cause of deforestation.

The world's riches: biodiversity and change (2)

EFFECTS OF DEFORESTATION

There are many effects of deforestation, including:

- disruption to the circulation and storage of nutrients
- surface erosion and compaction of soils
- sandification
- increased flood levels and sediment content of rivers
- climatic change
- loss of biodiversity.

Deforestation disrupts the closed system of nutrient cycling within tropical rainforests. Inorganic elements are released through burning and are quickly flushed out of the system by the high-intensity rains.

Soil erosion is also associated with deforestation. As a result of soil compaction, there is a decrease in infiltration, an increase in overland runoff and surface erosion.

Sandification is a process of selective erosion. Raindrop impact washes away the finer particles of clay and humus, leaving behind the coarser and heavier sand. Evidence of sandification dates back to the 1890s in Santarem, Rondonia.

As a result of the intense surface runoff and soil erosion, rivers have a higher flood peak and a shorter time lag. However, in the dry season river levels are lower, the rivers have greater turbidity (murkiness due to more sediment), an increased bed load, and carry more silt and clay in suspension.

Other changes relate to climate. As deforestation progresses, there is a reduction of water that is re-evaporated from the vegetation, hence the recycling of water must diminish. Evapotranspiration (EVT) rates from savanna grasslands are estimated to be only about one-third of those of the tropical rainforest. Thus mean annual rainfall is reduced, and the seasonality of rainfall increases.

AMAZON'S RESCUE REVERSED

Government satellite images show that at least 1249 square miles (3235 km^2) of rainforest were lost between August and December 2007, mainly because of soy planting and cattle ranching. The true figure could be as high as 2700 square miles (almost 7000 km^2).

Environmentalists say as much as 20% of the rainforest has already been destroyed, mostly since the 1970s. A further 40% could be lost by 2050 if that trend is not reversed, they estimate.

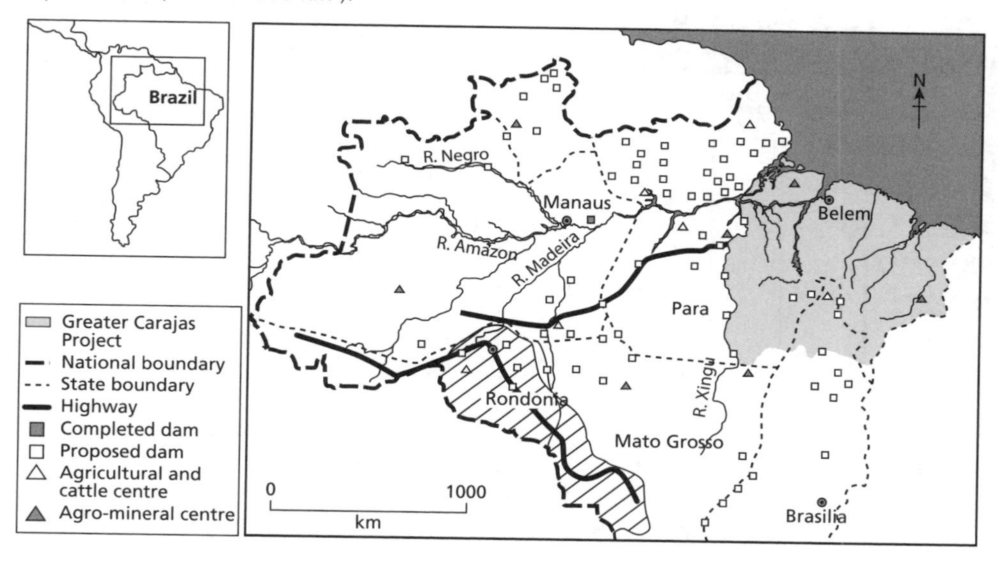

Economic development and deforestation in the Brazilian rainforest

THE COST OF ENVIRONMENTAL INACTION IN NIGERIA

The conventional constraint on government and private sector action has been concern about the costs of taking new environmental protection measures. This narrow preoccupation has overshadowed the equally important consideration of the mounting economic, social and ecological costs of not acting.

A recent World Bank study provides a stark assessment of the risks and enormous costs if no remedial action is taken. In sum, the long-term losses to Nigeria of not acting on growing environmental problems are estimated to be around $5000 million annually.

Soil degradation	3000
Water contamination	1000
Deforestation	750
Coastal erosion	150
Gully erosion	100
Fishery losses	50
Water hyacinth	50
Wildlife losses	10
Total	**5110**

Annual costs of inaction (US$ million/year)

Ecological footprints

CALCULATING ECOLOGICAL FOOTPRINT

Everything used for our daily needs and activities comes from natural resources. The **ecological footprint**, measured in acres or hectares (ha), calculates the amount of the earth's bioproductive space needed to keep a population at its current level of resource consumption. The calculation takes into account the following resources:

- **arable land:** the amount of land required for growing crops
- **pasture land:** the resources required for growing animals for meat, hides, milk, etc.
- **forests:** for fuel, furniture, housing, etc., also providing many ecosystem services such as climate stability, erosion prevention
- **oceans:** for fish and other marine products
- **infrastructure needs:** transportation, factories, housing, etc. based on the built-up land used for these needs
- **energy costs:** the land required for absorbing carbon dioxide emissions and other energy wastes.
- Species extinction, and toxic pollution of the air, water and land, are not yet taken into account in calculating the ecological footprint.

ECOLOGICAL FOOTPRINT – GLOBAL AND NATIONAL

The planet's biological productive capacity (biocapacity) is estimated at 1.9 ha per person. Currently, countries are using up 2.2 ha per person, living beyond the planet's biocapacity to sustain us by 15%, or by a deficit of 0.4 ha per person. This deficit is showing up as failing natural ecosystems – forests, oceans, fisheries, coral reefs, rivers, soil, water, and global warming.

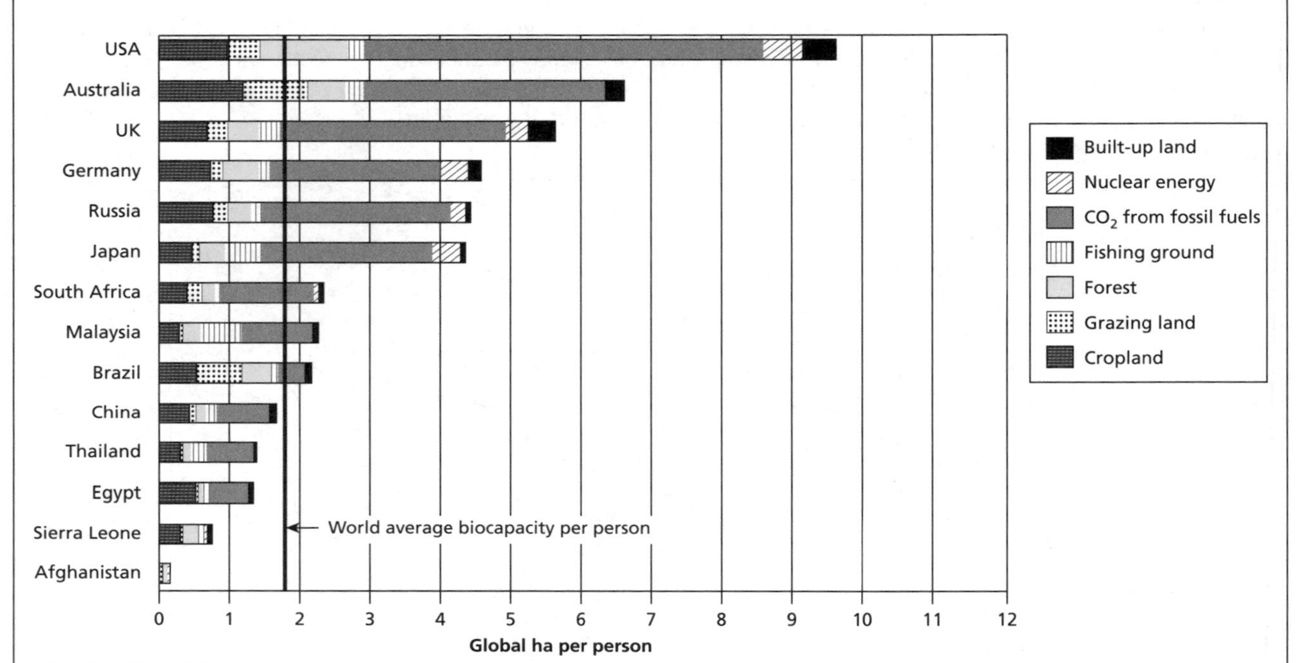

Global ecological footprints

Source: *WWF*

The planet's biocapacity is affected by the global population as well as the rate of consumption. Higher consumption depletes the planet's carrying, renewal and regeneration capacities. Estimates indicate that, if global population trends continue, the ecological footprint available to each person would be reduced to 1.5 ha per person by 2050 and, if consumption rates as prevalent in the rich western countries are adopted by the majority of humanity, then we would need four to five planets more to sustain ourselves.

The USA is the country with the largest per capita footprint in the world – a footprint of 9.57 ha. If everyone on the planet was to live like an average American, our current planet's biocapacity could support only about 1.2 billion people. On the other hand, if everyone lived like an average person in Bangladesh, where the per capita footprint is just 0.5 ha, the earth could support roughly 22 billion people.

The global ecological footprint grew from about 70% of the planet's biological capacity in 1961 to about 120% in 1999. Furthermore, future projections show that humanity's footprint is likely to grow to about 180% or even 220% of the earth's biological capacity by the year 2050.

Environmental sustainability

THE ENVIRONMENTAL SUSTAINABILITY INDEX

The environmental sustainability index (ESI) is produced by a team of environmental experts from Yale and Columbia Universities. Using 21 indicators and 76 measurements, including natural resource endowments, past and present pollution levels, and policy efforts, the report creates a "sustainability score" for each country, with higher scores indicating better environmental sustainability.

The 10 most sustainable countries, as ranked by the ESI, are dominated by wealthy, sparsely populated nations with an abundance of natural resources. Finland has been ranked first, with Norway, Sweden and Iceland all figuring in the top five. The only developing nations in the top 10 are Uruguay and Guyana, both of which have relatively low population densities and an abundance of natural resources. Conversely, the only densely populated countries that have received even above-average rankings are Japan, Germany, the Netherlands and Italy, some of the richest countries on the list.

ESI rank	Country	ESI score
1	Switzerland	95.5
2	Sweden	93.1
13	Germany	86.3
21	Japan	84.5
51	South Korea	79.4
95	Zimbabwe	69.3
105	China	65.1
120	India	60.3
125	Bangladesh	58.0
149	Niger	39.1

Environmental sustainability index of high population density countries, 2005

The table shows the overall ESI rankings only of countries and territories in which more than half the land area has a population density of over 100 people per km^2.

Environmental sustainability is essential for helping poor people. They are highly dependent on the environment and its resources (fresh water, crops, fish, etc.), which provide roughly two-thirds of household income for the rural poor.

Climate change is dramatically reshaping the environment on which poor people depend. The knock-on effects from climate change include increased rainfall variability (meaning more droughts and increased flooding), reduced food security, spread of disease, increased risk of accidents and damage to infrastructure. The poor are most vulnerable to these changes and have limited capability to respond to them.

The effects of climate change require a response at global, national and local levels. Most countries already fail to manage their environmental resources in a sustainable way. Climate change makes this an even more urgent priority.

Some facts and figures

- Overfishing has led to the collapse of many fisheries. One-quarter of global marine fish stocks are currently overexploited or significantly depleted.
- About 60% of the ecosystem services resources evaluated by the UN's Millennium Ecosystem Assessment (a measure of how ecosystems benefit people) are being degraded or used unsustainably.
- Between 10% and 30% of mammal, bird and amphibian species face extinction.
- Global timber production has increased by 60% in the past four decades. This means that roughly 40% of forest area has been lost, and deforestation continues at a rate of 13 million ha per annum.

CHALLENGES AND SOLUTIONS

Environmental concerns are fundamental to long-term sustainable development. Efforts must be made to improve understanding of the environmental impact of development strategies and to recognize the link between environmental degradation and poverty.

The poor, who are most dependent on natural resources and most affected by environmental degradation, lack the information or the access to participate in decision-making and policy development. In contrast, those who are most influential in policy development have little understanding of the costs and benefits associated with environmental policy. Economic growth and the environment are often still viewed as competing objectives. But investing in environmental management can be cost-effective, and it contributes to improving livelihoods.

MANAGING THE KORUP NATIONAL PARK

The Korup National Park was created in 1986 by the government of Cameroon with the support of the WWF. Under Cameroon law, human activity in the park is limited to tourism, research and recreation. The project is designed to "protect and manage the National Park and integrate it into the local economy and regional development plans".

One example of sustainable development in Korup is that of community forests. These are large areas of forest in which villagers obtain and manage part of the communal forest in a sustainable way. The project is reviewed regularly by the government and the WWF.

Management of Korup is very important – it contains over 400 species of trees, 425 species of birds, 120 species of fish and 100 mammal species. Over 60 species occur only in Korup, and 170 species are considered to be endangered or vulnerable.

www.mount-cameroon.org/korup/population_culture.html

Malthus, Boserup and the limits to growth

MALTHUS

In 1798 the Reverend Thomas Malthus produced his *Essay on the Principle of Population*. He believed that there was a finite optimum population size in relation to food supply, and that any increase in population beyond this point would lead to a decline in the standard of living and to "war, famine and disease". His theory was based on two principles:

1. In the absence of checks, population would grow at a geometric or exponential rate (1, 2, 4, 8, 16... etc.) and could double every 25 years.
2. Food supply at best only increases at an arithmetic rate (1, 2, 3, 4, 5... etc.).

Malthus suggested preventive and positive checks as two main ways by which population could be curbed once this ceiling had been reached. Preventive checks included abstinence from marriage, a delay in the time of marriage and abstinence from sex within marriage. Positive checks, such as lack of food, disease and war, directly affected mortality rates.

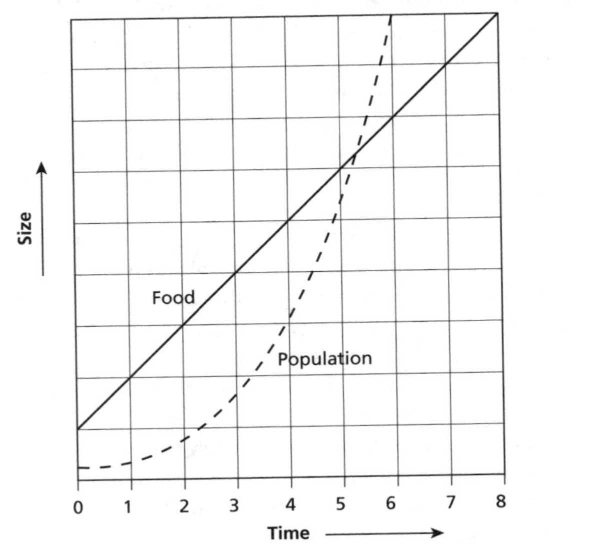

Relationship between population and food supply, after Malthus

INCREASING THE CARRYING CAPACITY: BOSERUP

A different view to that of Malthus is that of Esther Boserup (1910–99). She believed that people have the resources of knowledge and technology to increase food production and that when a need arises someone will find a solution.

Boserup suggested that in a pre-industrial society, an increase in population stimulated a change in agricultural techniques so that more food could be produced. Population growth thus enabled agricultural development to occur.

Boserup assumed that people knew of the technologies required by more intensive systems and used them when the population grew. If knowledge were not available, then the agricultural system would regulate the population size in a given area.

Increased food production

There have been many ways since Malthus's time in which people have increased food production. These include:

- draining marshlands
- extensification
- intensification
- reclaiming land from the sea
- cross-breeding of cattle
- high-yield varieties of plants
- terracing on steep slopes
- growing crops in greenhouses
- using more sophisticated irrigation techniques
- making new foods such as soy
- using artificial fertilizers and pesticides
- farming native species of crops and animals
- fish farming.

THE LIMITS TO GROWTH MODEL

This study examined the five basic factors that determine and therefore ultimately limit growth on the planet: population; agricultural production; natural resources; industrial production; and pollution.

Many of these factors were observed to grow at an exponential rate. Food production and population grew exponentially until the rapidly diminishing resource base forces a slowdown in industrial growth. Because of natural delays in the system, both population and pollution continue to increase for some time after the peak of industrialization. Population growth is finally halted by a rise in the death rate due to decreased food, water and medical services.

The team concluded that if the trends continued, the limits to growth would be reached by about 2070.

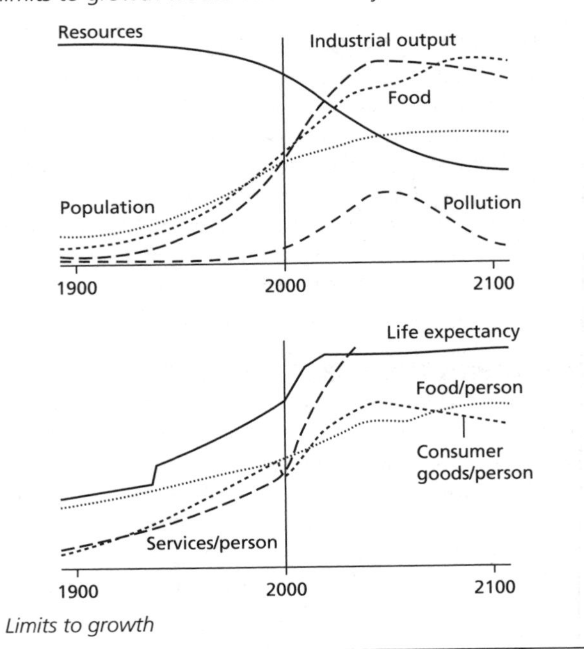

Limits to growth

Changing patterns of oil production and consumption

PRODUCTION

In 2003 global oil production was at 70 million barrels per day. Eight producers, Saudi Arabia, the USA, Russia, Iran, China, Venezuela, Mexico and Norway, accounted for over 50% of total production. Oil production is marginal or non-existent in many countries, notably Africa.

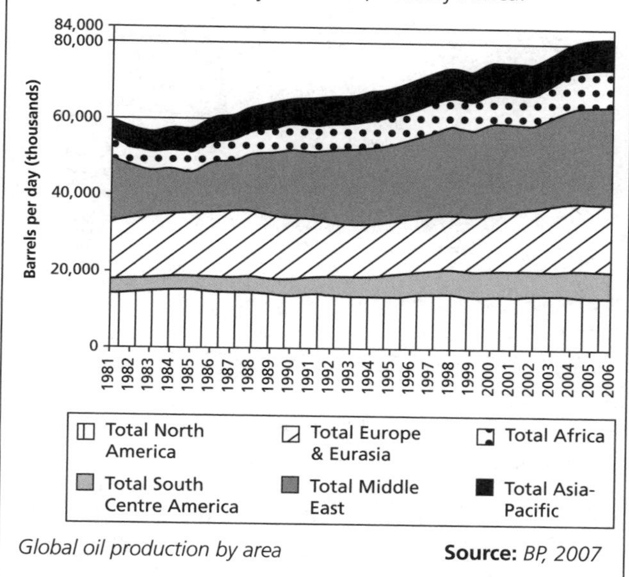

Global oil production by area **Source:** *BP, 2007*

CONSUMPTION

Seven countries, the USA, Japan, China, Germany, Russia, Italy and France, accounted for over 50% of global demand. Oil demand is roughly a function of population and level of development.

Oil consumption has nearly tripled since 1965. In 2006, demand was almost 84 million barrels per day. A significant share of the new oil demand is assumed by Pacific Asian nations going through rapid industrialization, particularly China, which has become the world's second largest importer after the USA.

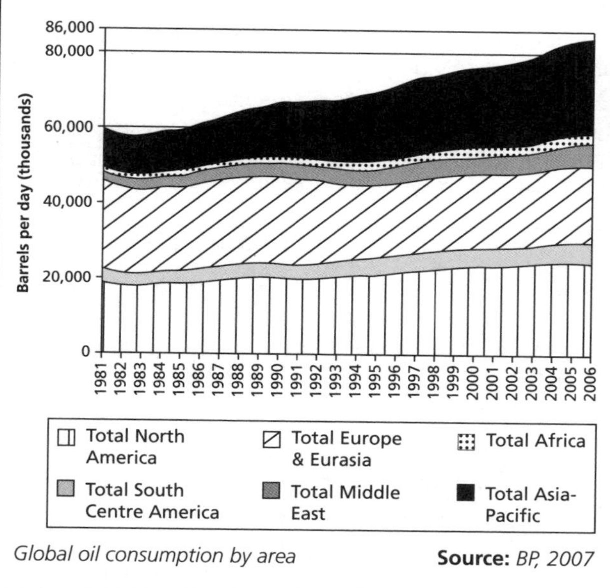

Global oil consumption by area **Source:** *BP, 2007*

OIL REFINING

Over 80% of oil refining now takes place in Europe, North America and Japan. However, the separation between production and refining causes problems. For example, oil was considered a cheap fuel and many countries became dependent on it. However, as a result of the oil price rise in 1973 many countries had to reassess their energy policy.

Oil reserves

Oil reserves are generally found in geological structures such as anticlines, fault traps and salt domes. At present rates of production and consumption, reserves could last for another 40 years. Nearly two-thirds of the world's reserves are found in the Middle East, followed by Latin America (12.5%) and then equally by the developed world, centrally planned economies (CPEs) and developing countries.

THE GEOGRAPHIC IMPLICATIONS OF MIDDLE EAST OIL

The importance of the Middle East as a supplier of oil is critical. Involvement in the Gulf War (1991) is a case in point. The Organization of Petroleum Exporting Countries (OPEC) controls the price of crude oil, and this has increased its economic and political power. It has also increased dependency on the Middle East by all other regions. This provides an incentive for rich countries to increase energy conservation or develop alternative forms of energy.

Countries therefore need to:

- maintain good political links with the Middle East and strive for political stability in the region
- involve the Middle East in economic cooperation
- reassess coal and nuclear power as energy options.

ENVIRONMENTAL IMPLICATIONS

The importance of oil as the world's leading fuel has had many negative effects on the natural environment. For example:

- oil slicks from tankers such as the *Torrey Canyon* (1967), *Exxon Valdez* (1989) and the *Braer* (1993)
- damage to coastlines, fish stocks and communities dependent on the sea
- water pollution caused by tankers illegally washing/cleaning out tankers in the North Sea
- Gulf War damage – storage of oil and oil wells can be targets for destruction causing immeasurable environmental damage.

Environmental disasters continue to plague the oil industry. In 1996, the *Sea Empress* ran aground off Milford Haven, Britain's largest oil terminal. Between 50,000 and 70,000 tonnes of oil escaped from the tanker. Oil slicks are a hazard to local wildlife. In places where the oil is spread thinly over the surface it kills off plankton and enters into the food chain. In addition, the disposal of "retired" platforms such as the Brent Spar is a major problem.

The changing importance of alternative energy sources

RENEWABLE RESOURCES

Types of renewable energy include hydroelectric power (HEP), solar, wind and tidal.

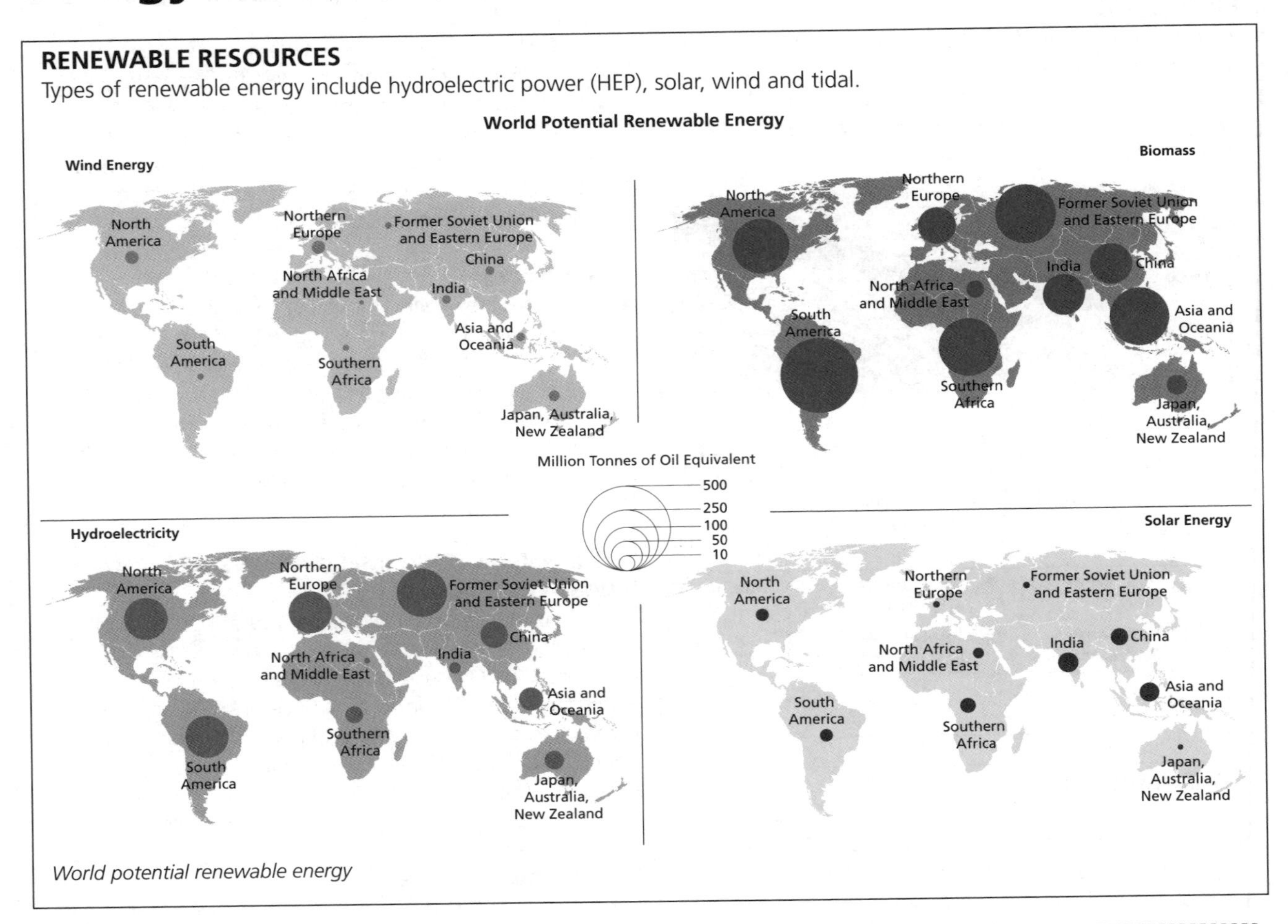

World potential renewable energy

EXTENSION

Visit
http://earthtrends.wri.org/images/renewable_energy_potential.jpg and
http://www.cleanedge.com/images/CleanEnergyProjected07.gif

TRENDS IN RENEWABLE ENERGY SOURCES

Globally, renewable energy is growing fast. The rates of development of renewable energy sources are far exceeding those of fossil fuels such as oil, coal and natural gas. In 2006, wind and solar development grew by 20% and 40% respectively. Renewable energy will become increasingly important as the world attempts to reduce greenhouse gas emissions to levels that are necessary to curb global warming.

The sixth annual *Clean Energy Trends Report* found that the market for renewable energy sources was about $55 billion worldwide in 2006, and forecast growth to $226 billion by 2016.

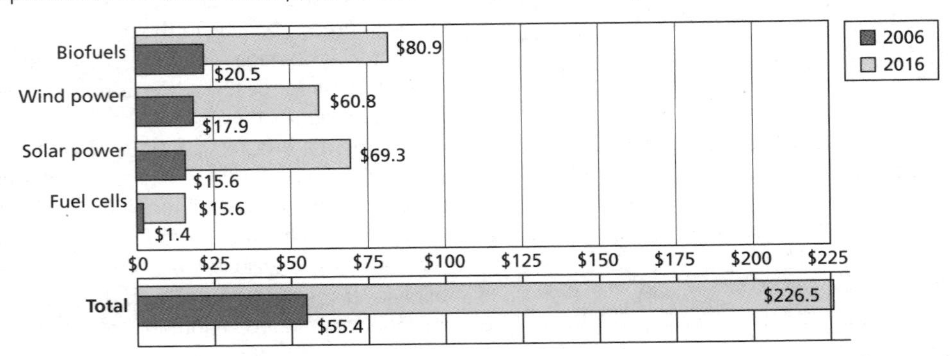

Clean energy projected growth, 2006–16 (US$ billions)

Source: *Clean Edge, 2007*

Alternative energy

SOLAR POWER

Energy from the sun is clean, renewable and so abundant that the amount of energy received by the earth in 30 minutes is the equivalent to all the power used by humans in one year. In the UK, solar energy falling on buildings could meet two-thirds of electricity needs.

The advantages

- No finite resources involved – less environmental damage
- No atmospheric pollution
- Suitable for small-scale production

The disadvantages

- Affected by cloud, seasons, night-time
- Not always possible when demand exists
- High costs

The high costs of solar power make it difficult for the industry to achieve its full potential. Each unit of electricity generated by solar energy costs 4–10 times as much as that derived from fossil fuels. At present it does not make a significant contribution to energy efficiency.

Although solar energy is increasing at a rate of 15–20% per year, it is from a tiny starting base and the annual production of photovoltaic (PV) cells is enough only to power one small city.

WIND POWER

Wind power is good for small-scale production. It needs an exposed site, such as a hillside, flat land or proximity to the coast. It also requires strong, reliable winds. Such conditions are found at Altamont Pass, California, for example.

The advantages

- No pollution of air, ground or water
- No finite resources involved
- Reduction in environmental damage elsewhere
- Suitable for small-scale production

The disadvantages

- Visual impact
- Noisy
- Winds may be unreliable

Large-scale development is hampered by the high cost of development, the large number of wind pumps needed, and the high cost of new transmission grids. Suitable locations for wind farms are normally quite distant from centres of demand.

TIDAL POWER

Tidal power is a renewable, clean energy source. It requires a funnel-shaped estuary, free of other developments, with a large tidal range. The River Rance in Brittany has the necessary physical conditions.

Large-scale production of tidal energy is limited for a number of reasons:

- high cost of development
- limited number of suitable sites
- environmental damage to estuarine sites
- long period of development
- possible effects on ports and industries upstream.

NUCLEAR POWER

Although nuclear power is not a renewable form of energy, it is often grouped with renewables since the amount of raw material (plutonium) needed to produce a large amount of energy is very small.

Advantages	Disadvantages
It is a cheap, reliable and abundant source of electricity	Uranium is a radioactive material and so the nuclear power industry is faced with the hazards of waste disposal and the problems of decommissioning old plants and reactors
Unlike coal and oil, which have reserves estimated to last 300 years and 50 years approximately, there is a plentiful supply of uranium – enough for it to be considered a renewable form of energy	Rising environmental fears concerning the safety of nuclear power and nuclear testing are based on experience: disasters such as Chernobyl, 1986
Uranium fuel is available from countries such as the USA, Canada, South Africa and Australia, so western Europe would not have to rely on potentially unstable regions such as the Middle East for its energy needs	Recession in the 1990s and 2000s has reduced the demand for energy – less energy development is now required
The EU is in favour of nuclear power and estimates that 40% of the EU's electricity will be provided by nuclear power (15% of total energy)	The EU, for example, has a diverse range of energy suppliers – the threat of disruption to any one source is therefore less worrying than it used to be

Hydroelectric power

HYDROELECTRIC POWER

HEP is a renewable form of energy that harnesses fast-flowing water with a sufficient head.

The **location** of HEP stations depends on:

- relief – namely a valley that can be dammed
- geology – a stable, impermeable bedrock
- river regime – a reliable supply of water
- climate – a reliable supply of water
- market demand – to be profitable
- transport facilities – to transport the energy.

The **site** depends on:

- local valley shape (narrow and deep)
- local geology (strong, impermeable rocks)
- lake potential (a large head of water)
- local land use (non-residential)
- local planning (lack of restrictions).

However, there are difficulties with HEP:

- HEP plants are very costly to build.
- Only a few places have a sufficient head of water.
- Markets are critical. This is because plants need to run at full capacity to be economical. In some cases a market is created: for example, aluminium smelters are often located close to HEP plants in order to use up the excess energy.

THE IMPACTS OF THE THREE GORGES DAM

The decision to build the Three Gorges Dam on the Yangtze in China highlighted some of the conflicts apparent in the way people use the river. The dam was completed in 2009.

The facts

- The Three Gorges Dam is over 2 km long and 100 m high.
- The lake is over 600 km long.
- Over 1 million people were moved to make way for the dam and the lake.
- The Yangtze provides 66% of China's rice and contains 400 million people.
- The Yangtze drains 1.8 million km^2 and discharges 700 km^3 of water annually.

The advantages

- The Three Gorges Dam will generate up to 18,000 megawatts, eight times more than Egypt's Aswan Dam and 50% more than the world's largest existing HEP dam, the Itaipu in Paraguay.
- It will enable China to reduce its dependency on coal.
- It will supply energy to Shanghai (population 13 million, one of the world's largest cities) and Chongqing (population 3 million, an area earmarked for economic development).
- It will protect 10 million people from flooding. (Over 300,000 people in China died as a result of flooding in the 20th century.)
- It will allow shipping above the Three Gorges: the dams have raised water levels by 90 m, and turned the rapids in the gorge into a lake.
- It has generated thousands of jobs.

Protests against the building of the dam

- Most floods in recent years have come from rivers which join the Yangtze below the Three Gorges Dam.
- The region is seismically active and landslides are frequent.
- The port at the head of the lake may become silted up as a result of increased deposition and the development of a delta at the head of the lake.
- Much of the land available for resettlement is over 800 m above sea level, and is colder, with infertile thin soils on relatively steep slopes.
- Dozens of towns, for example Wanxian and Fuling with 140,000 and 80,000 people respectively, had to be flooded.
- Up to 530 million tonnes of silt are carried through the Gorge annually: the first dam on the river lost its capacity within seven years and one on the Yellow River filled with silt within four years.
- To reduce the silt load, afforestation is needed, but the resettlement of people will cause greater pressure on the slopes above the dam.
- The dam interferes with aquatic life – the Siberian crane and the white flag dolphin are threatened with extinction.
- Archaeological treasures were drowned, including the Zhang Fei temple.
- It has cost as much as $70 billion.

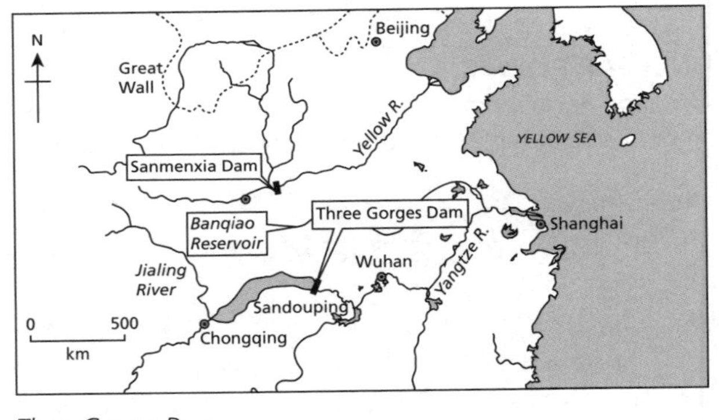

Three Gorges Dam

Conservation, waste reduction, recycling and substitution

DEFINITIONS

Recycling refers to the processing of industrial and household waste (such as paper, glass and some metals and plastics) so that materials can be reused. This saves scarce raw materials and helps reduce pollution. The UK, for example, has long lagged behind other EU countries with recycling mainly because there are many more landfill sites which are cheaper to use. The UK has a recycling target of 33% by 2015.

Reuse refers to the multiple use of a product by returning it to the manufacturer or processor each time. Reuse is usually more energy- and resource-efficient than recycling.

Reduction (or "reduce") refers to using less energy, for example turning off lights when not needed, or using only the amount of water needed when boiling a kettle.

Substitution refers to using one resource rather than another – the use of renewable resources rather than non-renewable resources would be a major benefit to the environment.

Landfill is the burying of waste in the ground, and then covering over the filled pit with soil and other material. Landfill may be cheap but it is not always healthy – and sites will eventually run out. Most landfill is domestic waste, but a small amount of hazardous waste is allowed on general sites.

Fly-tipping is when people or companies dump waste and old equipment. It is an increasing problem. There are many reasons for the increase, including:

- increased costs of landfill
- more goods such as TVs, computers and refrigerators classified as "hazardous" and subject to restrictions on how they are disposed of
- the introduction of strict new EU regulations mean that a high proportion of new products must be recycled – this can be costly to manufacturers and purchasers.

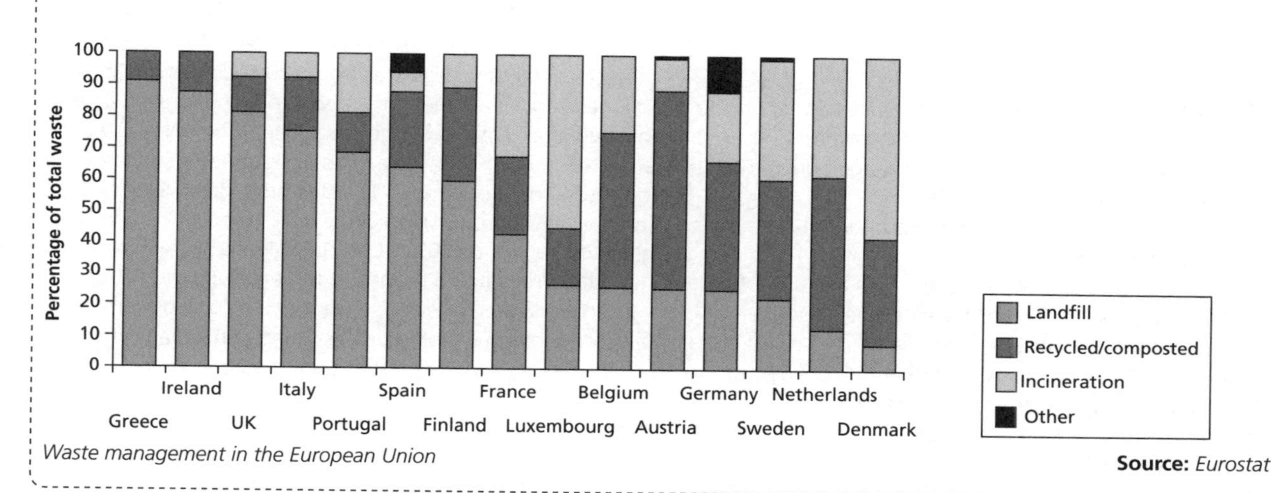

Waste management in the European Union

Source: *Eurostat*

WASTE IMPORTS IN CHINA

A fairly new environmental problem is the dumping of old computer equipment. To make a new PC requires at least 10 times its weight in fossil fuels and chemicals. This can be as high as 240 kg of fossil fuels, 22 kg of chemicals and 1500 kg of clean water. Old PCs are often shipped to LEDCs for recycling of small quantities of copper, gold and silver. PCs are placed in baths of acid to strip metals from the circuit boards, a process highly damaging to the environment and to the workers who carry it out.

China imports more than 3 million tonnes of waste plastic and 15 million tonnes of paper and cardboard each year. Containers arrive in the UK and other countries with goods exported from China, and load up with waste products for the journey back. A third of the UK's waste plastic and paper (200,000 tonnes of plastic rubbish and 500,000 tonnes of paper) is exported to China each year. Low wages and a large workforce mean that this waste can be sorted much more cheaply in China, despite the distance it has to be transported.

China is increasingly aware that this is not "responsible recycling" and that countries are exporting their pollution to them. They have begun to impose stricter laws on what types of waste can be imported.

EXTENSION

Visit

www.unescap.org for resource consumption and management in Asia and the Pacific region.

National and global initiatives

INTERNATIONAL POLICY TO PROTECT CLIMATE

The 1988 Toronto conference on climate change called for the reduction of CO_2 emissions by 20% of the 1988 levels by 2005. Also in 1988 the Intergovernmental Panel on Climate Change (IPCC) was established by the United Nations Environment Programme (UNEP) and the World Meteorological Organization.

The UN Conference on the Environment and Development (UNCED) was held in 1992 in Rio de Janeiro. It covered a range of subjects and there were a number of statements, including the Framework Convention on Climate Change (FCCC). This came into force in March 1994.

"The ultimate objective [of the convention]...is to achieve...stabilization of greenhouse gas concentrations in the atmosphere at a level that would prevent dangerous anthropogenic interference with the climate system."

The Kyoto Protocol

The Kyoto Protocol (1997) was an addition to the Rio Convention. It gave all MEDCs legally binding targets for cuts in emissions from the 1990 level by 2008–12. The EU agreed to cut emissions by 8%, Japan by 7% and the USA by 6%. Some countries found it easier to make cuts than others.

There are three main ways for countries to keep to the Kyoto target without cutting domestic emissions:

- Plant forests to absorb carbon or change agricultural practices (e.g. keep fewer cattle).
- Install clean technology in other countries and claim carbon credits for themselves.
- Buy carbon credits from countries such as Russia where traditional heavy industries have declined and the national carbon limits are underused.

Even if greenhouse gas production is cut by between 60% and 80% there is still enough greenhouse gas in the atmosphere to raise temperatures by 5 °C. The Kyoto agreement was only meant to be the beginning of a long-term process, not the end of one. It excludes, for example, carbon emissions from international flights and shipping, because they are classed as orphan emissions, not owned by any country.

Furthermore, the guidelines for measuring and cutting greenhouses gases were not finished in Kyoto. For example, it was not decided to what extent the planting of forests and carbon trading could be relied upon. George W Bush, then President of the USA, rejected the Kyoto Protocol since it would hurt the US economy and employment.

Although the rest of the world could proceed without the USA, that country emits about 25% of the world's GHGs. So without the USA, and LEDCs such as China and India, the reduction of carbon emissions would be seriously hampered. According to the Kyoto rules, 55 countries must ratify the agreement to make it legally binding worldwide, and 55% of the emissions being reduced must come from MEDCs. If the EU, eastern Europe, Japan and Russia agree, they could just make up 55% of the MEDCs' emissions. Without the USA (and Australia and Canada), it would be difficult to achieve this goal.

However, in November 2007 Australia joined the other MEDCs committed to tackling climate change by signing the Kyoto agreement to limit CO_2 emissions, at once distancing itself from the USA and ending a 10-year diplomatic exile on the issue. The decision took place on the first day of the UN conference in Bali (see below). The USA, which is responsible for 25% of the world's climate change emissions, was still backing voluntary targets to fight climate change.

Bali, 2007

The existing global treaty on greenhouse gases, agreed in Kyoto, expires in 2012. Thus, in November 2007, under the auspices of the UN, delegates from 180 countries met in Bali, Indonesia to set an agenda and start negotiations on a new international climate change agreement. The UN wanted an agreement to limit the earth's average temperature increase to no more than 2 °C above pre-industrial levels. This foresees emissions peaking in the next 10–15 years and then being cut rapidly by 50% of the 1990 levels by 2050. The negotiations included proposals for legally binding cuts in carbon emissions for rich countries and a contribution from large developing nations such as China and India.

The Bali agreement started two years of intense negotiations over how to prevent a possible 4 °C rise in global temperatures this century, which would threaten the food and water supplies of billions of people and drive thousands of species to extinction. It will commit countries to agree a new deal by 2009, which would come into force in 2013.

The Europeans wanted it to state clearly that rich countries needed to slash carbon emissions by 25–40% of 1990 levels by 2020. In the end they may have to settle for a 50% cut globally on 2000 levels by 2050. Developing countries such as China and India will not be set binding targets, but will probably be asked to adopt voluntary goals on energy conservation, and possibly on pollution from certain industries.

EXTENSION

Visit
www.panda.org/about_wwf/what_we_do/climate_change/index.cfm to see the WWF's site on climate change. Find out about the causes and potential solutions to climate change and see what you can do too.

EXAM QUESTIONS ON PAPER 1 – THE CORE

Key features

Timing: You have 1 hour 30 minutes

Choice: None in section A. In Section B answer one out of three questions.

Structure

Paper 1 – The Core consists of Sections A and B.

Section A has four questions, two of which are based on stimulus material. Each question relates to one of the four core topics. Lower level command terms such as *describe* and *explain* will be used in this section.

Section B requires an extended response to one out of three essay-style questions, with the emphasis on synthesis and evaluation. The questions may relate to one or more topics.

The questions in Section A below are organized under topic headings instead of in the exam format. A mock exam paper can be compiled by combining questions into four groups giving a total of 45 marks for Section A. In Section B, each of the three questions should have a different focus to avoid overlap with Section A. Each question is worth 15 marks.

The total mark for Paper 1 (sections A and B) is 60.

Section A

1 Populations in transition

The population pyramids show actual and predicted structural change for one Middle Eastern country between 2000 and 2025.

a) Describe three changes in this country's population between 2000 and 2025. [4]

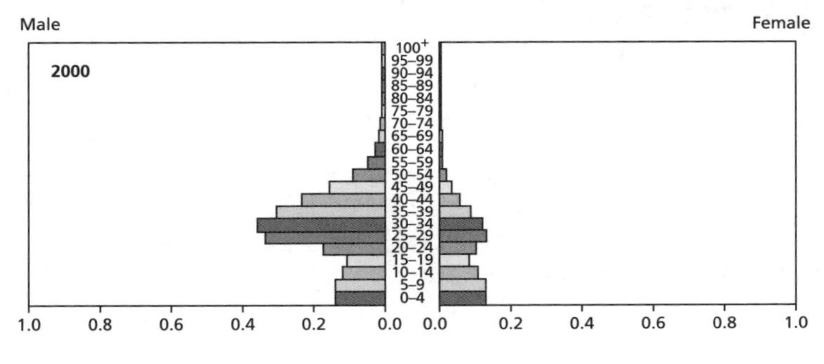

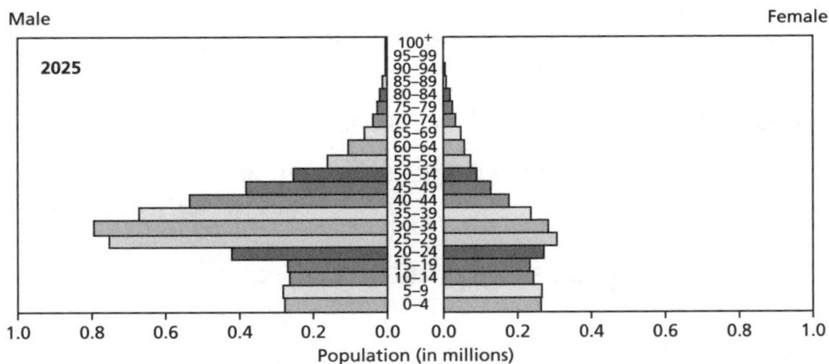

Source: *US Census Bureau, International Data Base*

b) Define population momentum. [3]

c) Explain the advantages and disadvantages for a country in having an ageing population. [6]

d) Draw a fully labelled diagram to show the process of natural increase in the population of a country over time. [5]

2 Disparities in wealth and development

a) Describe and briefly explain the relationship between the global pattern of wealth and trade. [5]

b) Explain the limitations of crude death rate as an international indicator of the standard of living. [4]

c) Describe two advantages and two disadvantages of international aid. [4]

d) Explain how gender inequality may limit a country's economic development. [6]

3 Patterns in environmental quality and sustainability

a) Describe two pollutants which are causes of unsafe drinking water. [2+2]

b) Explain two ways in which people can prevent the occurrence of soil degradation. [3+3]

c) Explain one management strategy designed to achieve environmental sustainability on the local scale. [5]

d) Draw a labelled diagram to show the radiation inputs and outputs in the atmosphere which result in global warming (the enhanced greenhouse effect). [6]

4 Patterns in resource consumption

a) Describe two ways in which the views of the neo-Malthusians have been realized. [5]

b) Select two global regions shown in the graph and explain the differences in their ecological footprints. [5]

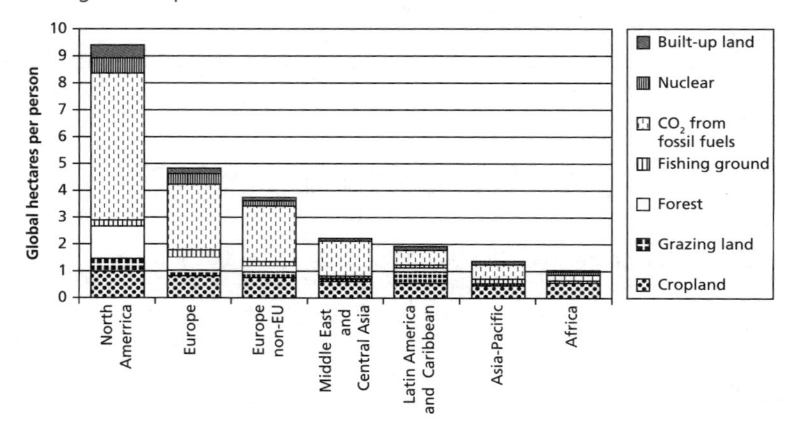

Ecological footprints of major global regions

c) Referring to examples, explain what is meant by overpopulation. [5]

d) Describe two advantages and two disadvantages of nuclear power. [4]

Section B

1 Referring to examples, discuss the extent to which migration is a response to the uneven distribution of resources. [15]

2 Discuss the relative importance of socio-economic factors as a cause of migration. [15]

3 Discuss the idea that sustainable development can be achieved through population control. [15]

4 Discuss the relationship between fertility and poverty. [15]

5 Describe the changing global pattern of economic development and evaluate the methods used to measure it. [15]

6 Describe the aims of Millennium Development Goals and assess the progress that has been made towards achieving them. [15]

7 Explain the likely effects of global warming and the international differences in response to it. [15]

8 "The consequences of global climate change are determined by poverty." Discuss this statement with reference to examples. [15]

9 Explain the importance of maintaining the biodiversity of tropical rainforests. [15]

10 Referring to examples, analyse the causes of water scarcity. [15]

11 Examine the reasons for the increasing use of renewable energy resources. [15]

12 Examine the methods adopted to reduce the consumption of one or more named resources. [15]

Drainage basin hydrology

BASIN HYDROLOGICAL CYCLE

In studying rivers, use is made of the basin hydrological cycle, in which the drainage basin, rather than the global system, is taken as the unit of study. The basin cycle is an open system: the main input is precipitation, which is regulated by various means of storage. The outputs include channel runoff, evapotranspiration and groundwater flow.

Water balance

The **water balance** shows the relationship between the inputs and outputs of a drainage basin. It is normally expressed as:

precipitation = Q (runoff/discharge) + E (evapotranspiration) +/− changes in storage (such as on the surface, in the soil and in the groundwater)

Throughput consists of the transfer of water through the system, from one storage to another, by means of the processes shown on the diagram as labelled arrows.

Interception

Interception refers to the capture of raindrops by plant cover that prevents direct contact with the soil. If rain is prolonged, the retaining capacity of leaves will be exceeded and water will drop to the ground (throughfall). Some will trickle along branches and down the stems or trunk (stemflow). Some is retained on the leaves and later evaporated.

Precipitation

Precipitation is the transfer of moisture to the earth's surface from the atmosphere. It includes dew, hail, rain, sleet and snow.

Soil moisture

The **zone of aeration** is a transitional zone in which water is passed upwards or downwards through the soil. Soil moisture varies with porosity (the number of pore spaces), and with permeability (the ability to transmit water).

Drainage basin hydrology

PRECIPITATION
Interception
Channel precipitation
1.VEGETATION
Stemflow & throughfall
2.SURFACE
Overland flow
Floods
Capillary rise
Infiltration
3. SOIL MOISTURE
Interflow
5. CHANNEL STORAGE
Percolation
Capillary rise
Evaporation
4. GROUND-WATER
Base flow
Recharge
Transpiration
EVAPO-TRANSPIRATION
Leakage
Transfer
RUNOFF
Output
Storage
Input

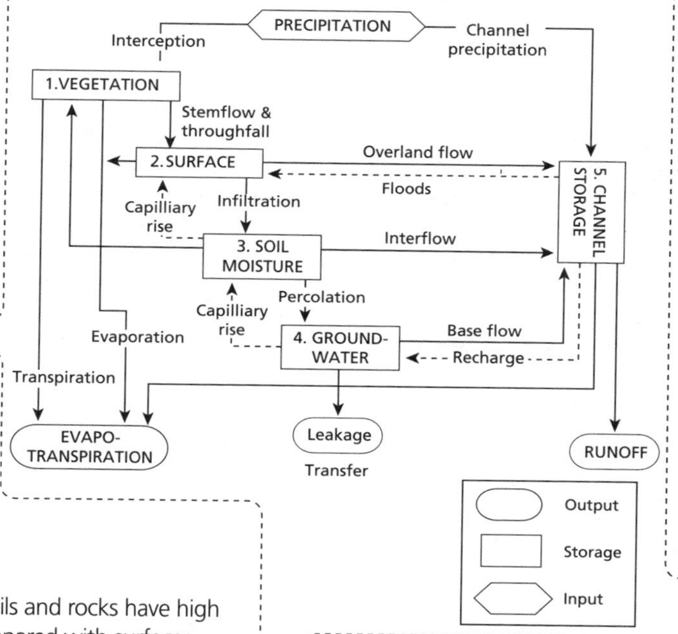

Infiltration

Infiltration is the process by which water sinks into the ground. **Infiltration capacity** refers to the amount of moisture that a soil can hold. **Infiltration rate** refers to the speed with which water can enter the soil. **Percolation** refers to water moving deep into the groundwater zone. **Overland runoff** occurs when precipitation intensity exceeds the infiltration rate, or when the infiltration capacity is reached and the soil is saturated.

Evaporation

Evaporation is the physical process by which a liquid becomes a gas. It is a function of:

- vapour pressure
- air temperature
- wind
- rock surface (e.g. bare soils and rocks have high rates of evaporation compared with surfaces which have a protective tilth where rates are low).

Evapotranspiration is the diffusion of water from vegetation and water surfaces to the atmosphere. **Potential evapotranspiration** is the rate of water loss from an area if there were no shortage of water.

Groundwater

The **groundwater zone** is normally divided into a zone of saturation, in which the underground water fills all the spaces in the rock, and a zone of aeration above it, in which the water does not fully saturate the pores. The **water table** divides one zone from the other.

Aquifers are rocks that hold water. They provide the most important store of water, regulate the hydrological cycle and maintain river flow.

EXTENSION

Visit
www.nwlg.org/pages/resources/geog/hydro_cycle/hydro/cycle.htm for animations on the hydrological cycle and **http://geography.about.com/cs/waterhydrology/** for links to some excellent sites on hydrology and rivers.

Discharge

Discharge refers to the volume of water passing a certain point per unit of time. It is usually expressed in cubic metres per second (cumecs). Normally, discharge increases downstream, as shown by the Bradshaw model.

BRADSHAW MODEL OF CHANNEL VARIABLES

Bradshaw's model shows changes to channel characteristics over the course of a river. Water velocity and discharge increase downstream, while channel bed roughness and load particle size decrease.

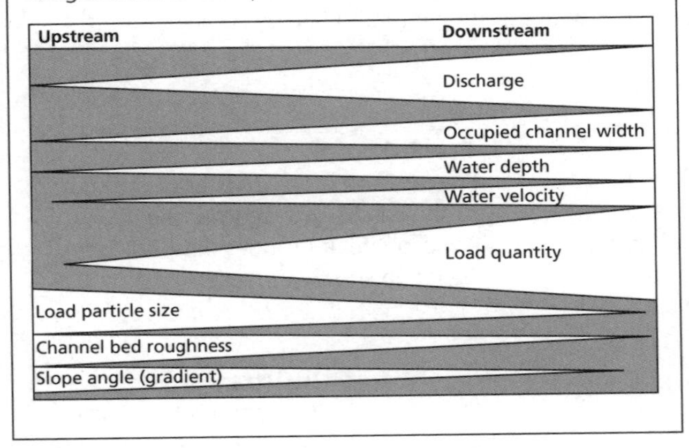

CHANGING CHANNEL CHARACTERISTICS

As a river travels downstream, changes can occur to its width, depth, velocity, discharge and efficiency. Efficiency is measured by the **hydraulic radius**, i.e. cross-sectional area/wetted perimeter (CSA/WP).

River level

③ Flood – high friction

② Bankfull – maximum efficiency (low friction)

① Below bankfull – high friction

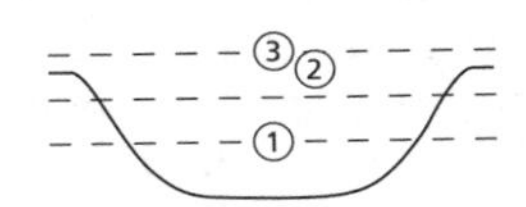

Shape

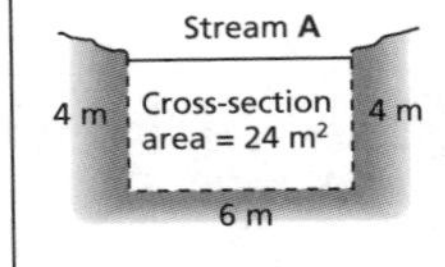

Very efficient (low relative friction)

Stream B

2 m | Cross-section area = 24 m² | 2 m

12 m

Inefficient (high relative friction)

--- wetted perimeter

Wetted perimeters	Hydraulic radius
Stream A: $4 + 4 + 6 = 14$ m	**Stream A:** $\frac{24}{14} = 1.71$ m
Stream B: $2 + 2 + 12 = 16$ m	**Stream B:** $\frac{24}{16} = 1.5$ m

THE LONG PROFILE

A number of processes, such as weathering and mass movement, interact to create variations in cross profiles and long profiles. **Cross profiles** are cross sections across the river valley, cutting the valley at right angles. **Long profiles** show changes in gradient along the river valley from the source to the mouth. Irregularities, or **knick points**, may be due to:

- geological structure/lithology (e.g. hard rocks erode slowly, which can result in the formation of waterfalls and rapids)
- variations in the load (e.g. when a tributary with a coarse load may lead to a steepening of the gradient of the main valley)
- sea level changes – a relative fall in sea level (isostatic recovery, eustatic fall, etc.) will lead to renewed downcutting, which enables the river to erode former floodplains and form new terraces and knick points.
- Rivers tend to achieve a condition of equilibrium, or grade, and erode the irregularities. There is a balance between erosion and deposition in which a river adjusts to its capacity and the amount of work being done. The main adjustments are in channel gradient, leading to a smooth concave profile.

The profile of the River Exe in England is typical of a graded river: concave and gradually decreasing towards the mouth of the river.

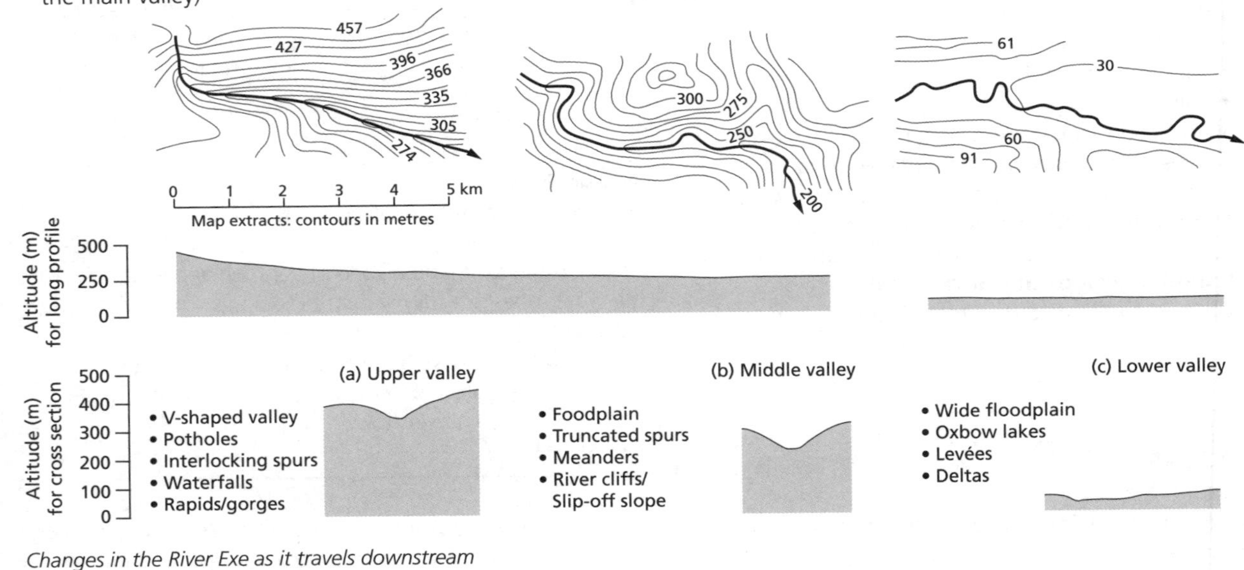

Changes in the River Exe as it travels downstream

Storm hydrographs

A **storm** or **flood hydrograph** shows how a river channel responds to the key processes of the hydrological cycle. It measures the speed at which rainfall falling on a drainage basin reaches the river channel. It is a graph on which river discharge during a storm or runoff event is plotted against time.

Discharge

Discharge (Q) is the volume of flow passing through a cross section of a river during a given period of time (usually measured in cumecs or m^3/sec).

Recessional limb

Recessional limb is influenced by geological composition and the behaviour of local aquifers. Larger catchments have less steep recessional limbs; likewise flatter areas.

Rising limb

Rising limb indicates the amount of discharge and the speed at which it is increasing. It is very steep in a flash flood or in small drainage basins where the response is rapid. It is generally steep in urbanized catchments.

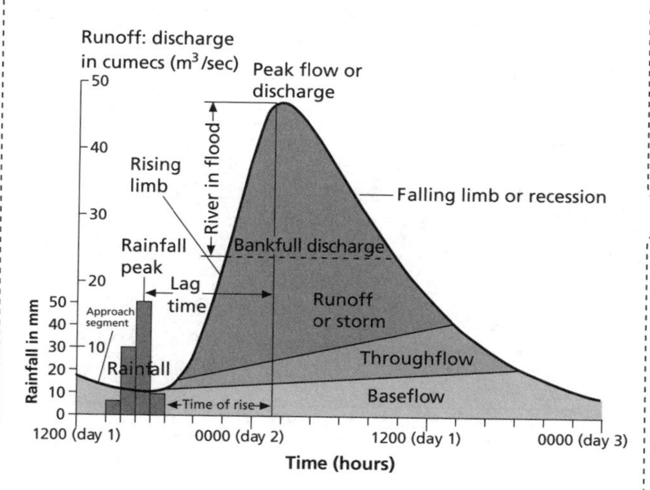

Hydrograph size (area under the graph)

- The higher the rainfall, the greater the discharge.
- The larger the basin size, the greater the discharge.

Peak flow or discharge

Peak flow or discharge is higher in larger basins. Steep catchments will have lower infiltration rates; flat catchments will have high infiltration rates, so more throughflow and lower peaks.

Lag time

Lag time is the time interval between peak rainfall and peak discharge. It is influenced by basin shape, steepness and stream order.

Runoff

The runoff curve reveals the relationship between overland flow and throughflow. Where infiltration is low, antecedent moisture high, surface impermeable and rainfall strong, overland flow will dominate.

Baseflow

Baseflow is the seepage of groundwater into the channel – very important where rocks have high pore space. A slow movement, it is the main, long-term supplier of the river's discharge.

VARIATION IN HYDROGRAPHS

A number of factors affect flood hydrographs:

- Climate (rainfall total, intensity, seasonality)
- Soils (impermeable clay soils create more flooding)
- Vegetation (vegetation intercepts rainfall and so flooding is less likely)
- Infiltration capacity (soils with a low infiltration capacity cause much overland flow)
- Rock type (permeable rocks will allow water to infiltrate, thereby reducing the flood peak)
- Slope angle (on steeper slopes there is greater runoff)
- Drainage density (the more stream channels there are, the more water that gets into rivers)
- Human impact (creating impermeable surfaces and additional drainage channels increases the risk of flooding; dams disrupt the flow of water; afforestation schemes increase interception)
- Basin size, shape and relief (small, steep basins reduce lag time, while basin shape influences where the bulk of the flood waters arrive)

URBAN HYDROLOGY AND THE STORM HYDROGRAPH

Urban hydrographs are different to rural ones. They have:

- a shorter lag time
- a steeper rising limb
- a higher peak flow (discharge)
- a steeper recessional limb.

This is because there are more impermeable surfaces in urban areas (roofs, pavements, roads, buildings), as well as more drainage channels (gutters, drains, sewers).

Flooding in Bangladesh

BANGLADESH AND HER RIVERS

Much of Bangladesh has been formed by deposition from three main rivers – the Brahmaputra, the Ganges and the Meghna. The sediment from these and over 50 other rivers forms one of the largest deltas in the world, and up to 80% of the country is located on the delta. As a result much of the country is just a few metres above sea level and is under threat from flooding and rising sea levels. To make matters worse, Bangladesh is a very densely populated country (over 900 people per km^2) and is experiencing rapid population growth (nearly 2.7% per annum).

Almost all of Bangladesh's rivers have their source outside the country. For example, the drainage basin of the Ganges and Brahmaputra covers 1.75 million km^2 and includes the Himalayas, the Tibetan Plateau and much of northern India. Total rainfall within the Brahmaputra-Ganges-Meghna catchment is very high and very seasonal: 75% of annual rainfall occurs in the monsoon between June and September. Moreover, the Ganges and the Brahmaputra carry snowmelt waters from the Himalayas. Peak discharges of the rivers are immense – up to 100,000 cumecs in the Brahmaputra, for example. In addition to water, the rivers carry vast quantities of sediment. This is deposited annually to form temporary islands and sandbanks.

FLOODING IN BANGLADESH

There are five main types of flooding in Bangladesh – river floods, overland runoff, flash floods, "back-flooding" and storm surges. Flooding in Bangladesh is due to a variety of factors. The combination of:

- discharge peaks of the big rivers
- high runoff from the Meghalaya Hills
- heavy rainfall
- high groundwater tables
- spring tides

creates particularly favourable conditions for large-scale flooding.

In addition, lateral river embankments and the disappearance of natural water storage areas in the lowlands seem to have a significant impact on the flooding processes.

CAUSES AND EFFECTS

Snowmelt in the Himalayas, combined with heavy monsoonal rain, causes peak discharges in all the major rivers during June and July. This leads to flooding and destruction of agricultural land. Outside the monsoon season, heavy rainfall causes extensive flooding (which may be advantageous to agricultural production, since it is a source of new nutrients). In addition, the effects of flash floods, caused by heavy rainfall in northern India, have been intensified by the destruction of forest, which reduces interception, decreases water retention and increases the rate of surface runoff.

Human activity in Bangladesh has increased the problem. Attempts to reduce flooding by building embankments and dikes have prevented the backflow of flood water into the river. This leads to a ponding of water (also known as "drainage congestion") and back-flooding. In this way, embankments have sometimes led to an increase in deposition in drainage channels, and this can cause large-scale deep flooding.

Bangladesh is also subject to coastal flooding. Storm surges caused by intense low-pressure systems are funnelled up the Bay of Bengal.

The effects
In the 1998 floods:

- 4750 people were killed
- 66% of Bangladesh was flooded
- 23 million people were made homeless
- 130,000 cattle were killed
- 660,000 ha of crops were damaged
- 400 factories were closed
- 11,000 km of roads were damaged
- 1000 schools were damaged or destroyed.

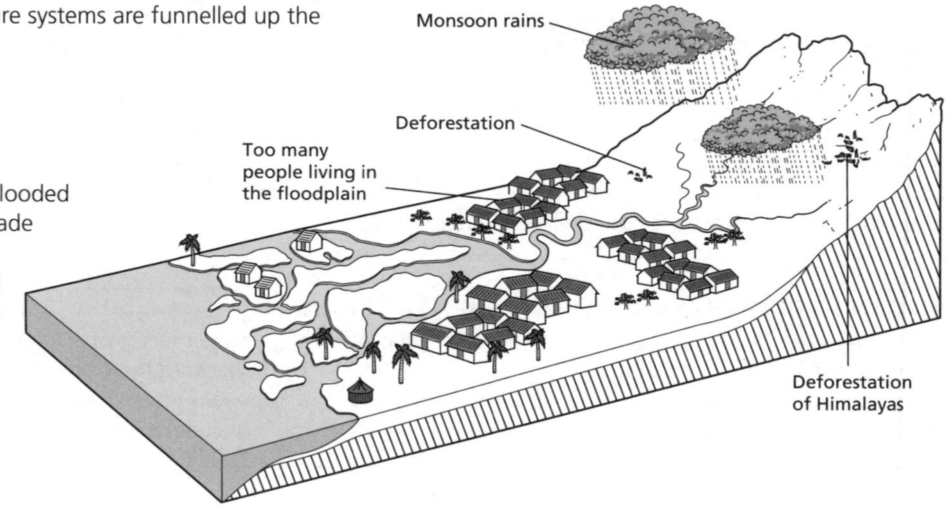

Monsoon rains
Deforestation
Too many people living in the floodplain
Deforestation of Himalayas

THE ADVANTAGES OF FLOODING

During the monsoon, between 30% and 50% of the entire country is flooded. The flood waters:

- replenish groundwater reserves
- provide nutrient-rich sediment for agriculture in the dry season
- provide fish (fish supply 75% of dietary protein and over 10% of annual export earnings)
- reduce the need for artificial fertilizers
- flush pollutants and pathogens away from domestic areas.

The effects of megadams

The number of large dams (more than 15 m high) being built is increasing rapidly and is reaching a level of almost two completions every day. Examples of such megadams include the Akosombo (Ghana), Tucurui (Brazil), Hoover (USA) and Kariba (Zimbabwe).

ADVANTAGES

The advantages of dams are numerous. In the case of the Aswan High Dam on the River Nile, Egypt, they include:

- **flood and drought control:** dams allow good crops in dry years as, for example, in Egypt in 1972 and 1973
- **irrigation:** 60% of water from the Aswan Dam is used for irrigation and up to 4000 km of the desert are irrigated
- **hydroelectric power:** this accounts for 7000 million kW hours each year
- improved **navigation**
- **recreation** and **tourism**.

It is estimated that the value of the Aswan High Dam to the Egyptian economy is about $500 million each year.

COSTS

On the other hand, there are numerous costs. For example, in the case of the Aswan High Dam:

- **water losses:** the dam provides less than half the amount of water expected
- **salinization:** crop yields have been reduced on up to one-third of the area irrigated by water from the dam due to salinization (see page 00)
- **groundwater changes:** seepage leads to increased groundwater levels and may cause secondary salinization
- **displacement of population:** up to 100,000 Nubian people have been removed from their ancestral homes
- **drowning of archaeological sites:** Rameses II and Nefertari at Abu Simbel had to be removed to safer locations; however, the increase in the humidity of the area has led to an increase in the weathering of ancient monuments
- **seismic stress:** the earthquake of November 1981 is believed to have been caused by the Aswan Dam; as water levels in the dam decrease, so too does seismic activity
- **deposition within the lake:** infilling is taking place at about 100 million tonnes each year
- **channel erosion** (clear water erosion) beneath the channel: lowering the channel by 25 mm over 18 years, a modest amount
- **erosion of the Nile Delta:** this is taking place at a rate of about 2.5 cm each year
- **loss of nutrients:** it is estimated that it costs $100 million to buy commercial fertilizers to make up for the lack of nutrients each year
- **decreased fish catches:** sardine yields are down 95% and 3000 jobs in Egyptian fisheries have been lost
- **spread of diseases** such as schistosomiasis (bilharzia) due to increased stagnant water.

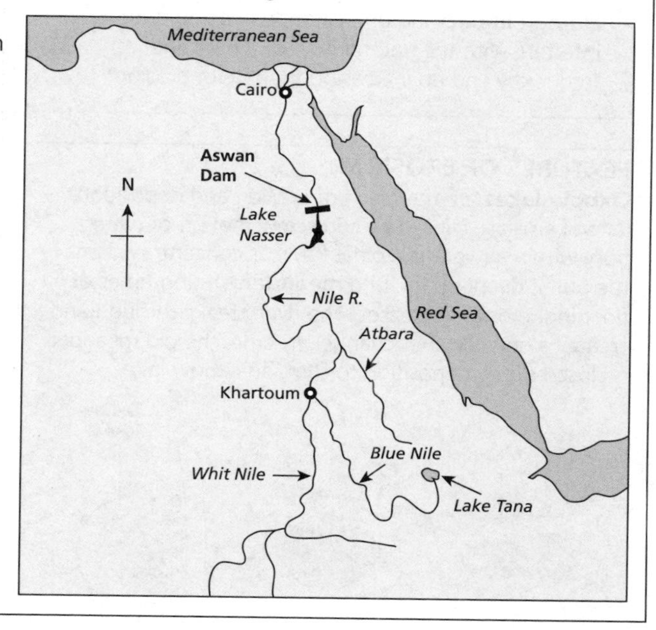

SUSTAINABLE USE OF WATER

All water is a resource common to all, the use of which should be subject to national control. There shall be no ownership of water but only a right to its use. The objective of managing the nation's water resources is to achieve optimum long term social and economic benefit for our society from their use, recognizing that water allocations may have to change over time. The water required to meet peoples' basic domestic needs should be reserved.

The development, apportionment and management of water resources should be carried out using the criteria of public interest, sustainability, equity and efficiency of use in a manner which reflects the value of water to society whilst ensuring that basic domestic needs, the requirements of the environment and international obligations are met. Responsibility should, where possible, be *delegated* to a catchment or regional level in such a manner as to enable interested parties to participate and reach consensus. The right of all citizens to have access to basic water services (the provision of potable water supply and the removal and disposal of human excreta and waste water) necessary to afford them a healthy environment on an equitable, economically and environmentally sustainable basis should be supported.

Source: *Department of Water Affairs and Forestry, South Africa, Water law principles*

THE MAIN TYPES OF EROSION

- **Abrasion** (or **corrasion**) is the wearing away of the bed and bank by the load carried by a river.
- **Attrition** is the wearing away of the load carried by a river. It creates smaller, rounder particles.
- **Hydraulic action** is the force of air and water on the sides of rivers and in cracks.
- **Solution** (or **corrosion**) is the removal of chemical ions, especially calcium, which causes rocks to dissolve.

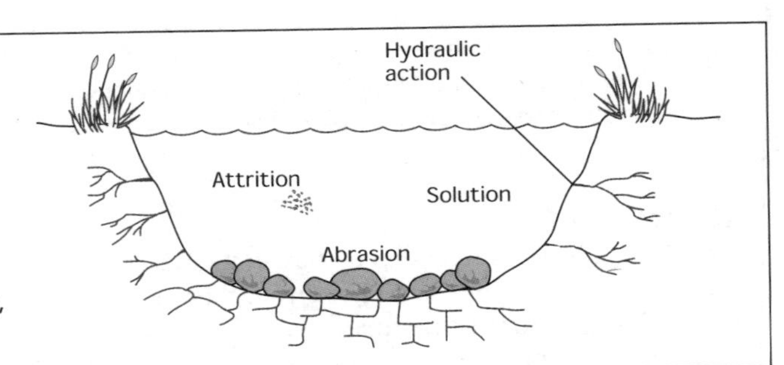

FACTORS AFFECTING EROSION

- **Load:** the heavier and sharper the load, the greater the potential for erosion.
- **Velocity and discharge:** the greater the velocity and discharge, the greater the potential for erosion.
- **Gradient:** increased gradient increases the rate of erosion.
- **Geology:** soft, unconsolidated rocks, such as sand and gravel, are easily eroded.
- **pH:** rates of solution are increased when the water is more acidic.
- **Human impact:** deforestation, dams and bridges interfere with the natural flow of a river and frequently end up increasing the rate of erosion.

THEORY OF RIVER CHANNEL LOAD

The **capacity** of a stream refers to the largest amount of debris that a stream can carry; its **competence** refers to the diameter of the largest particle that can be carried. The **critical erosion velocity** is the lowest velocity at which grains of a given size can be moved. The relationship between these variables is shown by means of a **Hjulström curve**.

There are three important features on Hjulström curves:

- The smallest and largest particles require high velocities to lift them.
- Higher velocities are required for entrainment than for transport.
- When velocity falls below a certain level (**settling** or **fall velocity**), particles are deposited.

FEATURES OF EROSION

Oxbow lakes are the result of erosion and deposition. Lateral erosion, caused by corkscrew motion of water flow within a river (helicoidal flow), is concentrated on the outer, deeper bank of a **meander**. During times of flooding, erosion increases. The river breaks through and creates a new, steeper channel. In time, the old meander is closed off by deposition to form an oxbow lake.

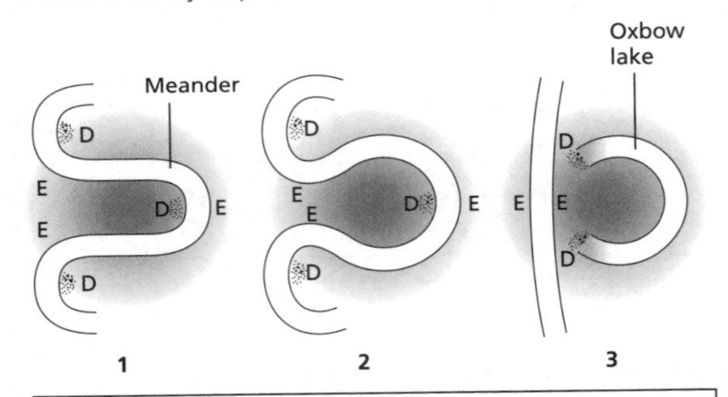

1 Eroson (E) and deposition (D) around a meander (a bend in a river).
2 Increased erosion during flood conditions. The meander becomes exaggerated.
3 The river breaks through during a flood. Further deposition causes the old meander to become an oxbow lake.

THE MAIN TYPES OF TRANSPORTATION

- **Suspension:** small particles are held up by turbulent flow in the river.
- **Saltation:** heavier particles are bounced or bumped along the bed of the river.
- **Solution:** the chemical load is carried dissolved in the water.
- **Traction:** the heaviest material is dragged or rolled along the bed of the river.
- **Flotation:** leaves and twigs are carried on the surface of the river.

WATERFALLS

Waterfalls frequently occur on horizontally bedded rocks. The soft rock is undercut by hydraulic action and abrasion. The weight of the water and the lack of support cause the waterfall to collapse and retreat. Over thousands of years the waterfall may retreat enough to form a gorge of recession.

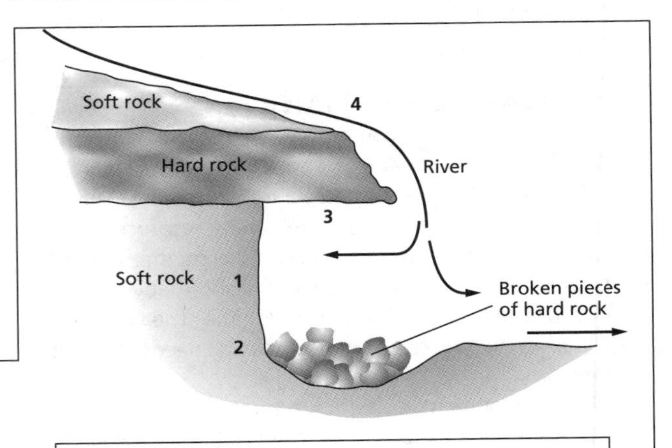

1 Hydraulic impact.
2 Abrasion of soft rock by hard fragments.
3 Lack of support by soft rock.
4 Weight of water causes unsupported hard rock to collapse.

Floodplain management: deposition

Deposition occurs as a river slows down and it loses its energy. Typically, this occurs as a river floods across a floodplain, or enters the sea behind a dam. It is also more likely during low-flow conditions (such as in a drought) than during high-flow (flood) conditions – as long as the river is carrying sediment. The larger, heavier particles are deposited first; the smaller, lighter ones later. Features of deposition include deltas, levées, slip-off slopes (point bars), oxbow lakes, braided channels and floodplains.

LEVÉES

When a river floods its speed is reduced; slowed down by friction caused by contact with the floodplain. As its velocity is reduced, the river has to deposit some of its load. It drops the coarser, heavier material first to form raised banks, or **levées**, at the edge of the river. This means that over centuries the levées are built up of coarse material, such as sand and gravel, while the floodplain consists of fine silt and clay.

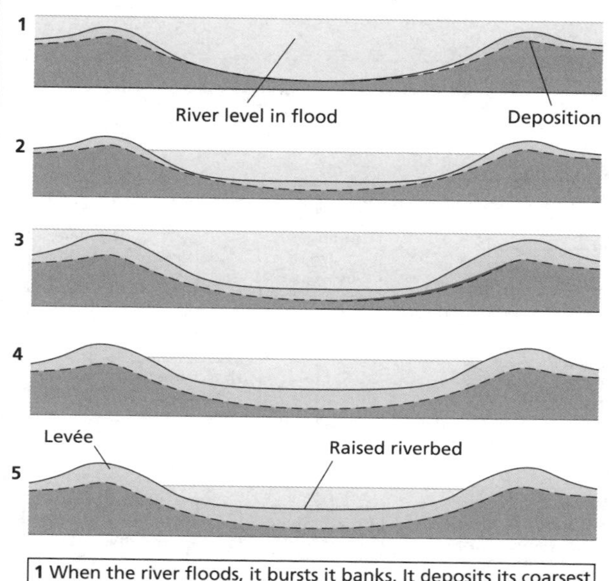

1 When the river floods, it bursts it banks. It deposits its coarsest load (gravel and sand) closer to the bank and the finer load (silt and clay) further away.
2, 3, 4. This continues over a long time, for centuries.
5 The river has built up raised banks called levées, consisting of coarse material, and a floodplain of fine material.

RIVER TERRACES

A river terrace is an eroded floodplain, generally separated from the new floodplain by a steep slope. It is formed due to changes in gradient, sediment load, climate change or human activity, or, indeed, any combination of these. It is the result of both deposition and erosion.

- Many terraces are formed by changes in base level (sea level).
- Changes in fluvial erosion and deposition, due to alternating cold and warm phases, are associated with the formation of terraces.
- Human activity can also lead to the formation of terraces. Deforestation for agricultural land reduces vegetation cover. As interception decreases, overland runoff increases, and there is accelerated erosion of part of the floodplain. This can lead to the formation of terraces upstream, as well as increased deposition downstream.

MEANDERS

Meandering is the normal behaviour of fluids and gases in motion. Meanders can occur on a variety of materials from ice to solid rock. Meander development occurs in conditions where channel slope, discharge and load combine to create a situation where meandering is the only way that the stream can use up the energy it possesses equally throughout the channel reach.

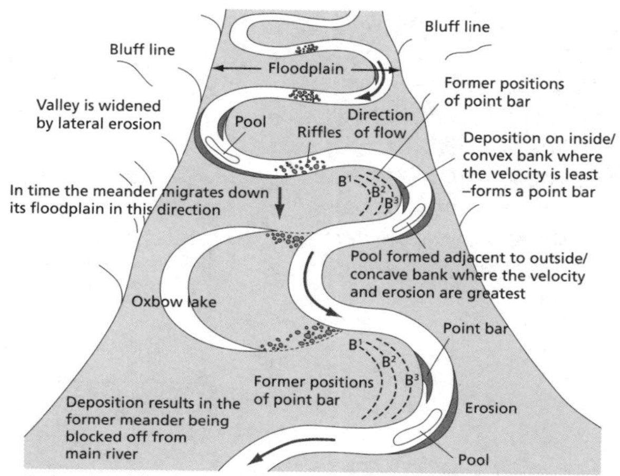

A river is said to be meandering when its **sinuosity** ratio exceeds 1.5. The **wavelength** of meanders is dependent on three major factors: channel width, discharge, and the nature of the bed and banks.

Development of a meander through time

(a) (b) (c)

Pool

Riffle

5 times the bedwidth

Original course

Sinuosity is: actual channel length / straight line distance

one wavelength – usually 10 times the bedwidth

Line of main current (THALWEG)

Sinuosity

uman modification of floodplains

URBAN HYDROLOGY

Storm-water sewers

- Reduce the distance that storm water must travel before reaching a channel
- Increase the velocity of flow because sewers are smoother than natural channels
- Reduce storage: sewers are designed to drain quickly away

Replacement of vegetated soils with impermeable surfaces

- Reduces storage and so increases runoff
- Decreases evapotranspiration because urban surfaces are usually dry
- Increases velocity of overland flow
- Reduces infiltration and percolation

Building activity

- Clears vegetation, which exposes soil and increases overland flow
- Disturbs and dumps the soil, increasing erodability
- Eventually protects the soil with an armour of concrete or tarmac

Encroachment on the river channel

- Embankments, reclamation and riverside roads
- Usually reduces channel width, leading to higher floods
- Bridges can restrict free discharge of floods and increase levels upstream

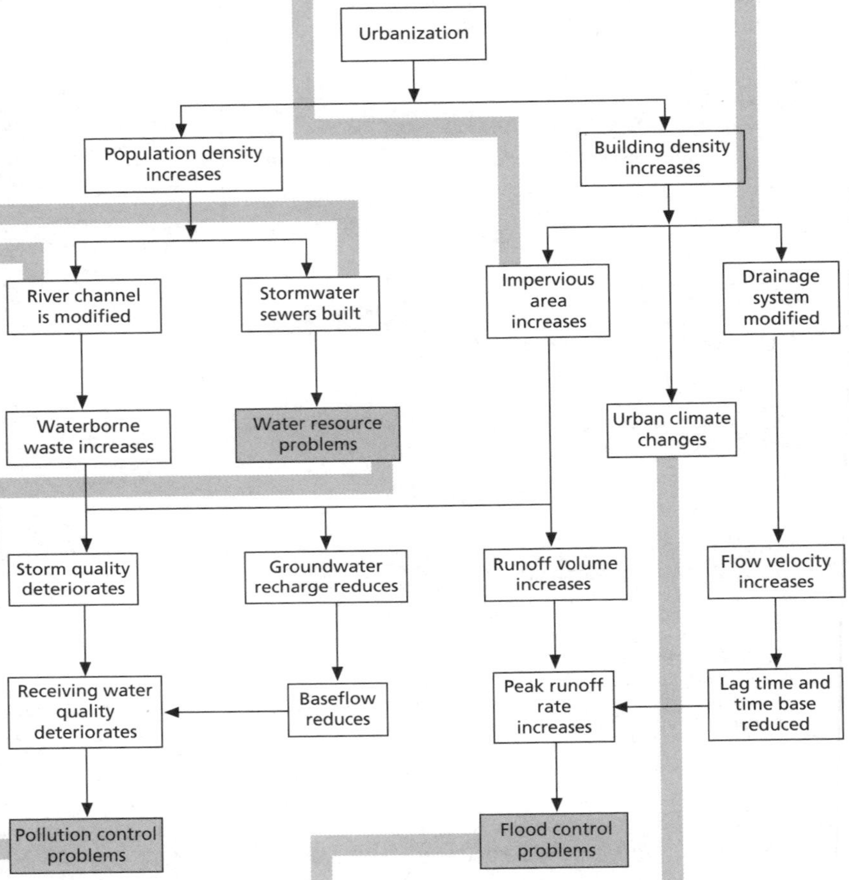

Water resource problems

- Groundwater recharge may be reduced because sewers bypass the mechanisms of percolation and seepage
- Groundwater abstraction through wells may also reduce the store locally
- Irrigation can draw on water resources, leading not only to depletion but also to pollution

Pollution control problems

- Storm water that washed off roads and roofs can contain heavy metals, volatile solids and organic chemicals
- Annual runoff from 1 km of the M1 Motorway in England included 1.5 tonnes of suspended sediment, 4 kg of lead, 126 kg of oil and 18 kg of aromatic hydrocarbons

Flood control problems

- Urbanization increases the peak of the mean annual flood, especially in moderate conditions
- A 243% increase in flood levels resulted from the building of Stevenage New Town in England
- However, during heavy prolonged rainfall, saturated soil behaves in a similar way to urban surfaces

Rainfall climatology of urban areas

- Greater aerodynamic roughness and urban heat island
- More rainfall, especially in summer
- Heavier and more frequent thunderstorms

Effect of urbanization on hydrological processes

Alternative stream management strategies

PERCEPTION AND RESPONSE

Perception of flooding is in part related to the frequency and the magnitude of floods. The responses to flooding are the result of knowledge, perception, money, technology, the characteristics of the flood and the success of the prediction. Responses include:

- bearing the loss
- emergency action
- flood-proofing
- flood control
- land-use zoning
- flood insurance.

Emergency action includes the removal of people and property, and flood-fighting techniques, such as sandbags. Much depends on the efficiency of forecasting and the time available to warn people and clear the area.

Flood-proofing includes sealing walls, sewer adjustment by the use of valves, covering buildings and machinery.

Land-use management is a further way of limiting the damage. However, there are practical problems, such as the difficulty of estimating the damage and use of potential land. Moreover, protection works may give a false sense of security.

Flood insurance is widely seen as a good alternative to floodplain management, but its lack of availability in many poor communities makes it of limited use.

The most effective way of controlling floods is through protective measures along flood channels. There are a variety of options (below).

FLOOD CONTROL – PROTECTIVE MEASURES ALONG FLOOD CHANNELS

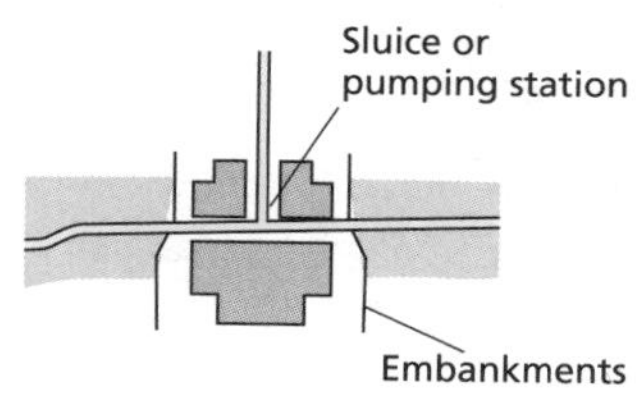

1 Flood embankments with sluice gates. The main problem with this is it may raise flood levels up- and downstream.

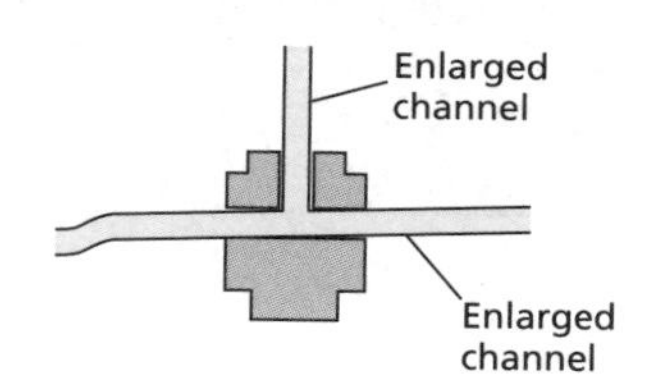

2 Channel enlargement to accommodate larger discharges. One problem with such schemes is that as the enlarged channel is only rarely used it becomes clogged with weed.

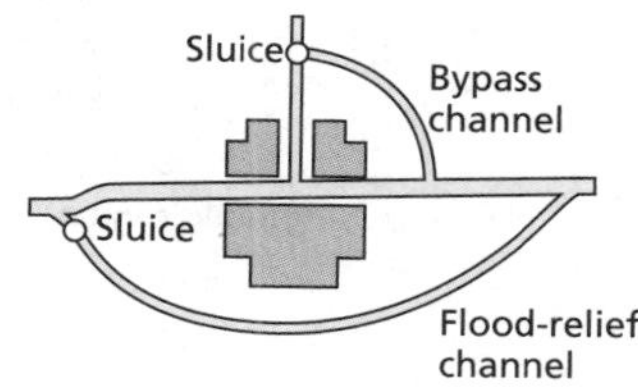

3 Flood relief channel. This is appropriate where it is impossible to modify the original channel as it tends to be rather expensive, e.g. the flood relief channels around Oxford, UK.

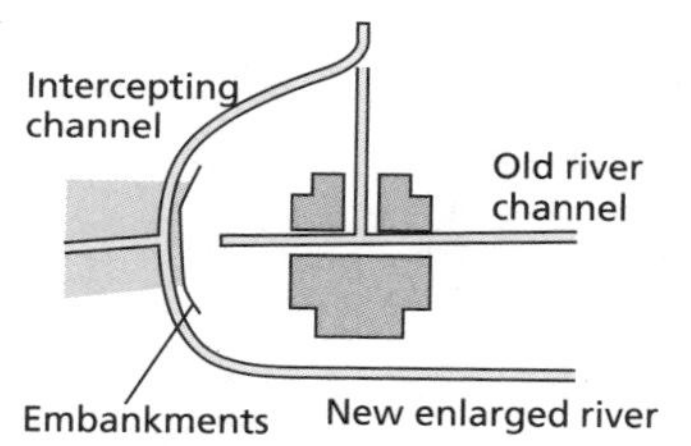

4 Intercepting channels. These divert only part of the flow away, allowing flow for town and agricultural use, e.g. the Great Ouse Protection Scheme in England's Fenlands.

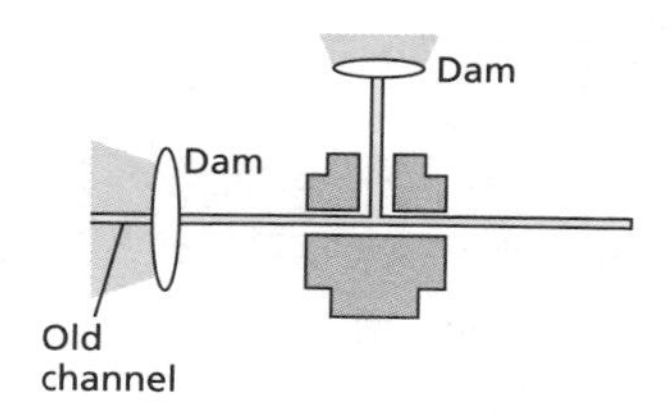

5 Flood storage reservoirs. This solution is widely used, especially as many reservoirs created for water supply purposes have a secondary flood control role, such as the intercepting channels along the Loughton Brook, UK.

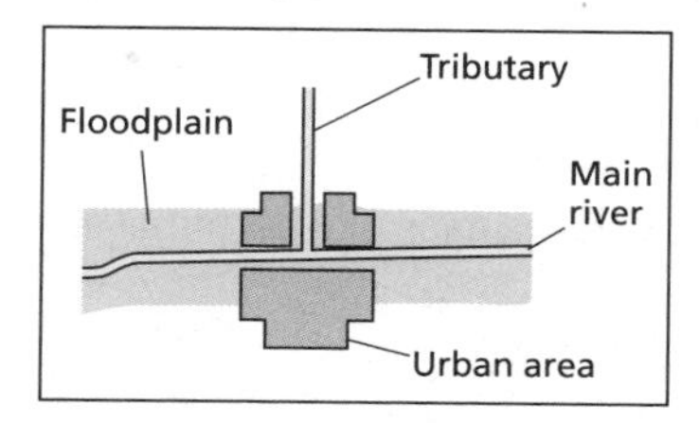

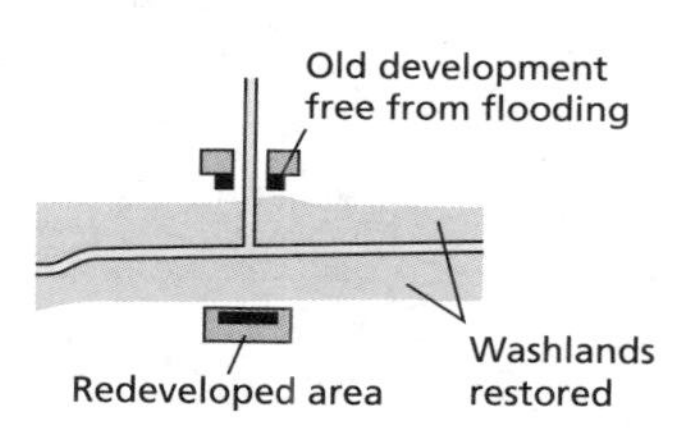

6 Removal of settlements. This is rarely used because of cost, although many communities, e.g. the village of Valmeyer, Illinois, USA were forced to leave following the 1993 Mississippi floods.

OTHER METHODS

Other measures include levées, removing boulders from riverbeds to riverbanks (reducing channel roughness and protecting banks from erosion), and raising the level of the floodplain.

Flood abatement (through the changing of land use in the drainage basin) tackles the problem by slowing down the rate at which water from storms reaches the river channel. This can be achieved through several means:

- Afforestation increases interception and evapotranspiration.
- Terracing of farmland enables overland flow to be controlled.
- Contour ploughing and strip cultivation enable control of overland flow.

Groundwater management (1)

- **Groundwater** refers to subsurface water. The permanently saturated zone within solid rocks and sediments is known as the **phreatic zone**, and here nearly all the pore spaces are filled with water. The upper layer of this is known as the **water table**. The water table varies seasonally – it is higher in winter following increased levels of precipitation. The zone that is seasonally wetted and seasonally dries out is known as the **aeration zone** or the **vadose zone**. Most groundwater is found within a few hundred metres of the surface, but it has been found at depths of up to 4 km beneath the surface.

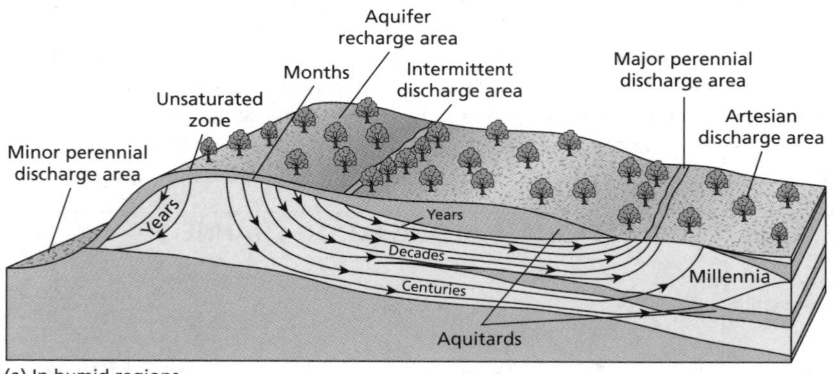

(a) In humid regions

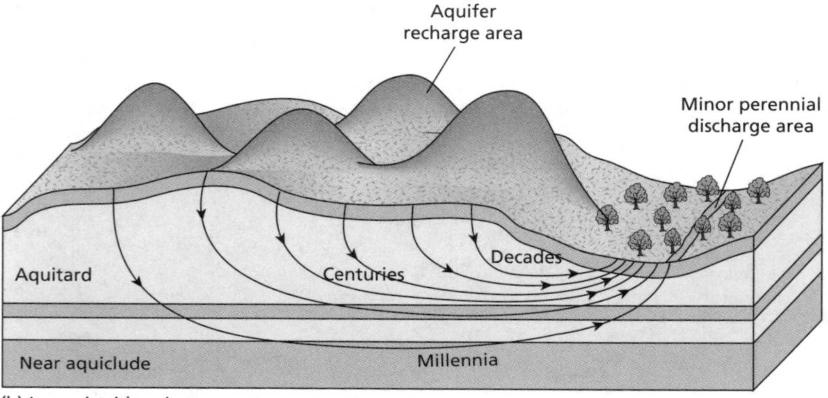

(b) In semi-arid regions

Groundwater

Groundwater may take as long as 20,000 years to be recycled. Hence, in some places, groundwater is considered a non-renewable resource.

Aquifers (rocks which contain significant quantities of water) provide a great reservoir of water. The water moves very slowly and acts as a natural regulator in the hydrological cycle by absorbing rainfall which otherwise would reach streams rapidly. In addition, aquifers maintain stream flow during long dry periods.

Aquifers are permeable rocks such as sandstones or limestones. A rock which will not hold water is known as an **aquiclude** or **aquifuge**. These are impermeable rocks, such as clay, which prevent large-scale storage and transmission of water.

An **aquitard** is a layer of rock which prevents the movement of water.

Groundwater recharge occurs as a result of:

- infiltration of part of the total precipitation at the ground surface
- seepage through the banks and bed of surface water bodies such as rivers, lakes and oceans
- groundwater leakage and inflow from adjacent aquicludes and aquifers
- artificial recharge from irrigation, reservoirs, etc.

- Losses of groundwater result from:
- evapotranspiration, particularly in low-lying areas where the water table is close to the ground surface
- natural discharge by means of spring flow and seepage into surface water bodies
- groundwater leakage and outflow through aquicludes and into adjacent aquifers
- artificial abstraction.

Groundwater management (2)

GROUNDWATER POLLUTION IN BANGLADESH

There has been an increase in the incidence of cancers in Bangladesh. This has been caused by naturally occurring arsenic in groundwater pumped up through tube wells. Estimates by the World Health Organization suggest that as many as 85 million of the country's 125 million population will be affected by arsenic-contaminated drinking water.

For 30 years, following the lead of Unicef, Bangladesh has sunk millions of tube wells, providing a convenient supply of drinking water free from the bacterial contamination of surface water that was killing 250,000 children a year. But the water from the wells was never tested for arsenic contamination, which occurs naturally in the groundwater. One in 10 people who drink the water containing arsenic will ultimately die of lung, bladder or skin cancer.

The first cases of arsenic-induced skin lesions were identified across the border in West Bengal, India, in 1983. Arsenic poisoning is a slow disease. Skin cancer typically occurs 20 years after people start ingesting the poison. The real danger is internal cancers, especially of the bladder and lungs, which are usually fatal. Bangladeshi doctors have been warned to expect an epidemic of cancers by 2010. The victims will be people in their 30s and 40s who have been drinking the water all their lives – people in their most productive years.

One solution to the problem is a concrete butt, collecting water by pipe from gutters. Another possible solution is a filter system. Neither is as convenient as the tube well it is designed to replace. Tube wells are easy to sink in the delta's soft alluvial soil, and for tens of millions of peasants the wells have revolutionized access to water.

WATER BALANCE IN AUSTRALIA

The water balance is based on studies in 51 catchments in Australia, ranging from the Great Artesian Basin and the Murray-Darling Basin to smaller basins such as Kangaroo Island. Many of the monitored areas are coastal, state capital cities, the eastern coastline, south and south-western coastlines.

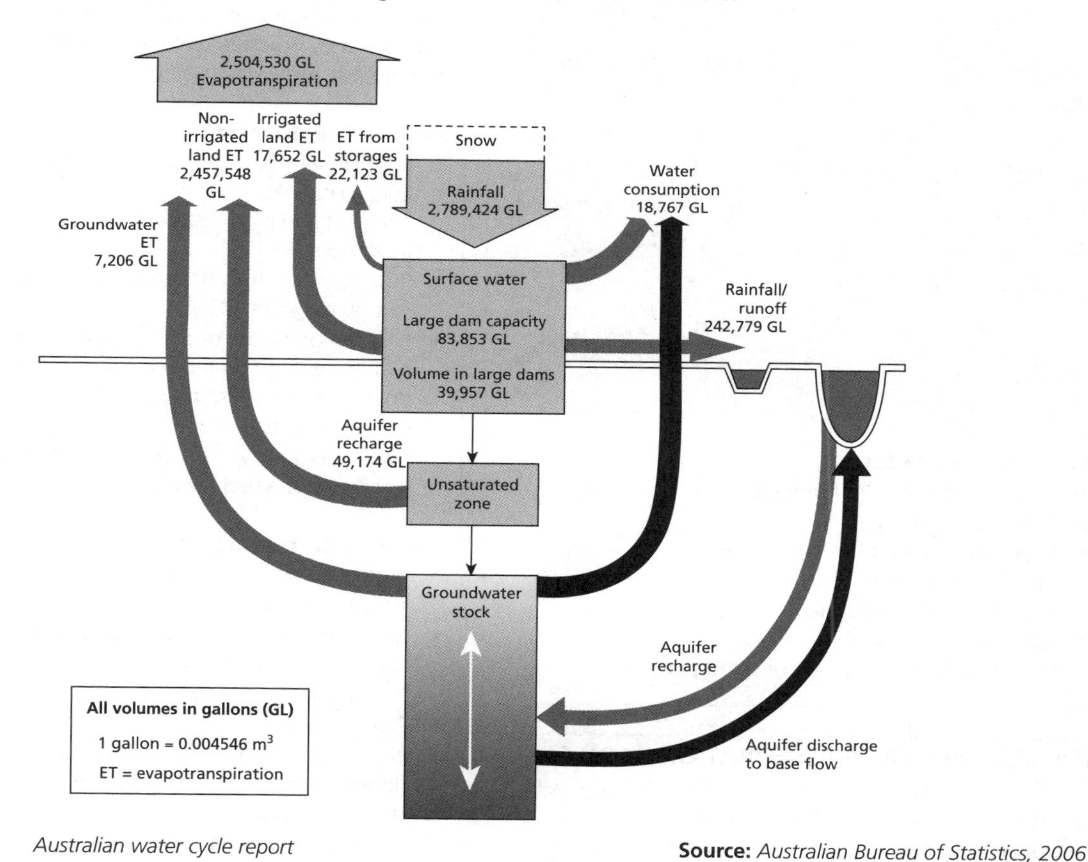

Australian water cycle report

Source: *Australian Bureau of Statistics, 2006*

Freshwater wetland management

WETLANDS

A wetland is defined as "land with soils that are permanently flooded". The Ramsar Convention, an international treaty to conserve wetlands, defines wetlands as "areas of marsh, fen, peatland or water, whether natural or artificial, permanent or temporary, with water that is static or flowing, fresh, brackish or salt." According to this classification, there are marine, coastal, inland and artificial types, subdivided into 30 categories of natural wetland and 9 human-made ones, such as reservoirs, barrages and gravel pits. Wetlands now represent only 6% of the earth's surface, of which 30% are bogs, 26% are fens, 20% are swamps, 15% are floodplains and 2% are lakes. It is estimated that there was twice as much wetland area in 1900 compared with 2000.

The value of wetlands

Wetlands provide many important social, economic and environmental benefits.

Functions	Products	Attributes
Flood control	Fisheries	Biological diversity
Functions	**Products**	**Attributes**
Sediment accretion and deposition	Game	Culture and heritage
Groundwater recharge	Forage	
Groundwater discharge	Timber	
Water purification	Water	
Storage of organic matter		
Food-chain support/cycling		
Water transport		
Tourism/recreation		

Loss and degradation

The loss and degradation of wetlands is caused by several factors, including:

- increased demand for agricultural land
- population growth
- infrastructure development
- river flow regulation
- invasion of non-native species and pollution.

CHANGING RIVER MANAGEMENT: THE KISSIMEE RIVER

Between 1962 and 1971 the 165 km meandering Kissimmee River and flanking floodplain in Florida, USA were **channelized** and transformed into a 90 km, 10 m deep drainage canal. The river was channelized to provide an outlet canal for draining floodwaters from the developing upper Kissimmee lakes basin, and to provide flood protection for land adjacent to the river.

Impacts of channelization

The channelization of the Kissimee River had several unintended impacts:

- the loss of 12,000–14,000 ha of wetlands
- a reduction in wading bird and waterfowl usage
- a continuing long-term decline in game fish populations.
- Concerns about the **sustainability** of existing ecosystems led to a state and federally supported restoration study. The result was a massive restoration project, on a scale unmatched elsewhere.

THE KISSIMEE RIVER RESTORATION PROJECT

The aim is to restore over 100 km² of river and associated floodplain wetlands. The project will benefit over 320 fish and wildlife species, including the endangered bald eagle, wood stork and snail kite. It will create over 11,000 ha of wetlands.

Restoration of the river and its associated natural resources requires **dechannelization**. This entails backfilling approximately half of the flood control channel and re-establishing the flow of water through the natural river channel. In residential areas the flood control channel will remain in place.

Costs of restoration

- It is estimated the project will cost $578 million (initial channelization cost $20 million). The bill is being shared by the state of Florida and the federal government.
- Restoration, which began in 1999, will not be completed until 2010.
- Restoration of the river's floodplain could result in higher losses of water due to evapotranspiration during wet periods. In extremely dry spells, navigation may be impeded in some sections of the restored river. It is, however, expected that navigable depths will be maintained for at least 90% of the time.

Benefits of restoration

- Higher water levels should ultimately support a natural river ecosystem again.
- Re-establishment of floodplain wetlands and the associated nutrient filtration function is expected to result in decreased nutrient loads to Lake Okeechobee.
- It is possible that restoration of the Kissimmee River floodplain could benefit populations of key avian species, such as wading birds and waterfowl, by providing increased feeding and breeding habitats.
- Potential revenue associated with increased recreational usage (such as hunting and fishing) and ecotourism on the restored river could significantly enhance local and regional economies.

Irrigation and agriculture

DEFINITION AND TYPES

Irrigation is the addition of water to areas where there is insufficient for adequate crop growth. Water can be taken from surface stores, such as lakes, dams, reservoirs and rivers, or from groundwater. Irrigation occurs in both rich and poor countries. For example, large parts of the USA and Australia are irrigated. There is evidence of irrigation in Egypt going back nearly 6000 years.

Types of irrigation range from total flooding, as in the case of paddy fields, to drip irrigation, where precise amounts are measured out to each individual plant.

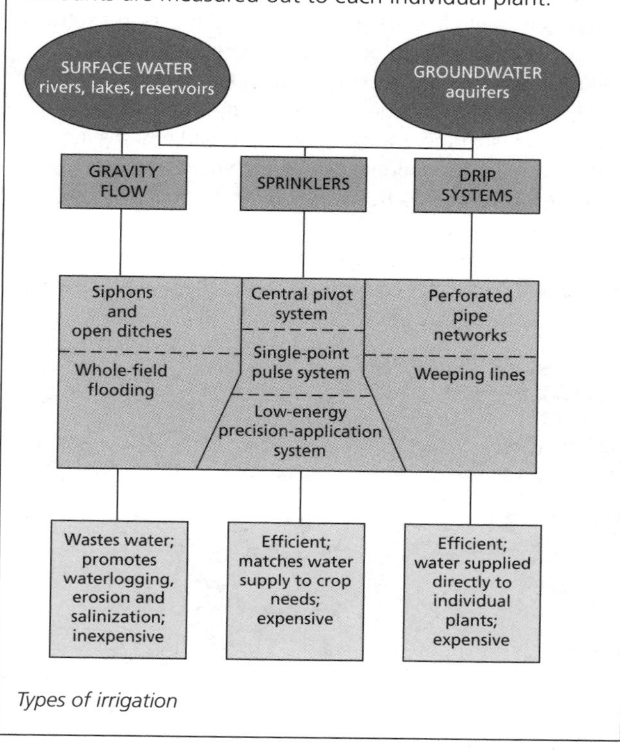

Types of irrigation

THE GANGA RIVER, INDIA

The Ganga River is over 2500 km long and drains an area of over 1 million km². Upwards of 250 million people live in the basin. Food production is very important in the Ganga Valley and irrigation is widely used. Water quality is low. The most common form of water pollution is organic matter from domestic sewage, municipal waste and agro-industrial effluent. The use of fertilizers and pesticides to feed high-yielding varieties of crops is increasing the trend. Water quality deteriorates during the dry season.

IMPACTS OF IRRIGATION

- In Texas, USA, irrigation has reduced the water table by as much as 50 m. By contrast, in the Indus Plain in Pakistan, irrigation has raised the water table by as much as 6 m since 1922, and caused widespread salinization (see below).
- Irrigation can reduce the earth's albedo (reflectivity) by as much as 10%. This is because a reflective sandy surface may be replaced by one with dark green crops.
- Irrigation can also cause changes in precipitation. Large-scale irrigation in semi-arid areas, such as the High Plains of Texas, has been linked with increased rainfall, hailstorms and tornadoes. Under natural conditions, semi-arid areas have sparse vegetation and dry soils in summer. However, when irrigated these areas have moist soils and a complete vegetation cover. Evapotranspiration rates increase, resulting in greater amounts of summer rainfall across Kansas, Nebraska, Colorado and the Texas Panhandle.
- Irrigation frequently leads to an increase in the amount of salt in the soil. This occurs when groundwater levels are close to the surface. In clay soils this may be within 3 m of the surface, whereas on sandy and silty soils it is less. Capillary forces bring water to the surface where it may be evaporated, leaving behind any soluble salts that it is carrying. This is known as **salinization**.
- Some irrigation, especially for paddy rice, requires huge amounts of water. As water evaporates in the hot sun, the salinity levels of the remaining water increase. This also occurs behind large dams.

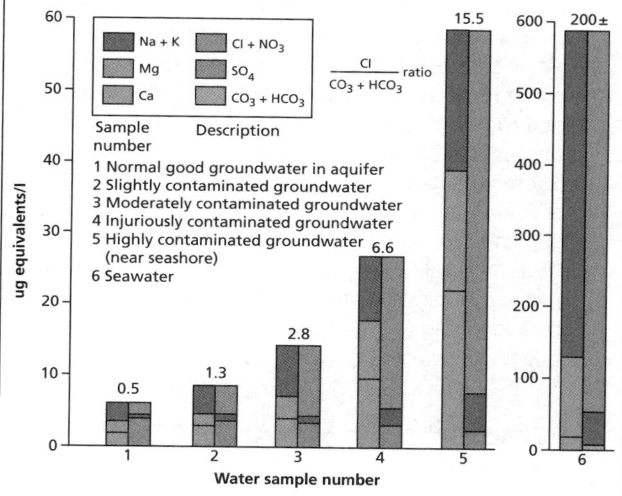

The amount and type of pollution varies with distance from the source of water

- Chemical changes are also important. In Salinas, California, salinization is characterized by an increase in dissolved salts and an increase in the ratio of chlorides to bicarbonates.
- **Eutrophication**, or nutrient enrichment, of water bodies has led to algal blooms, oxygen starvation and a decline in species diversity. This is most evident in poorly circulating waters, especially ponds and ditches. While there is a strong body of evidence to link increased eutrophication with increased use of nitrogen fertilizers, some scientists argue that increased phosphates from farm sewage are the cause.

Demand for water: local/national scale

MANAGING ISRAEL'S AQUIFERS

Water is one of the most sensitive and unsolvable problems in the Middle East. It has created great friction between Arabs and Jews; the example of Israeli-Palestinian tensions illustrates the problem clearly.

For decades Israel has obtained up to 80% of the 670 m^3 of water provided by the mountain aquifer. This aquifer is mostly located under the West Bank. The Israelis have occupied the West Bank since 1967 and have prevented the Palestinians from obtaining better access to the resource. The mountain aquifer is important for Israel as it provides:

- one-third of its water consumption
- 4% of its drinking water
- 50% of its agricultural water.

The 120,000 Jewish settlers in the West Bank use about 60 m^3 annually, compared with the 137 million m^3 used by 1.5 million West Bank Arabs. In addition, the West Bank settlers irrigate 70% of their cultivated land, compared with just 6% of Palestinian land.

The West Bank and Gaza are served by Israel's water carrier and, more importantly, the groundwater in the region's aquifers. The West Bank's aquifers, replenished by the rainfall on its hills, flow west, north and east from the central drainage divide (watershed). The eastern aquifer lies entirely within the West Bank, providing water for Palestinians and Israeli settlers; its usefulness is tempered by the fact that the main population centres are central or to the west.

Israel's 5½ million people consume three or four times as much water per head as the 2 million Palestinians. Forbidden to dig new wells or deepen old ones, Palestinians were kept very short, particularly for their crops (industry, under military occupation, barely existed).

What is bad in the West Bank is usually worse in Gaza; and water, or rather the lack of it, is no exception. The Gazans, like the West Bankers, get a little domestic water from Israel's national carrier, but most of their meagre supplies come from an aquifer that has been grossly exploited and is in a badly dilapidated state. The Gazans pump out about twice as much as can be safely withdrawn. Seawater creeps in, making the water so saline that it kills the citrus trees.

The problem of water supply is widespread throughout the Middle East region, with Jordan, Israel and Palestine suffering the most acute shortages. As part of the Israeli-Jordanian peace process in 1995, Israel agreed to provide Jordan with 150 million m^3 of water per annum. This will be supplied by:

- diverting water
- building new dams
- desalinization.
- Other possibilities include using the Litani and Awali rivers in Lebanon, cutting back on agriculture, and creating a regional water market whereby people pay for the water they use.

EXTENSION

Sketch maps

Sketch maps such as the one here showing ground water flows in Israel's mountain aquifer need to be clear. They should contain:

- orientation – a north arrow
- scale – km or m
- a key – here groundwater flow and the watershed are shown
- labels – such as the Dead Sea and names of towns.

Too much information and the sketch map becomes too cluttered – too little and means reader is left wondering what it means! Here we can see the different aquifers supplying different regions and towns.

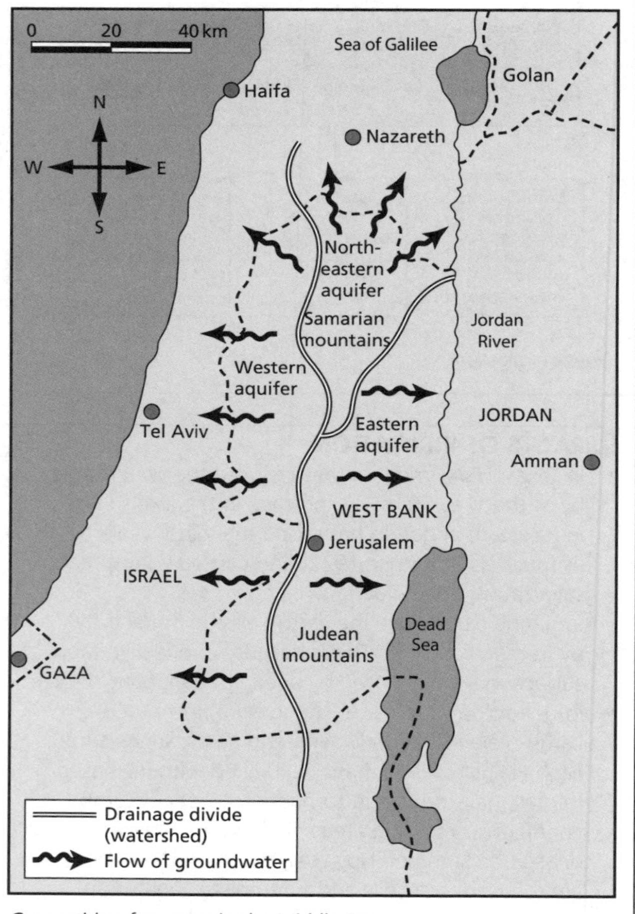

Competition for water in the Middle East

Demand for water: international scale

THE MEKONG

The Mekong is South-East Asia's largest river. It is the world's 12th longest and 21st in the size of its basin. The largest expanse of flat, well-watered and fertile land in the basin lies around Tonle Sap lake, but the devastating annual flood makes intensive agriculture difficult there. The surface area of the lake can swell to up to 10 times its normal size during the monsoon.

Unusually for such a large river in the heart of Asia, the population along the course of the Mekong is scanty. The largest city, Phnom Penh, has just 1.1 million inhabitants. This makes the river unusual in another respect: the pressures of a burgeoning population and fast economic growth are only just beginning to make their mark.

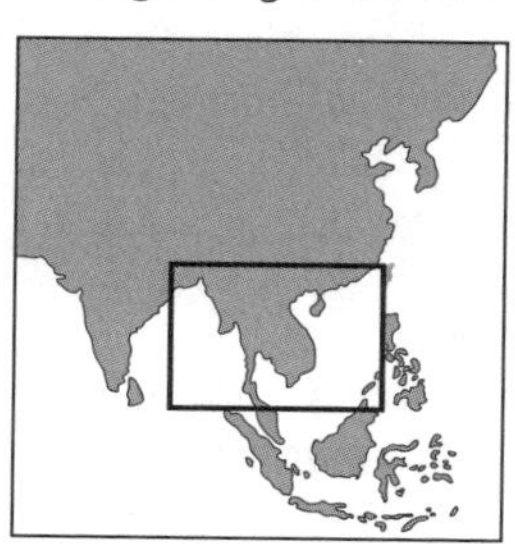

Until relatively recently, this huge river has been an economic backwater, remaining almost untouched until the 1990s. The first dam on the river, at Man Wan, in China, was not completed until 1993. The first bridge across the Lower Mekong (i.e. outside China) was built in 1994. However, population growth and economic growth are now creating a strain on the Mekong.

	Population growth, 2007 (%)	Economic growth, 2007 (%)
Cambodia	1.75	9.1
China	0.63	11.4
Laos	2.34	7.0
Burma (Myanmar)	0.8	5.5
Thailand	0.64	4.5
Vietnam	0.99	8.5

Population growth and economic growth in the Mekong region, 2007

DAMS ALONG THE MEKONG

The hydroelectric potential of the Mekong and its tributaries is considerable and largely untapped. Early plans to develop the river failed to materialize due to war and civil unrest. So far only 5% (1600 MWs) of the lower basin's hydroelectric potential of approximately 30,000 MW have been developed, and the few projects have all been on the tributaries.

Of the total potential of 30,000 MW in the Lower Mekong basin, approximately 13,000 are on the Mekong, the rest on its tributaries. There is 13,000 MW potential in Laos, 2200 MW on tributaries in Cambodia and 2000 MW on tributaries in Vietnam. In contrast, in the Upper Mekong basin in Yunnan Province, China, there is 23,000 MW potential.

The dams generate valuable electricity, aid irrigation and regulate flooding. However, in the process they have caused irreparable damage to what was, until recently, the Mekong's most valuable resource: its fisheries. The Mekong and its tributaries yield more fish than any other river system. The annual harvest, including fish farms, amounts to about 2 million tonnes. The Mekong is home to over 1200 different species of fish – more than any other river, save the Amazon and the Congo.

CAMBODIA

The cumulative impacts of the Mekong dams are likely to affect Cambodia significantly, where the river's annual floods create the world's fourth largest catch of freshwater fish and employment for 1.5 million people. Cambodia catches 400,000 tonnes of freshwater fish a year, ranking it only behind China, India and Bangladesh, but annual river levels are thought to have dropped at least 12% since the dams and irrigation works started upstream.

The situation could worsen rapidly if the proposed $4 billion Sambor dam is built. This is expected to flood nearly 800 km^2, displacing 60,000 people and affecting fishing. Meanwhile dams built by Vietnam on the Se San River, a major Mekong tributary, have been particularly damaging in Cambodia. Se San fishers have complained that there are fewer fish and that the river's erratic flows often wash away their nets.

About 80% of rice production in the Lower Mekong basin depends on water, silt and nutrients provided by the flooding of the Mekong. Dams on the Upper Mekong could mean less frequent floods, adversely affecting farming and fishing.

Distribution of oceans and ocean currents

DISTRIBUTION AND IMPORTANCE OF OCEANS

Oceans cover about 70% of the earth's surface, and are of great importance to humans in a number of ways. These include regulating global climates and as a source of economic materials. In addition, oceans are important for leisure and recreation.

Oceans cover about 50% of the earth's surface in the northern hemisphere and about 90% in the southern hemisphere. This is not always clear when looking at world maps.

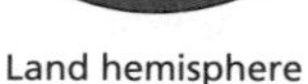

Land hemisphere

Sea hemisphere

Ocean or sea	Area in 1000 km²
Pacific Ocean	166,229
Atlantic Ocean	86,551
Indian Ocean	73,442
Arctic Ocean	13,223
South China Sea	2,975
Caribbean Sea	2,516
Mediterranean Sea	2,509
Bering Sea	2,261

The world's largest oceans and seas

DISTRIBUTION OF OCEAN CURRENTS

Warm ocean currents move water away from the equator, whereas cold ocean currents move water away from cold regions towards the equator. The major currents move huge masses of water over long distances. The warm Gulf Stream, for instance, transports 55 million m^3 per second. Without it, the "temperate" lands of north-western Europe would be more like the sub-Arctic. The cold Peru Current and the Benguela Current of south-west Africa bring in nutrient-rich waters dragged to the surface by offshore winds.

In addition, there is the **Great Ocean Conveyor Belt** (see page 68). This deep, grand-scale circulation of the ocean's waters effectively transfers heat from the tropics to colder regions, such as northern Europe.

EXTENSION

Map projections

Map projections convey a "message". The two globes show very different hemispheres – a land north and a sea south. The Mercartor Scale over-emphasizes the importance of the nothern hemisphere – for example it suggests that Greenland is roughly the same size as the African continent – and locates the British Isles close to the centre of the map. In contrast, the Peters Projection is more realistic for tropical regions but squeezes termperate regions.

Map projections must be treated with caution!

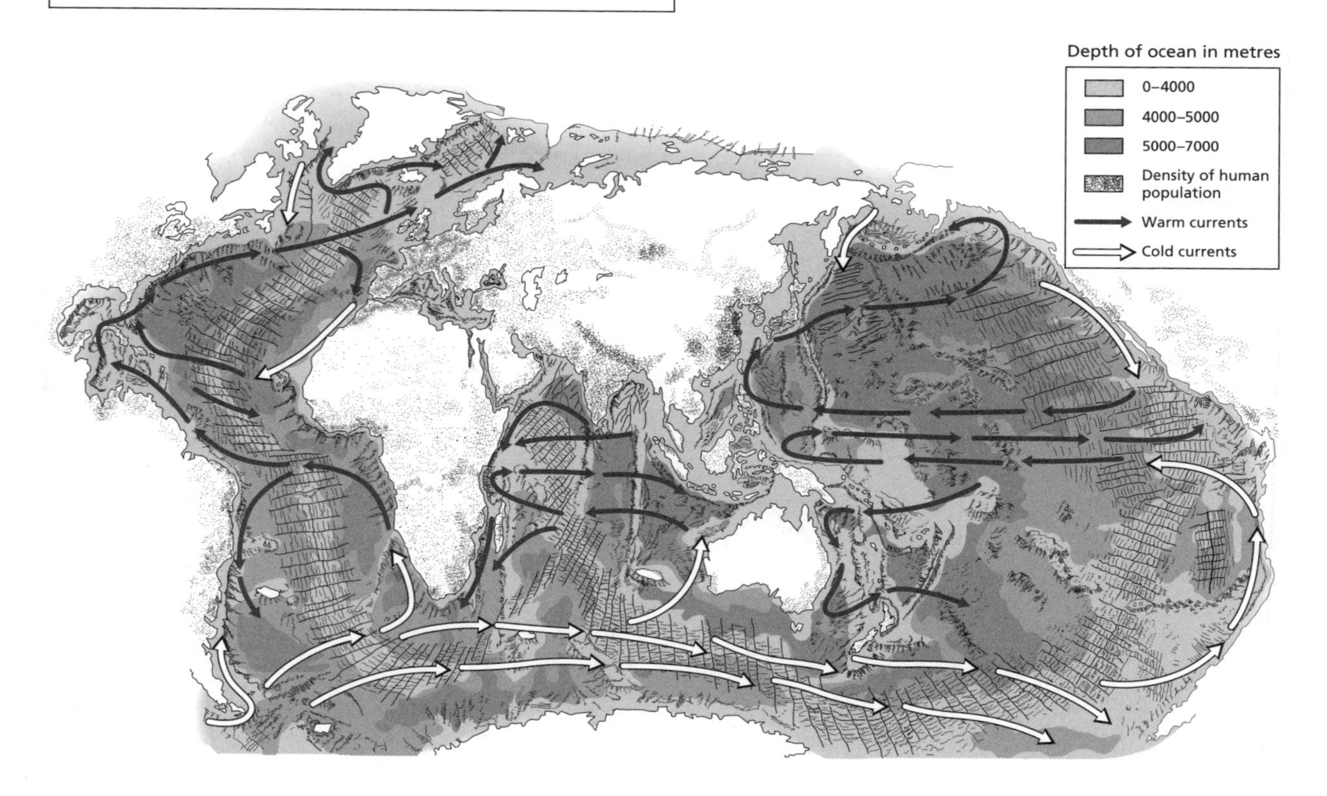

Ocean morphology

FEATURES OF THE OCEAN FLOOR

The ocean floor consists of many features such as deep sea trenches, mid-ocean ridges, transform faults, rift valleys, deep abyssal plains, continental slopes and continental shelves. Smaller features include submarine canyons and submarine volcanoes or seamounts.

Seamounts are extinct volcanic cones that lie below the surface. A **guyot** is a flat-topped volcano that once reached the surface but later subsided.

The **abyssal plain** is at the edge of the continental slope. These plains cover large areas of the sea floor at depths of between 4000 and 6000 m. They are generally flat and featureless.

The **continental slope** is the steeply sloping area of the seabed that stretches from the continental shelf to the abyssal plain. The continental slope may contain **submarine canyons** eroded by fast-flowing currents of water and sediment. In addition, sediment slumps down the canyon to form a steep, narrow valley on the continental slope. Submarine canyons are often located close to the point where a large river flows into the sea.

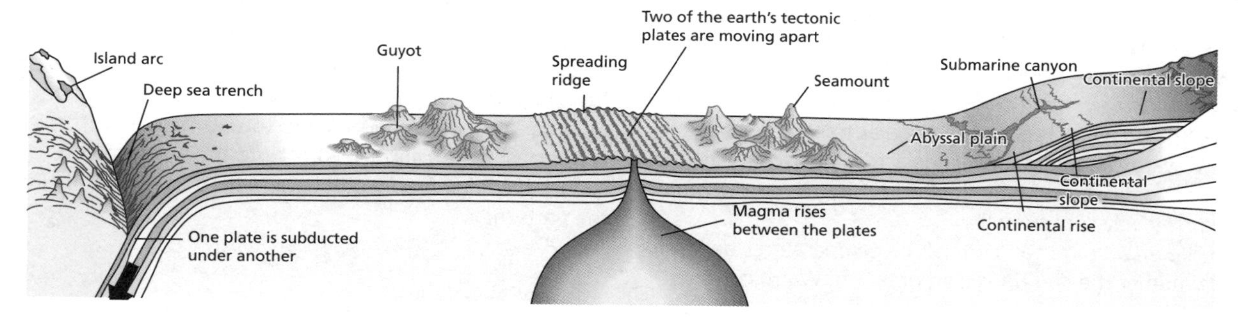

Ocean trenches are the deepest parts of the oceans. These are arc-shaped depressions, formed at subduction zones where one tectonic plate (usually an oceanic one) plunges under a less dense continental one. The Mariana Trench in the Pacific Ocean is over 11,000 m deep.

Mid-ocean ridges are the largest feature of the ocean floor. They are essentially a linear belt of submarine mountains. They occur at divergent (spreading or constructive) plate boundaries. New magma forces its way up between two plates and pushes them apart. In slow-spreading ridges, such as in the Mid-Atlantic, the rate of spreading is up to 5 cm/year. The ridges are characterized by a wide **rift valley** at their centre. This rift valley can be up to 20 km wide. In contrast, where the rate of spreading is rapid, as in the case of the East Pacific Rise, which spreads at a rate of about 17–18 cm/year, there are no rift valleys.

The **continental shelf** is a relatively flat area of seabed, stretching from the land to the edge of the continental slope. The continental shelf is less than 250 m deep and may be up to 70 km wide.

OCEANIC WATER

Oceanic water varies in its salinity and temperature. Average salinity is about 35 parts per thousand (ppt). Concentrations of salt are higher in warm seas, due to the high rates of evaporation of water. In tropical seas, salinity decreases sharply with depth. In contrast, in polar seas where there is an input of fresh water from rivers, salinity levels are low. Salinity levels increase with depth.

The predominant minerals in seawater are chloride (54.3%) and sodium (30.2%), which combine to form salt. Other important minerals in the sea include magnesium and sulphate ions.

Temperature

Temperature varies considerably at the surface of the ocean, but there is little variation at depth. In tropical and subtropical areas, sea surface temperatures in excess of 25 °C are caused by insolation (**in**coming **sol**ar radi**ation**). From about 300 m to 1000 m the temperature declines steeply to about 8–10 °C. Below 1000 m the temperature decreases to a more uniform 2 °C in the ocean depths.

The temperature profile is similar in the mid-latitudes (40–50 °N and S), although there are clear seasonal variations. Summer temperatures may reach 17 °C, whereas winter sea temperatures are closer to 10 °C. There is a more gradual decrease in temperature with depth (thermocline). In high latitudes and polar oceans, sea surface temperatures range between 0 °C and 5 °C. In some cases the temperature may be below freezing, but the water does not freeze because of its salinity. Below the surface, it reaches the uniform temperature of 2 °C in the deep ocean.

Oceans and climate (1)

SEA CURRENTS

Surface ocean currents are caused by the influence of prevailing winds blowing steadily across the sea. The dominant pattern of surface ocean currents (known as **gyres**) is roughly circular flow; the pattern of these currents is clockwise in the northern hemisphere and anticlockwise in the southern hemisphere. The main exception is the circumpolar current that flows around Antarctica from west to east. There is no equivalent current in the northern hemisphere because of the distribution of land and sea. Within the circulation of the gyres, water piles up into a dome. The effect of the rotation of the earth is to cause water in the oceans to push westward; this piles up water on the western edge of ocean basins, rather like water slopping in a bucket. The return flow is often narrow, fast-flowing currents such as the Gulf Stream. The Gulf Stream in particular transports heat northwards and then eastwards across the North Atlantic; it is the main reason why the British Isles have mild winters and relatively cool summers.

The effect of ocean currents on temperatures depends on whether the current is cold or warm. Warm currents from equatorial regions raise the temperatures of polar areas (with the aid of prevailing westerly winds). However, the effect is only noticeable in winter. For example, the North Atlantic Drift raises the winter temperatures of north-west Europe. By contrast, there are other areas which are made colder by ocean currents. Cold currents such as the Labrador Current off the north-east coast of North America may reduce summer temperatures, but only if the wind blows from the sea to the land.

In the Pacific Ocean there are two main atmospheric states – the first is warm surface water in the west with cold surface water in the east; the other is warm surface water in the east with cold in the west. In whichever case, the warm surface causes low pressure. As air blows from high pressure to low pressure, there is a movement of water from the colder area to the warmer area. These winds push warm surface water into the warm region, exposing colder deep water behind them and maintaining the pattern.

THE GREAT OCEAN CONVEYOR BELT

In addition to the transfer of energy by wind and the transfer of energy by ocean currents, there is also a transfer of energy by deep sea currents. Oceanic convection occurs from the polar regions, where cold, salty water sinks into the depths and makes its way towards the equator. The densest water is found in the Antarctic area; here seawater freezes to form ice at a temperature of around about −2 °C. The ice is fresh water, hence the seawater left behind is much saltier and therefore denser. This cold, dense water sweeps round Antarctica at a depth of about 4 km. It then spreads into the deep basins of the Atlantic, the Pacific and the Indian Oceans. Surface currents bring warm water to the North Atlantic from the Indian and Pacific Oceans. These waters give up their heat to cold winds which blow from Canada across the North Atlantic. This water then sinks and starts the reverse convection of the deep ocean current. The amount of heat given up is about a third of the energy that is received from the sun. Because the conveyor operates in this way, the North Atlantic is warmer than the North Pacific, so there is proportionally more evaporation there. The water left behind by evaporation is saltier and therefore much denser, which causes it to sink. Eventually the water is transported into the Pacific where it picks up more water and its density is reduced.

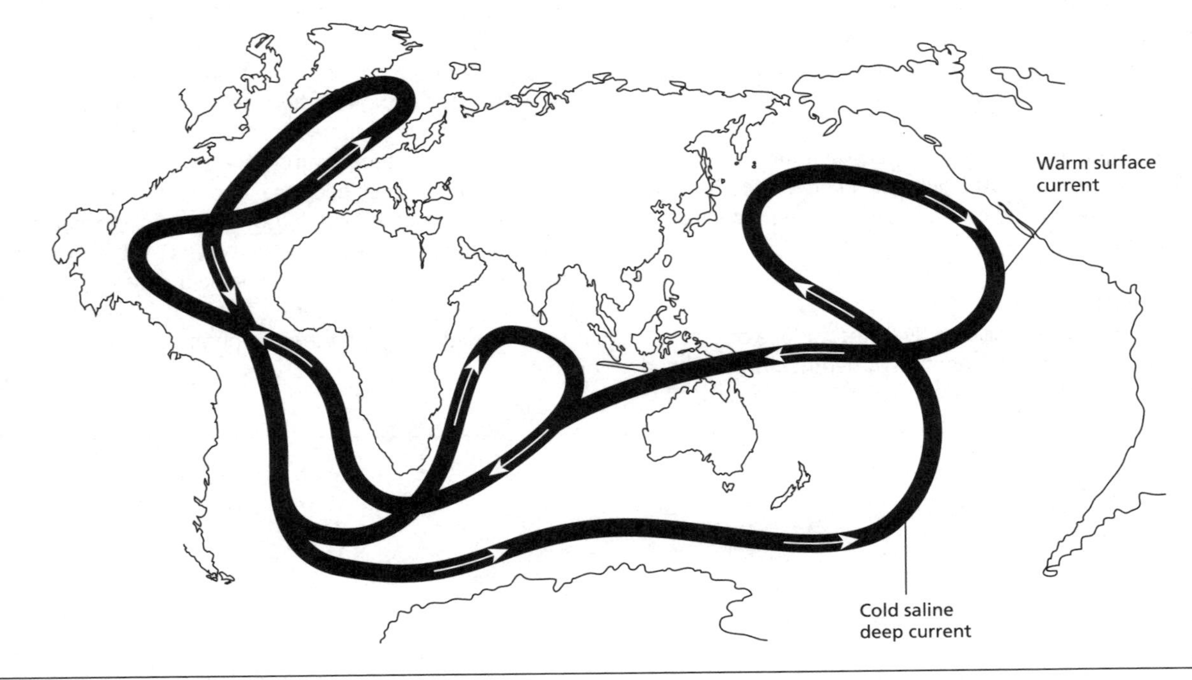

Oceans and climate (2)

EL NIÑO

El Niño – the "Christ Child" – is a warming of the eastern Pacific that occurs at intervals of between two and ten years, and lasting for up to two years. Originally, El Niño referred to a warm current that appeared off the coast of Peru, but it is now realized that this current is part of a much larger system.

Normal conditions in the Pacific Ocean

The Walker circulation is the east–west circulation that occurs in low latitudes. Near South America, winds blow offshore, causing upwelling of the cold, rich waters. By contrast, warm surface water is pushed into the western Pacific. Normally, sea surface temperatures (SSTs) in the western Pacific are over 28 °C, causing an area of low pressure and producing high rainfall. Over coastal South America, however, SSTs are lower, high pressure exists and conditions are dry.

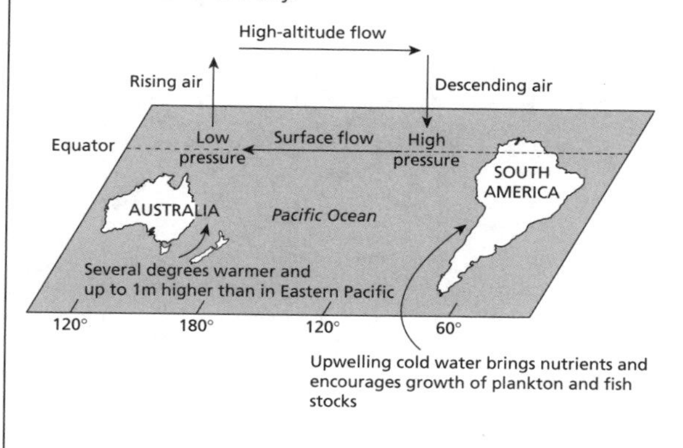

El Niño conditions in the Pacific Ocean

During El Niño episodes, the pattern is reversed. Water temperatures in the eastern Pacific rise as warm water from the western Pacific flows into the east Pacific. During ENSO (El Niño Southern Oscillation) events, SSTs of over 28 °C extend much further across the Pacific. Low pressure develops over the eastern Pacific, high pressure over the west. Consequently, heavy rainfall occurs over coastal South America, whereas Indonesia and the western Pacific experience warm, dry conditions. Some of these events can be disastrous.

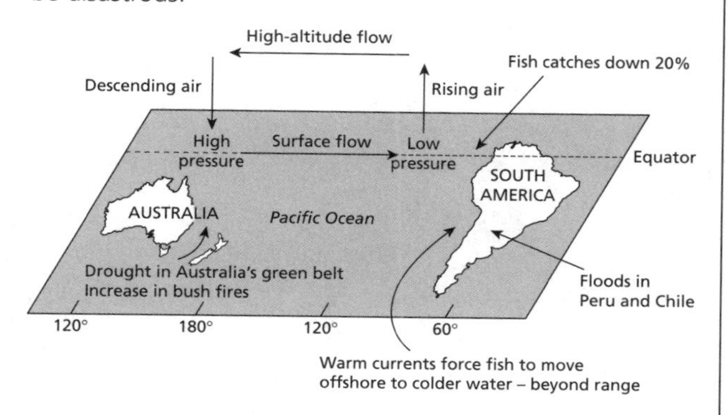

LA NIÑA

La Niña is an intermittent cold current that flows from the east across the equatorial Pacific Ocean. It is an intensification of normal conditions, whereby strong easterly winds push cold, upwelling water off the coast of South America into the western Pacific. Its impact extends beyond the Pacific and has been linked with unusual rainfall patterns in Africa's Sahel region and in India, and with unusual temperature patterns in Canada.

THE IMPACTS OF EL NIÑO AND LA NIÑA

Managing the impacts that these events cause is difficult for many reasons.

- They cannot be predicted with much accuracy.
- They affect large parts of the globe, not just the Pacific.
- Some of the countries affected do not have the resources to cope.
- There are indirect impacts on other parts of the world though trade and aid (teleconnections).

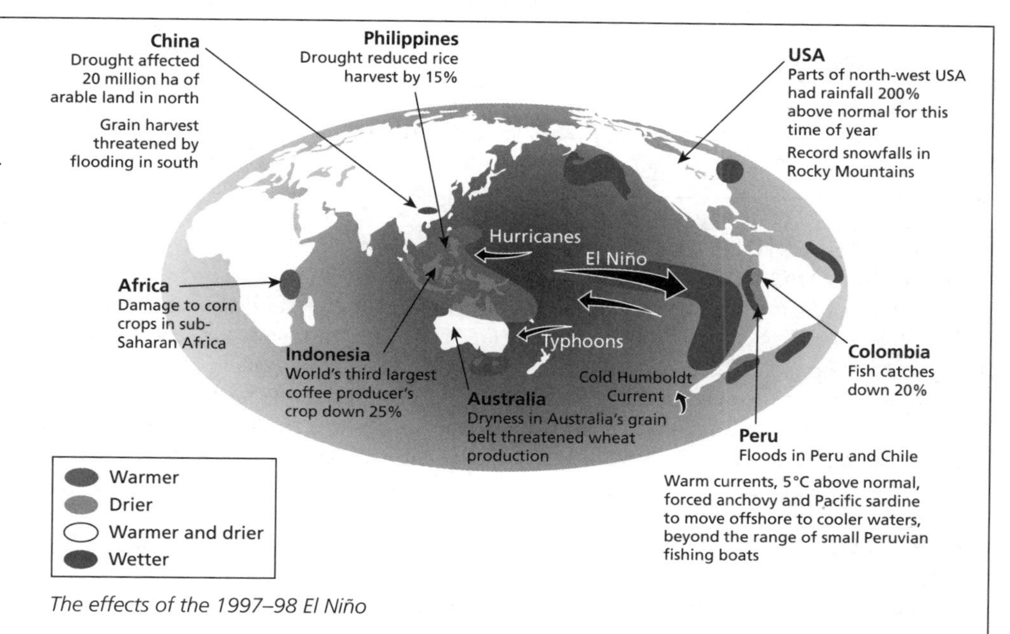

The effects of the 1997–98 El Niño

Oceans and resources (1)

Oceans are a rich source of resources.

Saltwater contains nutrients and minerals.

Oil and gas deposits are found in the continental shelf. The Persian Gulf accounts for 66% of the world's proven oil reserves and 33% of the world's proven gas reserves. The continental shelf area of the Gulf of Mexico has been explored and developed since the 1940s.

The continental shelf contains sediments such as gravel, sand and mud. These come from the erosion of rocks and are transported by rivers to the sea.

Diamonds can be found in the continental shelf areas off Africa and Indonesia.

Near ocean ridges and rift valleys there are rich deposits of sulphur – some associated with hydrothermal vents ("black smokers").

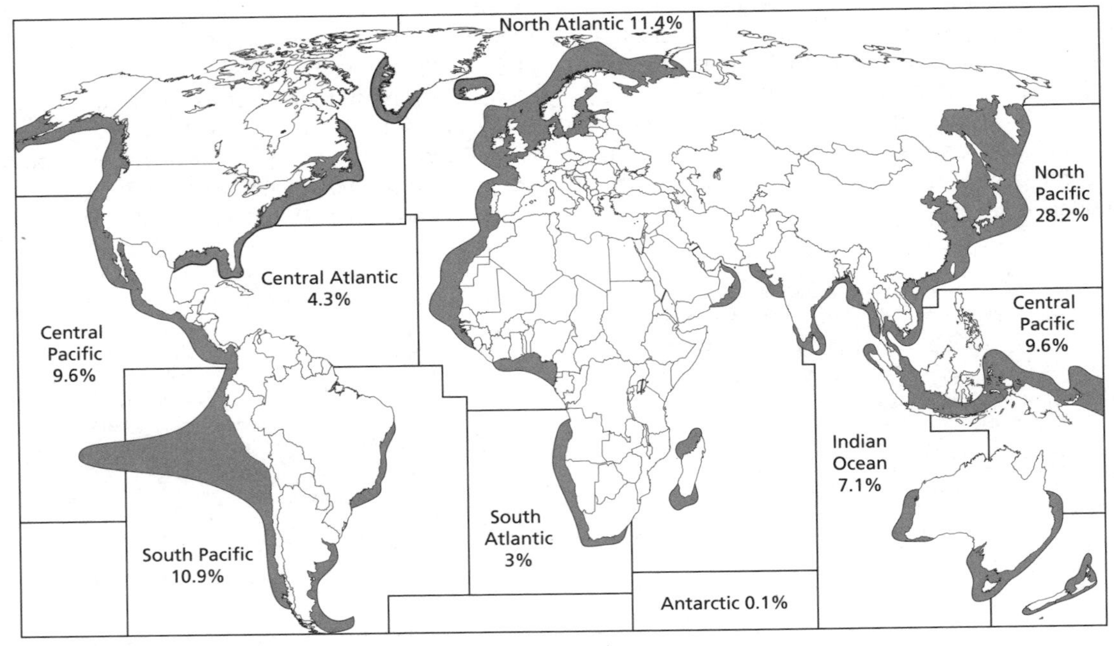

Main fishing grounds, showing percentage of world's catch (2001)

Gold and manganese are found on the ocean floor. Ocean floor sediments are formed of sand, mud and silt. Deep ocean floors are covered in ooze. Ocean sediments have a varied source. Some are fine silts carried by turbidity currents. Others come from sands and dust blown by wind off the continents. Some heavier material is carried by icebergs, and deposited as the ice gradually melts.

Authigenic sediments are precipitates of chemicals, such as iron oxide, from seawater, in forms such as manganese nodules. Manganese nodules are fist-sized and located on the abyssal plain. To date, no economic way has been developed for mining these chemicals. Biogenic ooze is the skeletal remains of microscopic organisms that once lived in the ocean.

The oceans provide a valuable supply of fish. The worldwide harvest of fish was 5 m tonnes in 1900 and about 90 m tonnes in 2000. Fish account for about 10% of the protein eaten by people. It is the only major food source still gathered from the wild.

Oceans vary in their ecological productivity. Net primary productivity (NPP) varies from 120 g/m^2/year in the open oceans to 360 g/m^2/year in the continental shelves. In contrast, estuaries have an NPP of 1500 g/m^2/year. The Gulf of Mexico has a very large fishing industry, especially shrimp and red snapper.

OCEANS AS A STORE AND SOURCE OF CARBON DIOXIDE

The major reservoirs of carbon dioxide are fossil fuels ($10{,}000 \times 10^{12}$ kg of carbon), the atmosphere (750×10^{12} kg of carbon) and the oceans ($38{,}000 \times 10^{12}$ kg of carbon). Oceans play a key role in the carbon cycle. Photosynthesis by plankton generates organic compounds of carbon dioxide. Some of this material passes through the food chain and sinks to the ocean floor, where it is decomposed into sediments. Eventually it is destroyed at subduction zones, where ocean crusts are subducted beneath the continental plates. Carbon dioxide is later released during volcanic activity. The transfer of carbon dioxide from ocean to atmosphere involves a very long time-scale.

EXTENSION

Visit
www.ozcoasts.org.au/glossary/images/carbon_cyclefig1.jpg for a diagram of the carbon cycle.

Oceans and resources (2)

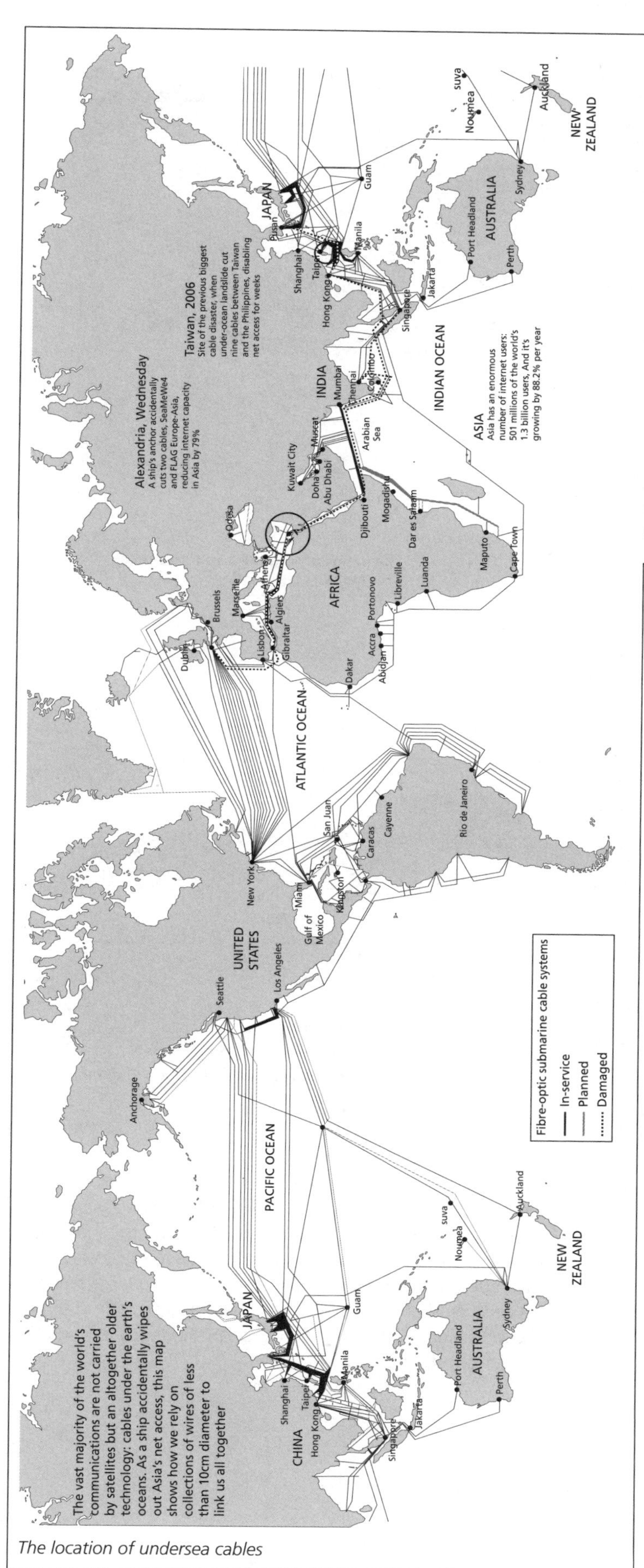

The location of undersea cables

UNDERWATER CABLES

The continental shelf and open ocean have also been used for the laying of cables. An internet blackout in January 2008, which left 75 million people with only limited access, was caused by a single ship that tried to moor off the coast of Egypt in bad weather. Telephone and internet traffic was severely reduced across a huge swath of the region, including India, Egypt and Dubai.

The incident highlighted the fragility of a global communications network. The impact of the blackout spread wide, with economies across Asia and the Middle East struggling to cope.

Despite the clean, hi-tech image of the online world, much of the planet remains totally reliant on real-world connections put in place through massive physical effort. The expensive fibre-optic cables are laid at great cost in huge lines around the globe, directing traffic backwards and forwards across continents and streaming millions of conversations simultaneously from one country to another.

EXTENSION

Flow lines

Flow lines show the volume of movement between places. The thickness of the line indicates the volume, and the direction can be shown by an arrow – or suggest two-way movement as shown on the diagram above. In many cases absolute vaules are plotted, although relative values can be shown.

When using this technique, remember:

- avoid clutter by keeping the background as simple as possible
- choose an appropriate scale so that extreme values can be shown without any loss of clarity
- provide a key.

Overfishing (1)

THE PROBLEM

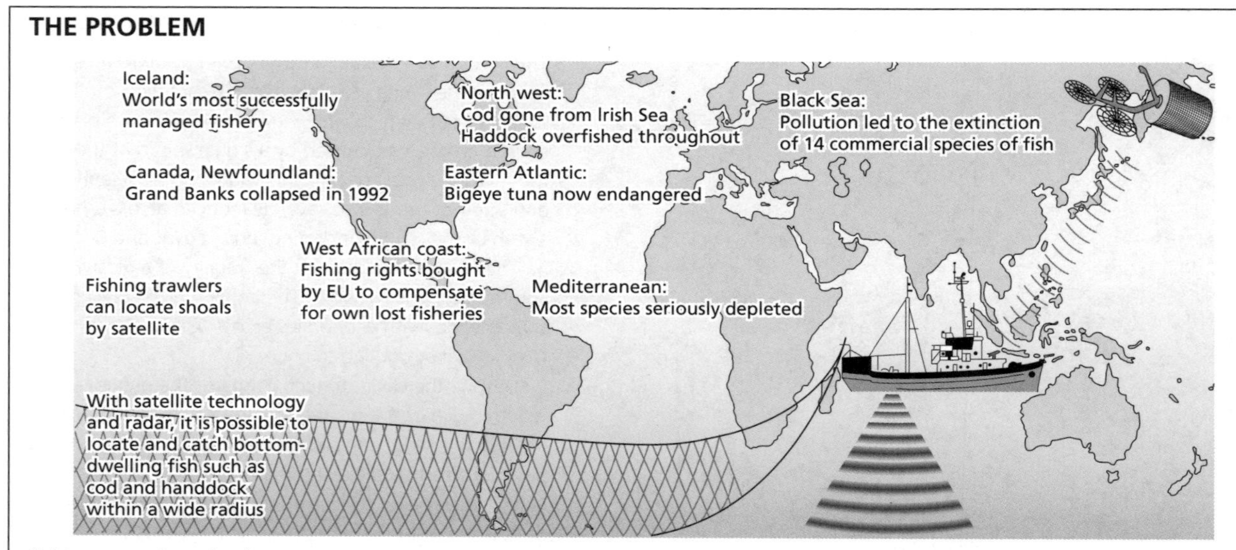

Fishing grounds under threat

Fishing fleets now catch fewer large, predatory fish, but more smaller fish further down the food chain. The most prized food fish, e.g. cod, which tend to be top-level predators, are declining, leaving smaller, less desirable fish. This not only affects the type of fish available for human consumption, it could change marine ecosystems for ever.

Larger, predatory fish need to eat large quantities of smaller fish. As their numbers fall, the numbers of smaller fish increase. This is why, despite overfishing of cod and other important species, total fish catches have remained high. However, the type of fish being caught is changing.

Even with larger boats and better technology, fish catches of species such as cod are falling. World fish stocks have declined rapidly – some species have become extinct. More and more ships are chasing fewer fish, and prices have risen quickly. Despite many attempts to save the fishing industry, e.g. through quotas and bans, there has been little success.

Nearly 70% of the world's stocks are in need of management. Cod stocks in the North Sea are less than 10% of 1970 levels. Fishing boats from the EU now regularly fish in other parts of the world, e.g. Africa and South America, to make up for the shortage of fish in EU waters. More than half of the fish consumed in Europe is now imported.

1890s	2
1900s	1
1910s	0
1920s	4
1930s	2
1940s	3
1950s	4
1960s	1
1970s	8
1980s	53
1990s	3

Fish extinctions

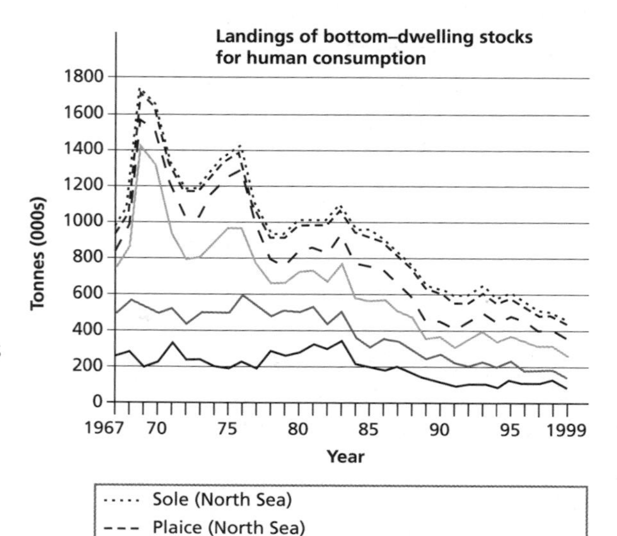

The decline in North Sea fisheries

Grand Banks

Once a fish stock is overfished, it is very difficult for it to recover. The Grand Banks area off Newfoundland was once the world's richest fishery. In 1992 it had to be closed to allow stocks to recover. It was expected to be closed for three years, but fish numbers, especially cod, have not yet recovered and it is still closed. The cod's niche in the ecosystem has been taken by other species, such as shrimp and langoustines.

Overfishing (2)

TOO MANY FISHERMEN, TOO FEW FISH

Many argue that measures such as quotas, bans and the closing of fishing areas still fail to address the real problems of the European fishing industry: too many fishermen are chasing too few fish and too many immature fish are being caught. For the fisheries to be protected and for the industry to be competitive on a world scale, the number of boats and the number of people employed in fishing must be reduced. At the same time, the efficiencies which come from improved technology must be embraced. A World Bank and FAO report in 2008 showed that up to $50bn per year is lost in poor management, inefficiency and overfishing in world fisheries. The report puts the total loss over the last 30 years at $2.2 trillion. The industry's fishing capacity continues to increase. The number of vessels is increasing slowly. However, each boat has greater capacity due to improved technology. Due to over-capacity, much of the investment in new technology is wasted. The amount of fish caught at sea has barely changed in the last decade. Fish stocks are depleted so the effort to catch the ones remaining is higher than it needs to be.

STRATEGIES FOR THE EUROPEAN FISHING INDUSTRY

The table suggests some possible strategies for the future, but there are no simple solutions to the problems associated with such a politically, economically and environmentally sensitive industry.

Action	Type of measure	Objectives
Conservation of resources		
Technical measures	Small meshed nets, minimum landing sizes, boxes	To protect juveniles and encourage breeding; to discourage marketing of illegal catches
Restriction of catches	TACs (total allowable catches) and quotas	To match supply to demand; to plan quota uptake throughout the season; to protect sensitive stocks
Limiting numbers of vessels	Fishing permits (which could be traded inter- or intra-nationally)	System applicable to EU vessels and other countries' vessels fishing in EU waters
Surveillance		
Checking of landings by EU and third-country vessels	Log books, computer/satellite surveillance	To apply penalties to overfishing and illegal landings
Structural		
	Structural aid to the fleet	To finance investment in fleet modernization (although commissioning of new vessels must be closely controlled), while providing reimbursement for scrapping, transfer and conversion
Reduction of unemployment, leading to an increase in productivity	Inclusion of zones dependent on fishing in Objectives 1, 2 and 5b of Structural Funds	To facilitate restructuring of the industry, to finance alternative local development initiatives to encourage voluntary/early retirement schemes
Markets		
Tariff policy	Minimum import prices, restrictions on imports	To ensure EU preference (although still bound under World Trade Organization)
Other measures		
Restrict number of vessels	Fishing licences	To discourage small, inefficient boats through large licence fees
Increase the accountability of fishermen	Rights to fisheries	Where fish stay put (e.g. shellfish), sections of the seabed can be auctioned off
		Where a whole fishery is controlled, quotas could be traded which would allow some fishermen to cash in and leave the sea

EXTENSION

Line graphs

Line graphs are quite simple graphs which show change over time. Line graphs use continuous data and they show trends. The changes can be relative or absolute. Line graphs can be simple – showing one feature, or multiple – showing many features, such as the graph on page 72 showing the change in landings of a number of North Sea fish species. In all line graphs there is an independent variable and a dependent variable. In this example the year is the independent variable (plotted on the x, or horizontal, axis) and the dependent variable is the fish catch (plotted on the y, or vertical, axis).

Pollution

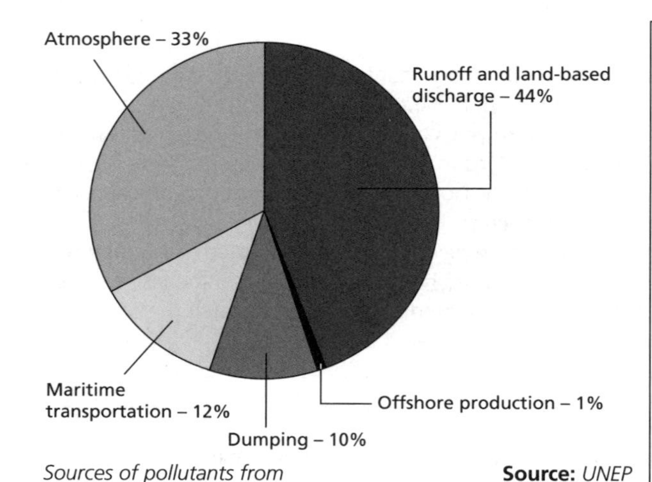

Sources of pollutants from human activities entering the sea

Source: *UNEP*

THREATS TO THE MARINE ENVIRONMENT

Less than a tenth of the sea floor has ever been explored; even so, the human hand is increasingly evident. Oil exploration is a major activity in regions such as the Gulf of Mexico, the South China Sea and the North Sea. The threats vary. There is growing evidence of widespread toxic effects on benthic communities on the floor of the North Sea in the vicinity of the 500+ oil production platforms in British and Norwegian waters. (*Benthic* means relating to the lowest layer of the ocean.) Meanwhile, oil exploration in the deep waters of the North Atlantic, north-west of Scotland, threatens endangered deep-sea corals. There is evidence, too, that acoustic prospecting for hydrocarbons in these waters may deter or disorientate some marine mammals.

In the future, the biological riches of the "black smokers" face threats from deep-sea mining. The mid-ocean hot springs spew out potentially valuable metal sulfides, such as gold, silver and copper. In the cold water, they are deposited in thick crusts, attracting exploitation. Rights have already been given to one company to prospect for metals on 4000 km^2 of the bed of the Bismarck Sea north of Papua New Guinea.

Role and importance of oceans

The oceans, like the atmosphere, are fundamental to the health of the planet. They dominate many of its cycling processes, as well as being the ultimate sink for a variety of pollutants. They absorb about 2 billion tonnes of carbon – in the form of carbon dioxide (CO_2) – and disperse an estimated 3 million tonnes of oil spilt annually from ships and, predominantly, from sources on land.

The oceans store a thousand times more heat than the atmosphere and transport enormous amounts of it around the globe. In consequence, they are largely responsible for determining climate on land. The warm Gulf Stream washing up from the tropics in the Atlantic Ocean keeps Europe many degrees warmer in winter than Hudson Bay on the opposite shore. The oscillation between El Niño and La Niña currents in the tropical Pacific Ocean fundamentally changes the weather across the ocean, flipping Indonesia, Australia and coastal South America into and out of droughts and floods.

All these processes now face disruption from the global scale of human activity, particularly climate change. Currently, the oceans moderate climate change by absorbing a third of the CO_2 emitted into the air by human activity. But several studies suggest that global warming will stratify the oceans and reduce their capacity to act as a CO_2 "sink" by 10–20% over the next century, accelerating warming.

RESPONSE TO THREATS

There have been some successes in the international handling of the marine environment. The International Whaling Commission's moratorium introduced in the mid-1980s has helped revive whale stocks. The United Nations Convention on the Law of the Sea, signed in 1982 but only entering into force in 1994, established a framework of law for the oceans, including rules for deep-sea mining and economic exclusion zones extending 200 nautical miles around nation states.

A series of international laws have effectively eliminated the discharge of toxic materials – from drums of radioactive waste to sewage sludge and air pollution from incinerator ships – into the waters around Europe. International public pressure in the mid-1990s forced the reversal by a major oil company of plans to scuttle the Brent Spar, a large structure from the North Sea offshore oil industry, into deep water west of Scotland. European agreements since then have indicated that all production platforms and other structures should be removed from the oilfields at the end of their lives wherever possible.

EXTENSION

Pie charts

Pie charts and proportional pie charts are frequently used on maps to show variations in the size and composition of a feature: in this example only the composition of pollution is shown.

Every 3.6° on the pie chart represents 1% of the circle. To plot vaules, convert them into percentages and multiply by 3.6 to work out the number of degrees to plot.

The advantges of pie charts include:

- they are easy to construct
- they are a striking visual techniqe
- they are relatively easy to read.

Disadvantages include:

- the over-emphasis of lage values
- they require time, care and patience to draw.

The geopolitics of oceans

EXCLUSIVE ECONOMIC ZONES

Exclusive economic zones (EEZs) have a profound impact on the management and conservation of ocean resources, since they recognize the right of coastal states to control over 98 million km^2 of ocean space. Coastal states are free to "exploit, develop and manage and conserve all resources – fish or oil, gas or gravel, nodules or sulphur – to be found in the waters, on the ocean floor and in the subsoil of an area, extending almost 200 nautical miles from its shore." Almost 90% of all known oil reserves under the sea fall under some country's EEZ. So too do the rich fishing areas – up to 98% of the world's fishing regions fall within an EEZ.

ASCENSION – A BRITISH EEZ IN THE SOUTH ATLANTIC?

The UK has claimed 200,000 km^2 of the Atlantic seabed surrounding Ascension Island, as the international race to establish sovereignty over underwater territories gains momentum. The mountainous ocean floor, up to 560 km from the isolated island in the South Atlantic, is believed to contain extensive mineral deposits. With no near neighbours, other states are unlikely to challenge the claim.

Ascension Island has a land area of around 100 km^2 but, due to its isolated location, it generates an EEZ with an area of more than 440,000 km^2. As mineral and energy prices have soared, there has been growing international interest in exploring the seabed for increasingly scarce reserves. The first deep-sea mining project – operating at depths of over 1600 m and aiming to extract gold, silver, copper and zinc from extinct volcanic vents – was due to start operating in the waters off Papua New Guinea in 2009.

The waters around Ascension Island are generally deeper than the Pacific and probably beyond current technological limits for extraction. The mid-Atlantic ridge does contain, however, similar volcanic black smoker vents that help concentrate valuable minerals. Britain has lodged, or is preparing, claims to underwater territories around Antarctica, the Falklands, Rockall in the north Atlantic and in the Bay of Biscay.

EXTENSION

Visit

http://www.geocities.com/aipsg/proc21-geo.html
for an article on the geopolitics of South Asia and the threat of war – it examines the changing role of the Indian Ocean.

http://www.lse.ac.uk/collections/alcoa/pdfs/berkmanpresentation.pdf
for a presentation on the Arctic Ocean geopolitics.

AN ARCTIC SCRAMBLE

In 2007, Russia claimed the North Pole by planting an underwater flag. In 2008, Canada, Denmark, Norway, Russia and the USA met in Greenland to discuss how to divide up the resources of the Arctic Ocean.

According to the US Geographical Survey, the Arctic could hold a quarter of the world's undiscovered gas and oil reserves. The five countries are racing to establish the limits of their territory, stretching far beyond their land borders. Climate change is a fact of the Arctic. The ice is melting and transport routes and natural resources which used to be inaccessible are opening up.

Environmental groups have criticized the scramble for the Arctic, saying it will damage unique animal habitats, and have called for a treaty similar to that regulating the Antarctic, which bans military activity and mineral mining.

Countries around the Arctic Ocean are rushing to stake claims on the Polar Basin seabed and its oil and gas reserves, made more tempting by rising energy prices. Resolving territorial disputes in the Arctic has gained urgency because scientists believe rising temperatures could leave most of the Arctic ice-free in summer months in a few decades' time. This would improve drilling access and open up the North-West Passage, a route through the Arctic Ocean linking the Atlantic and Pacific that would reduce the sea journey from New York to Singapore by thousands of miles.

Under the 1982 UN Law of the Sea Convention, coastal states own the seabed beyond existing 200 nautical mile (370 km) zones if it is part of a continental shelf of shallower waters. While the rules aim to fix shelves' outer limits on a clear geological basis, they have created a tangle of overlapping Arctic claims.

Coastal margins

PHYSICAL CHARACTERISTICS

Coastal environments are influenced by many factors, including physical and human processes. As a result, there is a great variety in coastal landscapes.

- **Geology properties** (rock): hard rocks such as granite and basalt give rugged landscapes, e.g. the Giant's Causeway in Northern Ireland, whereas soft rocks such as sands and gravels produce low, flat landscapes, e.g. around Poole Harbour on the south coast of England.
- **Geological structure: concordant** or **accordant** (Pacific) coastlines occur where the geological strata lie parallel to the coastline, e.g. the south coast of Ireland, whereas **discordant** (Atlantic-type) coastlines occur where the geological strata are at right angles to the shoreline, e.g. the south-west coast of Ireland.
- **Processes:** erosional landscapes, e.g. the east coast of England, contain many rapidly retreating cliffs, whereas areas of rapid deposition, e.g. the Netherlands, contain many sand dunes and coastal flats.
- **Sea-level changes** interact with erosional and depositional processes to produce advancing coasts (those growing due to either deposition and/or a relative fall in sea level) or retreating coasts (those being eroded and/or drowned by a relative rise in sea level).
- **Human impacts** are increasingly common – some coasts, e.g. in Florida, are extensively modified, whereas others are more natural, e.g. south-west Ireland.
- **Ecosystem type**, such as mangrove, coral, sand dune, salt marsh and rocky shore, adds further variety to the coastline.

WAVE REFRACTION

Wave refraction occurs when waves approach an irregular coastline or at an oblique angle (a). Refraction reduces wave velocity and, if complete, causes wave fronts to break parallel to the shore. Wave refraction concentrates energy on the flanks of headlands and dissipates energy in bays (b).

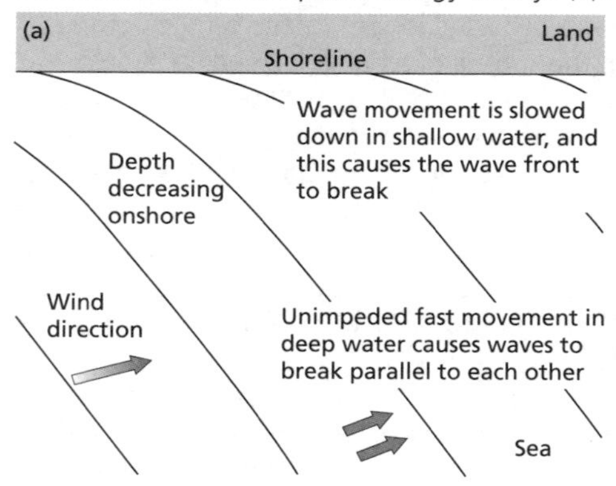

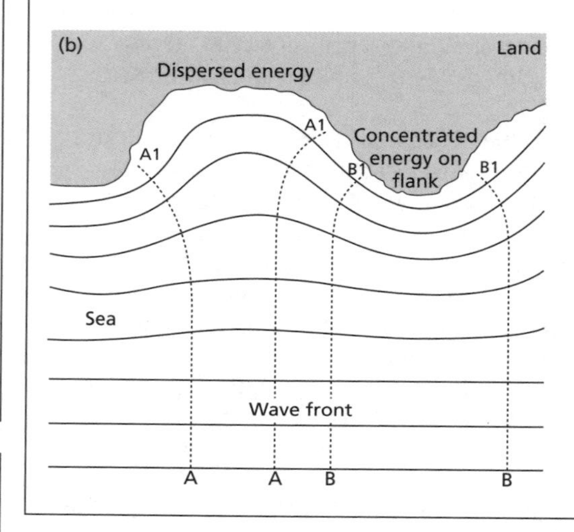

CONSTRUCTIVE AND DESTRUCTIVE WAVES

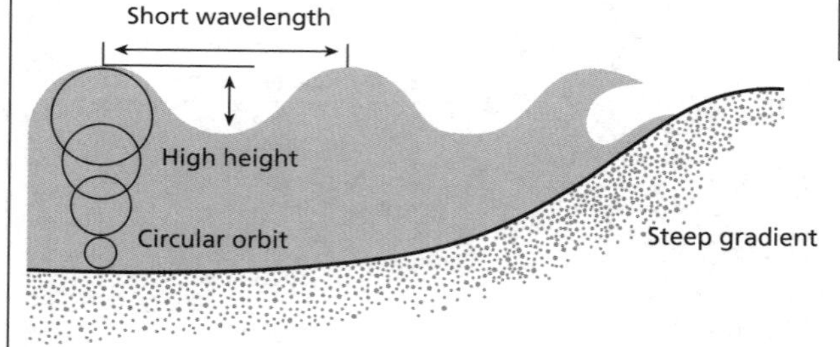

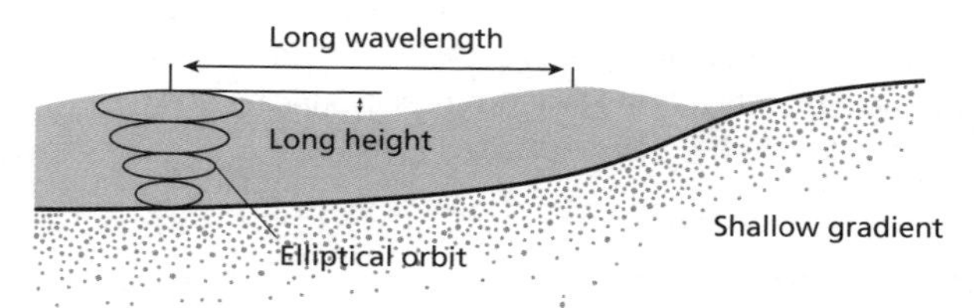

Destructive waves

- Erosional waves
- Also called "surging", "storm" or "plunging waves"
- Short wavelength, high height
- High frequency (10–12 per minute)
- Low period (one every 5–6 seconds)
- Backwash greater than swash
- Steep gradient
- High energy

Constructive waves

- Depositional waves
- Also called "spilling" or "swell" waves
- Long wavelength, low height
- Low frequency (6–8 per minute)
- High period (one every 8–10 seconds)
- Swash greater than backwash
- Low gradient
- Low energy

Coastal processes and landforms

EROSION

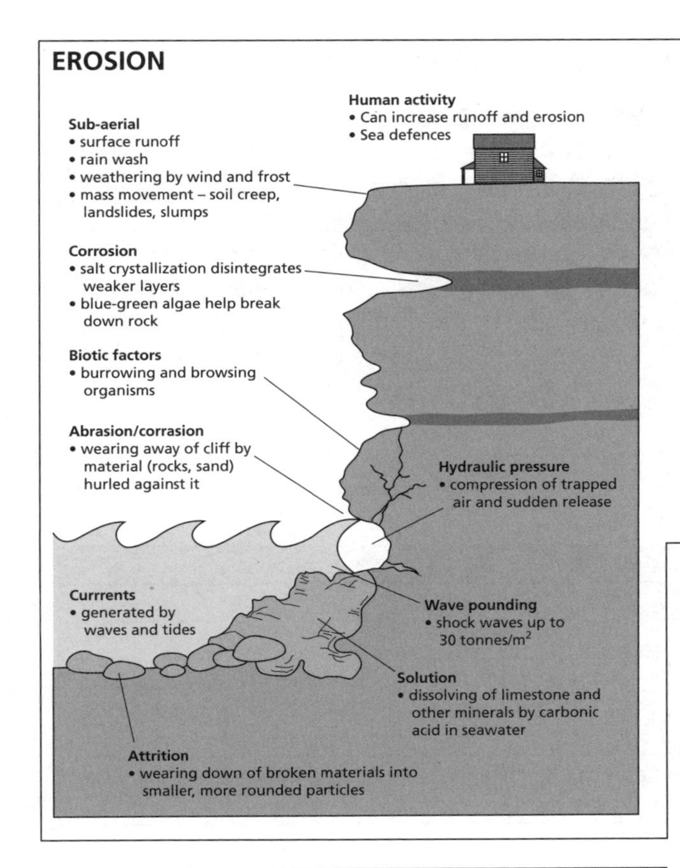

Coasts are shaped by the interplay of **marine** and **sub-aerial processes**. Marine, or cliff-foot, processes include:

- **abrasion**
- **hydraulic impact** or **quarrying**
- **solution**
- **attrition**.

Sub-aerial, or cliff-face, processes include:

- **salt weathering**: the process by which sodium and magnesium compounds expand in joints and cracks, thereby weakening rock structures
- **freeze–thaw weathering**: the process whereby water freezes, expands and degrades jointed rocks
- **biological weathering**: carried out by molluscs, sponges and urchins. It is very important in low-energy coasts.

SHORE PLATFORMS

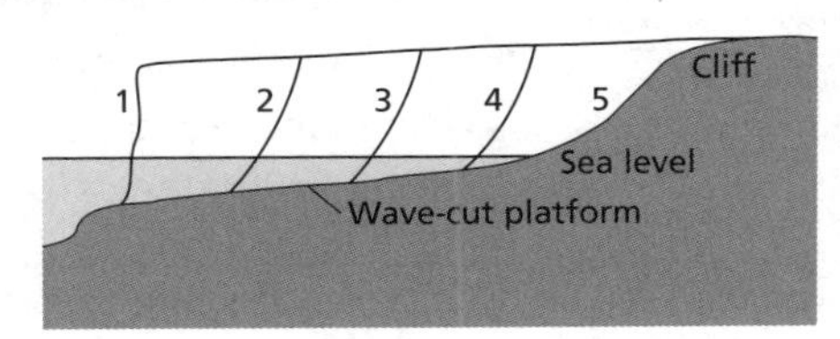

Shore platforms include **intertidal platforms (wave-cut platforms)**, **high-tide platforms** and **low-tide platforms**. Wave-cut platforms are most frequently found in high-energy environments and are typically less than 500 m wide with an angle of about 1°. Steep cliffs (1) are replaced by a lengthening platform and lower-angle cliffs (5), subjected to sub-aerial processes rather than marine forces. Alternatively, platforms might have been formed by frost action, salt weathering or biological action during lower sea levels and different climates.

LONGSHORE DRIFT

Refraction is rarely complete and consequently **longshore** or **littoral drift occurs**.

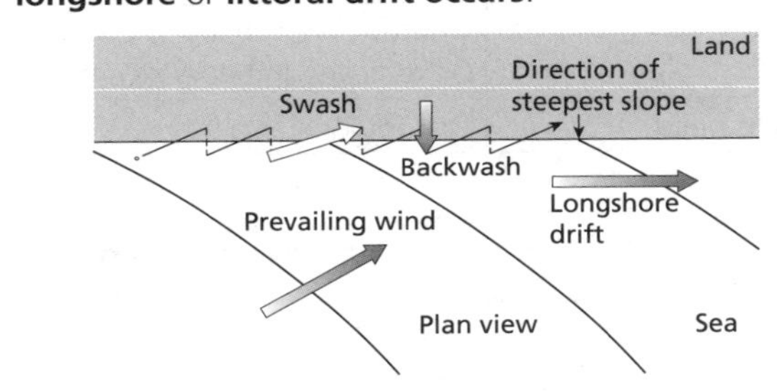

FEATURES OF DEPOSITION

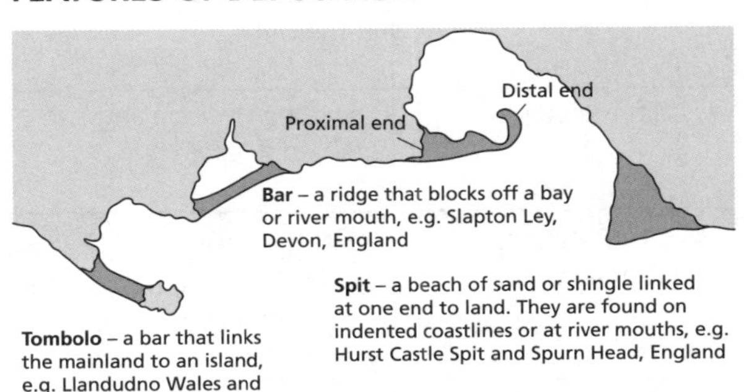

Essential requirements include:

- a large supply of material
- longshore drift
- an irregular, indented coastline, e.g. river mouths.

BEACH PROFILE

Storm beach – a noticeable, semi-permanent ridge, found at the level of the highest spring tides

Berms – small-scale beach ridges built up by successive levels of tides or storms

Cusps – semi-circular embayments found in the shingle or at the shingle–sand interface

Conflicts and management strategies

RELATIONSHIPS BETWEEN HUMAN ACTIVITIES AND COASTAL ZONE PROBLEMS

Human activity	Agents/consequences	Coastal zone problems
Urbanization and transport	Land-use changes; congestion; dredging of sediments	Loss of habitats and species diversity; lowering of groundwater table; saltwater intrusion
Tourism, recreation and hunting	Development and land-use changes (e.g. golf courses); ports and marinas	Loss of habitats and species diversity; disturbance; lowering of groundwater table; saltwater intrusion in aquifers
Fisheries and aquaculture	Port construction; fish processing facilities; fish farm effluents	Overfishing; impacts on non-target species; litter and oil on beaches; water pollution
Industry (including energy production)	Land-use changes; power stations; extraction of natural resources	Loss of habitats and species diversity; water pollution; eutrophication; thermal pollution

COASTAL MANAGEMENT

Type of management	Aims/methods	Strengths	Weaknesses
Hard engineering	**To control natural processes**		
Cliff-base management	*To stop cliff or beach erosion*		
• Sea walls	Large-scale concrete curved walls designed to reflect wave energy	Easily made; good in areas of high density	Expensive; lifespan about 30–40 years; foundations may be undermined
• Revetments	Porous design to absorb wave energy	Easily made; cheaper than sea walls	Lifespan limited
• Gabions	Rocks held in wire cages absorb wave energy	Cheaper than sea walls and revetments	Small scale
• Groynes	To prevent longshore drift	Relatively low costs; easily repaired	Cause erosion on downdrift side; interrupt sediment flow
• Rock armour	Large rocks at base of cliff to absorb wave energy	Cheap	Unattractive; small scale; may be removed in heavy storms
• Offshore breakwaters	To reduce wave power offshore	Cheap to build	Disrupt local ecology
• Rock strongpoints	To reduce longshore drift	Relatively low costs; easily repaired	Disrupt longshore drift; erosion downdrift
Cliff-face strategies	*To reduce the impacts of sub-aerial processes*		
• Cliff drainage	Removal of water from rocks in the cliff	Cost-effective	Drains may become new lines of weakness; dry cliffs may produce rockfalls
• Cliff regarding	Lowering of slope angle to make cliff safer	Useful on clay (most other measures are not)	Uses large amounts of land – impractical in heavily populated areas
Soft engineering	**Working with nature**		
• Offshore reefs	Waste materials, e.g. old tyres weighted down, to reduce speed of incoming waves	Low technology and relatively cost-effective	Long-term impacts unknown
• Beach nourishment	Sand pumped from seabed to replace eroded sand	Looks natural	Expensive; short-term solution
• Managed retreat	Coastline allowed to retreat in certain places	Cost-effective; maintains a natural coastline	Unpopular; political implications
• "Do nothing"	Accept that nature will win	Cost-effective!	Unpopular; political implications
• Red-lining	Planning permission withdrawn; new line of defences set back from existing coastline	Cost-effective	Unpopular; political implications

Coral reefs and mangroves

CORAL REEFS

Coral reefs are often described as the "rainforests of the sea" on account of their rich biodiversity and their vulnerability to destruction. Some coral is believed to be 2 million years old, although most is less than 10,000 years old. Coral reefs contain nearly a million species of plants and animals, and about 25% of the world's sea fish breed, grow, spawn and evade predators in coral reefs. Some of the world's best coral reefs include Australia's Great Barrier Reef, much of the reefs around the Philippines and Indonesia, Tanzania and the Comoros, and the Lesser Antilles in the Caribbean.

Pressures on coral reefs

Nearly two-thirds of the world's coral reefs are currently at risk from human activity. Destruction takes many forms:

- Construction of roads increases runoff, which can carry sediment from land-clearing areas, high levels of nutrients from agricultural areas, as well as many pollutants such as petroleum products or insecticides.
- Large sections of coral reefs are destroyed by boats dropping anchor or grounding. Fuel leakage is also damaging.
- Demand for souvenirs increases commercial exploitation of reefs.
- Fishing now uses dynamite to flush out fish and cyanide solution to catch live fish.
- Other destructive activities include the collection of specimens, mining for building and the cement industry, trampling.

In addition, indirect pressures include sedimentation from rivers and waste disposal from urban areas. Coastal development, especially for tourism, is taking its toll too.

Dust storms from the Sahara have introduced bacteria into Caribbean coral, while global warming may cause coral bleaching. Bleaching occurs when high temperatures expel the algae in coral, removing their colour – hence the coral appears bleached. Many areas of coral in the Indian Ocean were destroyed by the 2004 tsunami.

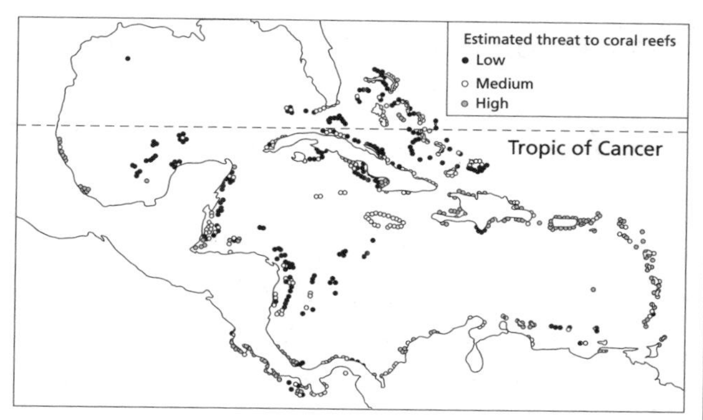

Tropic of Cancer

The value of coral

Coral reefs are of major biological and economic importance. Countries such as Barbados, the Seychelles and the Maldives rely on tourism. Florida's reefs attract tourism worth US$1.6 billion annually. The global value of coral reefs in terms of fisheries, tourism and coastal protection is estimated to be US$375 billion! Occupying less than 0.25% of the marine environment, they nevertheless shelter more than 25% of all known fish species.

MANGROVES

Mangroves are salt-tolerant forests of trees and shrubs that grow in the tidal estuaries and coastal zones of tropical areas. The muddy waters, rich in nutrients from decaying leaves and wood, are home to a great variety of sponges, worms, crustaceans, molluscs and algae. Mangroves cover about 25% of the tropical coastline, the largest being the 570,000 ha mangrove forest in the Sundarbans in Bangladesh.

The value of mangroves

Mangroves have many uses, such as providing large quantities of food and fuel, building materials and medicine. One hectare of mangrove in the Philippines can yield 400 kg of fish and 75 kg of shrimp. Mangroves also protect coastlines by absorbing the force of hurricanes and storms. They also act as natural filters, absorbing nutrients from farming and sewage disposal.

Pressures on mangroves

Despite their value, many mangrove areas have been lost to rice paddies and shrimp farms. As population growth in coastal areas is set to increase, the fate of mangroves looks bleak. Already most Caribbean and South Pacific mangroves have disappeared, while India, West Africa and South-East Asia have lost half of theirs.

Thailand	185,000 ha (1960–91) to shrimp ponds
Malaysia	235,000 ha (1980 and 1990) to shrimp ponds and farming
Indonesia	269,000 (1960–90) to shrimp ponds
Vietnam	104,000 (1960–74) due to US army
Philippines	170,000 ha (1967–76) mostly to shrimp ponds
Bangladesh	74,000 ha (since 1975) largely to shrimp ponds
Guatemala	9,500 ha (1965–84) to shrimp ponds and salt farming

Mangrove losses

Global distribution of extreme environments (1)

Extreme environments include, among others:

- cold and high-altitude environments (polar, glacial areas, periglacial areas; high mountains in non-tropical areas)
- hot, arid environments (hot deserts and semi-arid areas).

These areas are relatively inaccessible and tend to be viewed as inhospitable to human habitation. Despite this, they provide numerous opportunities for settlement and economic activity.

DISTRIBUTION OF EXTREME ENVIRONMENTS

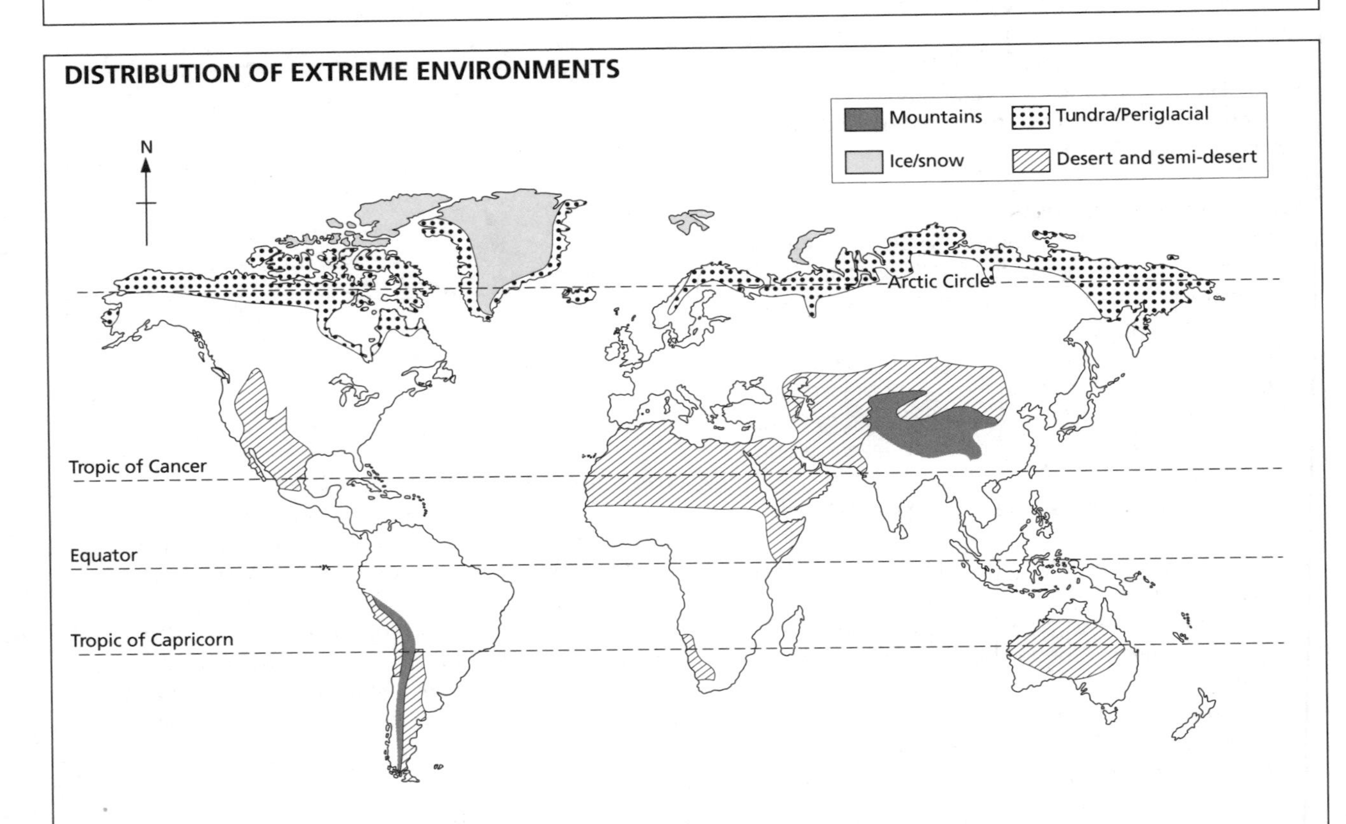

Cold and high-altitude environments

The distribution of cold environments is very uneven. Polar environments are located towards the North Pole and the South Pole, where levels of insolation are very low. In the northern hemisphere, there is a belt of periglacial environments (*periglacial* means "on the edge of glacial"). This zone is generally not found in the southern hemisphere except in small areas, given the relative lack of land mass at around 60–65°S.

Other cold environments are associated with high mountains. There are extensive areas of high ground in Asia, associated with the Himalayas; other high-altitude areas include the Andes and the Rockies. The mountains were formed as a result of tectonic activity:

- the Himalayas with the collision boundary between the Indian plate and the Eurasian plate
- the Andes with the collision and subduction of the oceanic Nazca plate under the South American plate
- much of the Rockies with the collision and subduction of the Juan de Fuca plate under the North American plate.

Desert and semi-arid environments

Desert and semi-desert areas cover as much as one-third of the earth's surface. They are generally located around the tropics and are associated with permanent high pressure systems which limit rain formation. There are four main factors which determine the location of the world's main deserts. They include:

- the presence of stable, high-pressure conditions at the tropics, e.g. the Sahara and the Great Australian deserts
- large distance from the sea (known as continentality), such as the central parts of the Sahara and Australia, and parts of the south-west USA
- rain-shadow effects as in Patagonia (South America) and the Gobi Desert in central Asia
- proximity to cold upwelling currents, which limit the amount of moisture held in the air, e.g. off the west coast of South America, helping to form the Atracama Desert, and off the west coast of southern Africa, helping to form the Namib Desert.

Global distribution of extreme environments (2)

CONDITIONS IN EXTREME ENVIRONMENTS

Cold and high-altitude environments

Cold environments are very varied in their characteristics. Mountain environments can be characterized by warm days and very cold nights. They may also receive large amounts of rainfall due to relief rain. Other mountain areas are in a rain-shadow area and receive low rainfall. Polar areas generally receive low rainfall. They are, in effect, cold deserts.

Owing to their steep nature, mountains are difficult areas to build on, and they act as barriers to transport. Soils are often thin, and suffer from high rates of overland runoff and erosion. In contrast, in periglacial areas – or tundra regions – the low temperatures produce low rates of evaporation and soils are frequently waterlogged. The growing season is relatively short – temperatures are above 6 °C for only a few months of the year.

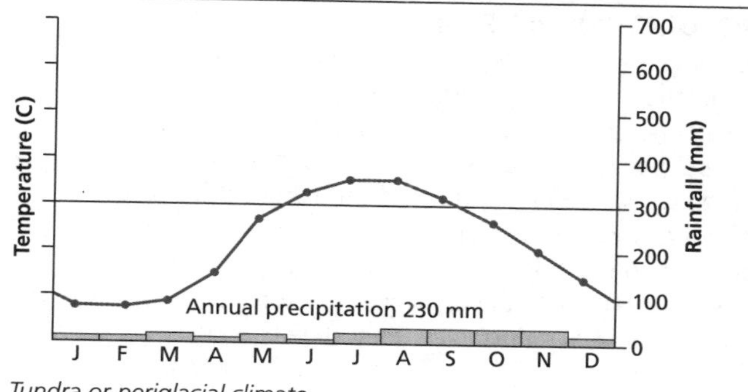

Tundra or periglacial climate

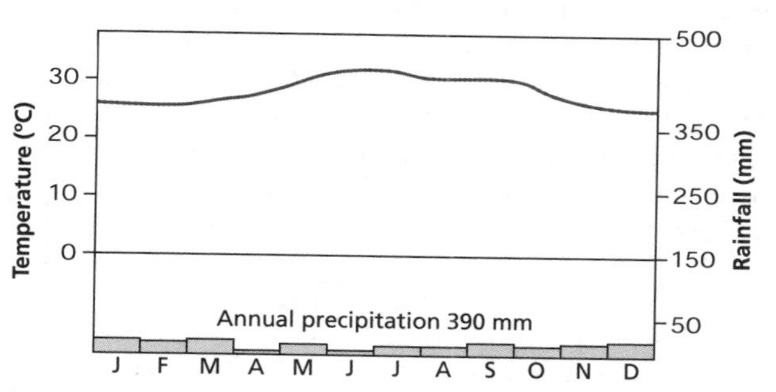

Desert climate

Desert and semi-arid environments

In desert areas, such as Aden, the lack of water acts as a major constraint for development. Temperatures are hot throughout the year but, in the absence of fresh water, farming, for example, is almost impossible. In semi-arid areas, annual rainfall varies between 250 and 500 mm, so there is some possibility for farming, especially where water conservation methods are used. On the other hand, the guarantee of warm, dry conditions could be excellent for tourism developments, especially in coastal areas, such as the Red Sea coast of Egypt.

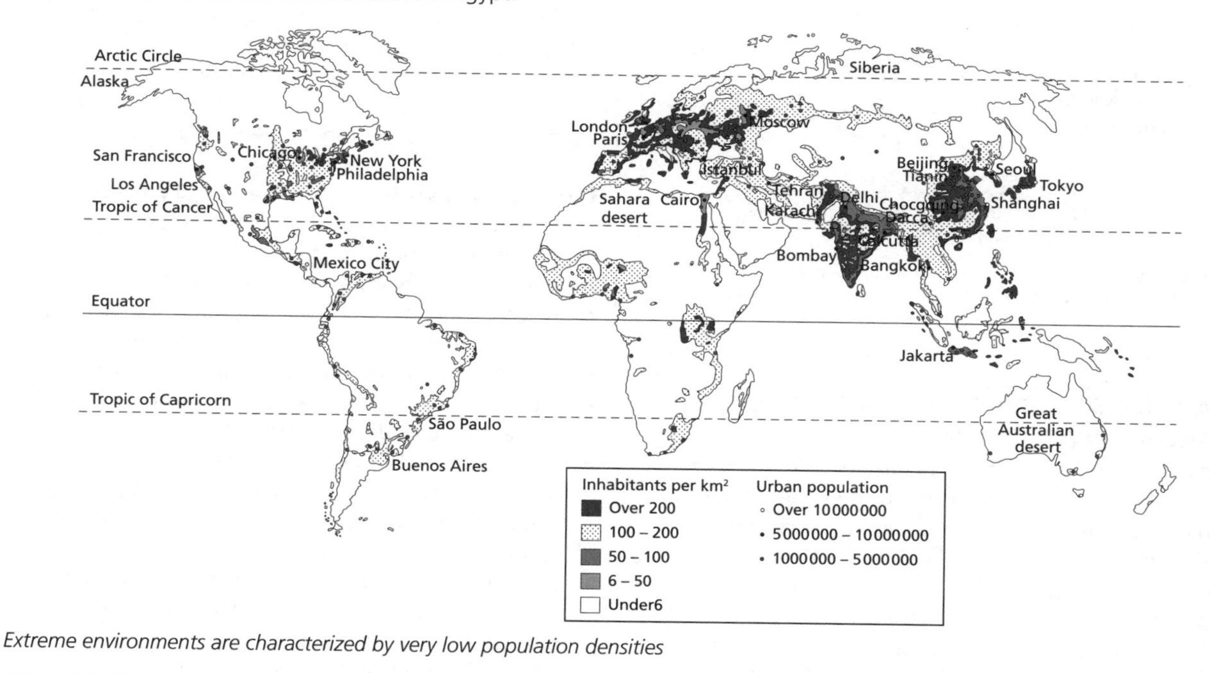

Extreme environments are characterized by very low population densities

People in extreme environments

POPULATION DENSITIES

Extreme environments are characterized by low population densities. Examples include densities of three people per km² in Australia, Iceland and Canada; two people per km² in Namibia; and just one person per km² in the western Sahara. Much of this can be put down to the extremes of climate: insufficient heat in Iceland and Canada, and insufficient water in the other three areas, are largely to blame. None of these environments is particularly "comfortable"; they all fall a long way outside the recognized "comfort zones for human habitation".

Other factors are important, too. Iceland is relatively remote and isolated. This makes communications costly, if not difficult. It also increases the cost of materials which have to be imported, such as timber for building. Similarly, Namibia is a long way from the economic core of southern Africa, and this increases the costs of imports and exports. Coastal areas are better off than inland areas but are still relatively undeveloped.

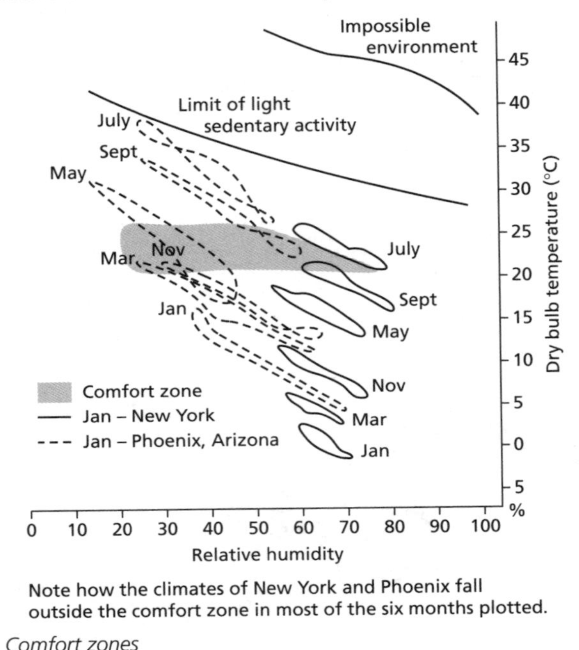

Comfort zones

TRADITIONAL COPING MECHANISMS IN EXTREME ENVIRONMENTS

Traditionally, periglacial pastures have been used by Inuit for herding or hunting caribou. The Inuit tend to be migratory, moving north into the tundra during the brief months of summer and heading southwards to the forest margins in winter. The Sami of Scandinavia also follow this pattern. To make up for the lack of decent pasture on land, many indigenous peoples have turned to rivers and the oceans. Fishing is extremely important in periglacial environments. For the Nenetsky of the Yamal Peninsula in Siberia, it is an important supplement to their diet. At the other end of the scale, fishing and fish-related products accounted for up to 70% of Iceland's GDP (2006). To cope with the cold conditions, Inuit populations have evolved a layer of fat which protects them from the extreme cold.

Desert inhabitants are also migratory. The Bedouin and the Fulani are excellent examples. To cope with the extreme temperatures in the daytime they avoid the direct sun and take a rest. They tend to travel in early morning and late afternoon. Their clothing – loose fitting garments – also helps them to cope with high temperatures. It reduces sweating and allows them to remain reasonably fresh.

Coping in the Sahel

The indigenous people of the Sahel in North Africa have adapted to these environmental conditions by a combination of strategies. As pastoralists, they make use of the limited resources of the Sahel and combat overgrazing by migrating to areas of seasonal growth while there is an opportunity. In doing so, they tend to leave vegetation around more permanent water sources for times when they will need it later. Such migration patterns also utilize arid areas that are not suitable for cultivation. The livestock herds are diversified – cattle are kept for income in the meat market, sheep and goats for milk and meat for internal consumption. Herd diversification also allows pastoralists to make use of a greater variety of the available vegetation resources because the animals have different grazing patterns. The diet of the indigenous people varies with conditions. More milk is consumed in the wetter periods, with meat being more common in the drier periods. Their animals are bartered with sedentary farmers for grain.

Coping with water shortages in dry areas

Some solutions are "natural" and require farmers to adapt to the natural environment. Adaptations to water shortages, both directly and indirectly, include:

- increased mobility (the traditional way of dealing with insufficient amounts of rainfall and pasture)
- management of size and composition of herds
- exchange of livestock and livestock products
- increased use of drought-tolerant species
- utilization of wild species and tree crops
- windbreaks to reduce wind erosion of bare soil
- irrigating with silt-laden river water to restore soil in badly eroded areas
- dune stabilization using straw checkerboards and planted xerophytes (plants which can withstand prolonged water shortage)
- land enclosure to reduce wind erosion.

Glacial environments

GLACIAL SYSTEMS

A glacial system is the balance between inputs, storage and outputs. Inputs include **accumulation** of snow, avalanches, debris, heat and meltwater. The main store is that of ice, but the glacier also carries debris, **moraine** and meltwater. The outputs are the losses due to **ablation**, the melting of snow and ice, and sublimation of ice to vapour, as well as sediment.

The **regime** of the glacier refers to whether the glacier is advancing or retreating:

- If accumulation > ablation, the glacier advances.
- If accumulation < ablation, the glacier retreats.
- If accumulation = ablation, the glacier is steady.

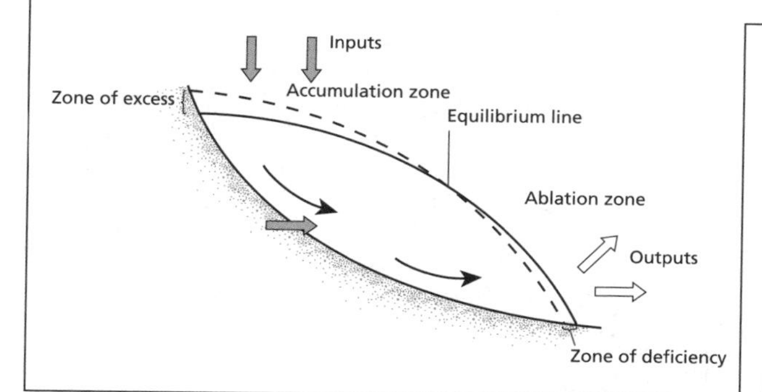

- Glacial systems can be studied on an annual basis or on a much longer time-scale. The size of a glacier depends on its regime, i.e. the balance between the rate and amount of supply of ice and the amount and rate of ice loss. The glacier will have a positive regime when the supply is greater than loss by ablation (melting, evaporation, calving, wind erosion, avalanche, etc.) and so the glacier will thicken and advance. A negative regime will occur when the wasting is greater than the supply (e.g. the Rhone glacier today) and so the glacier will thin and retreat. Any glacier can be divided into two sections: an area of accumulation at high altitudes generally, and an area of ablation at the snout.

GLACIAL EROSION

The amount and rate of erosion depends on the local geology, the velocity of the glacier, the weight and thickness of the ice, and the amount and character of the load carried. The methods of glacial erosion include plucking and abrasion.

Plucking

This occurs mostly at the base of the glacier and to an extent at the side. It is most effective in jointed rocks or those weakened by freeze–thaw. As the ice moves, meltwater seeps into the joints and freezes onto the rock, which is then ripped out by the moving glacier.

Abrasion

The debris carried by the glacier scrapes and scratches the rock, leaving **striations**.

Other mechanisms

Other mechanisms include meltwater, freeze–thaw weathering and pressure release. Although not strictly glacial nor erosional, these processes are crucial in the development of glacial scenery.

LANDFORMS PRODUCED BY GLACIAL EROSION

Cirques

In the northern hemisphere, these are generally found on north- or east-facing slopes where accumulation is highest and ablation is lowest. They are formed in stages:

1. A preglacial hollow is enlarged by **nivation** (freeze–thaw and removal by snow melt).
2. Ice accumulates in the hollow.
3. Having reached a critical weight and depth, the ice moves out in a rotational manner, eroding the floor by plucking and abrasion.
4. Meltwater trickles down the bergschrund, allowing the cirque to grow by freeze–thaw. (A *bergschrund* is a crevasse that forms when the moving glacier ice separates from the non-moving ice above.)

After glaciation, an armchair-shaped hollow remains, frequently filled with a lake, e.g. Blue Lake cirque, New South Wales, Australia.

Arêtes, peaks, troughs, basins and hanging valleys

Other features of glacial erosion include **arêtes** and **pyramidal peaks** (horns) caused by the headward recession (cutting back) of two or more cirques. Glacial **troughs** (or U-shaped valleys) have steep sides and flat floors. In plan view they are straight, since they have truncated the interlocking spurs of the preglacial valley. The ice may also carve deep **rock basins** frequently filled with **ribbon lakes**. **Hanging valleys** are formed by tributary glaciers which, unlike rivers, do not cut down to the level of the main valley, but are left suspended above, e.g. Stickle Beck in the Lake District, UK. They are usually marked by waterfalls.

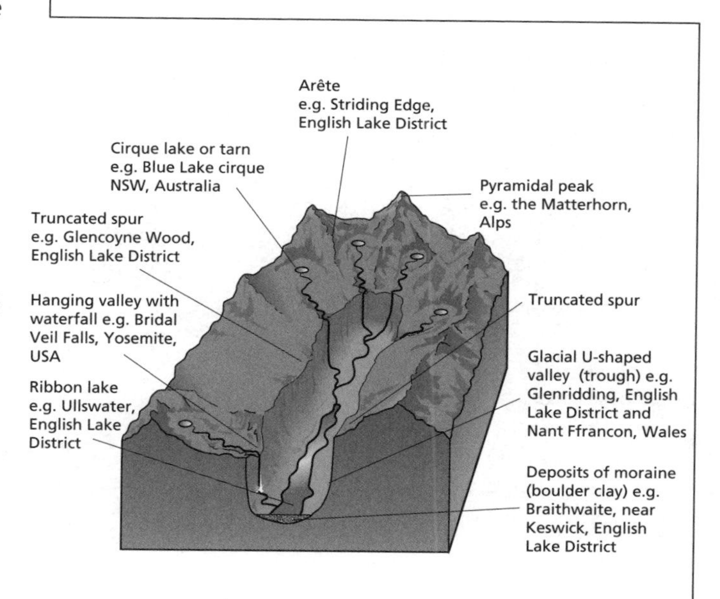

Glacial deposition

DEPOSITIONAL FEATURES

The term **drift** refers to all glacial and fluvioglacial deposits left after the ice has melted. Glacial deposits, or **till**, are angular and unsorted, and include erratics, drumlins and moraines. Till is often subdivided into **lodgement till**, material dropped by actively moving glaciers, and **ablation till**, deposits dropped by stagnant or retreating ice.

Characteristics of till

- Poor sorting – till contains a large range of grain sizes, e.g. boulders, pebbles, clay
- Poor stratification – no regular sorting by size
- Mixture of rock types – from a variety of sources
- Many particles have striations
- Long axis orientated in the direction of glacier flow
- Some compaction of deposits
- Mostly subangular particles

Erratics

Erratics are large boulders foreign to the local geology, e.g. the Madison Boulder in New Hampshire, USA, which is estimated to weigh over 4600 tonnes.

Moraines

Moraines are lines of loose rocks, weathered from the valley sides and carried by the glaciers. At the snout of the glacier is a crescent-shaped mound of **terminal moraine**. Its character is determined by the load the glacier was carrying, the speed of movement and the rate of retreat. The **ice-contact slope** (up-valley) is always steeper than the down-valley slope. Cape Cod in Massachusetts, USA, is a fine example of a terminal moraine.

Lateral moraines are ridges of materials found on the edge of a glacier. The lateral moraines on the Gorner Glacier in Switzerland are good examples. Where two glaciers merge and the two touching lateral moraines flow in the middle of the enlarged glacier, they are known as medial moraines. Again, the Gorner Glacier contains many examples of medial moraines.

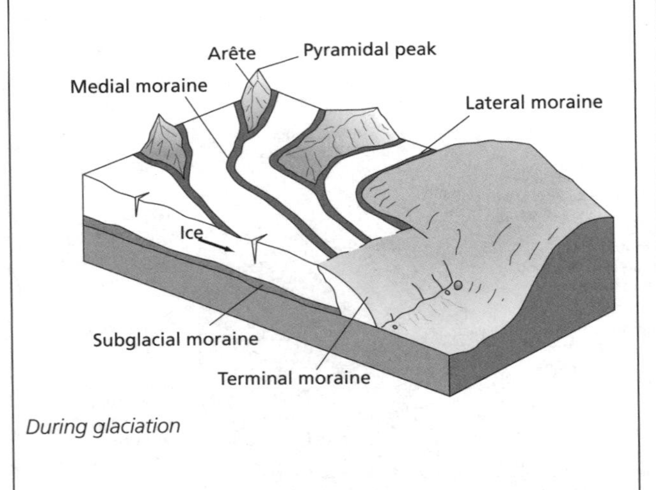

During glaciation

Drumlins

Drumlins are small oval mounds up to 1.5 km long and 100 m high, e.g. the drowned drumlins of Clew Bay in County Mayo, Ireland. One of the largest concentrations is in New York state, where there are over 10,000 drumlins. They are deposited due to friction between the ice and the underlying geology, causing the glacier to drop its load. As the glacier continues to advance, it streamlines the mounds.

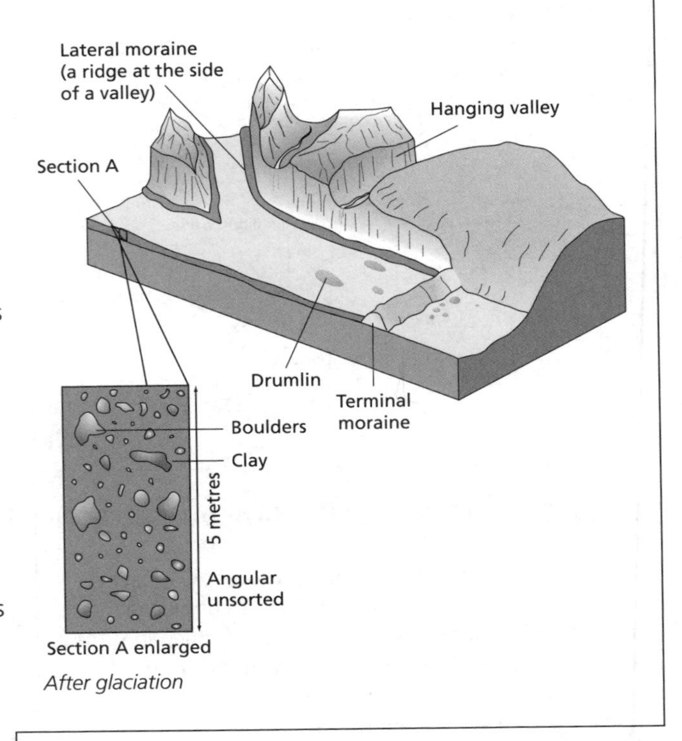

After glaciation

DIRECTION OF GLACIER MOVEMENT

These features can be used to determine the direction of glacier movement. Erratics pinpoint the origin of the material; drumlins and the long axes of pebbles in glacial till are orientated in the direction of glacier movement.

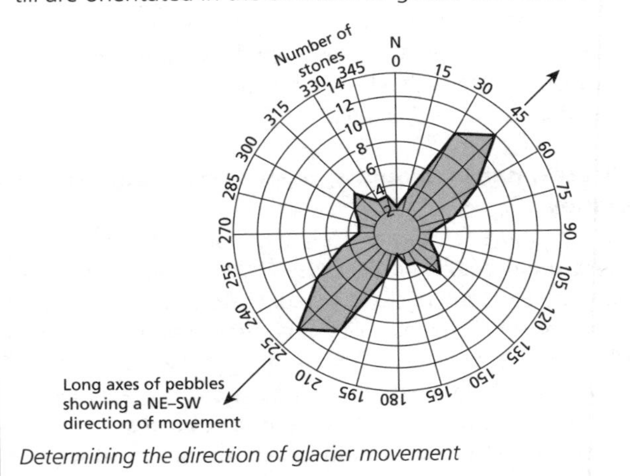

Determining the direction of glacier movement

Periglacial environments

Periglacial areas are found on the edge of glaciers or ice masses and are characterized by permafrost, impermeable permanently frozen ground, and freeze–thaw action. Summer temperatures rise above freezing, so ice melts.

Three types of periglacial region can be identified: Arctic continental, Alpine and Arctic maritime. These vary in terms of mean annual temperature and therefore the frequency and intensity with which processes operate.

PERMAFROST

Approximately 20% of the world's surface is underlain by permafrost, in places up to 700 m deep. Three types of permafrost exist: continuous, discontinuous and sporadic; these are associated with mean annual temperatures of −5° to −50 °C, −1.5° to −5 °C and 0° to −1.5 °C respectively. Above the permafrost is the active layer, a highly mobile layer which seasonally thaws out and is associated with intense mass movements. The depth of the active layer depends on the amount of heat it receives, and varies in Siberia from 0.2–1.6 m at 70 °N to between 0.7 and 4 m at 50 °N.

SOLIFLUCTION

Solifluction literally means flowing soil. In winter, water freezes in the soil causing expansion of the soil and segregation of individual soil particles. In spring, the ice melts and water flows downhill. It cannot infiltrate into the soil because of the impermeable permafrost. As it moves over the permafrost it carries segregated soil particles (peds) and deposits them further downslope as a solifluction lobe or terracette.

PATTERNED GROUND

Patterned ground is a general term describing the stone circles, polygons and stripes that are found in soils subjected to intense frost action, e.g. on the slopes of Kerio crater, southern Iceland. On steeper slopes, stone stripes replace stone circles and polygons. Their exact mode of formation is unclear, although ice sorting, differential frost heave, solifluction and the effect of vegetation are widely held to be responsible.

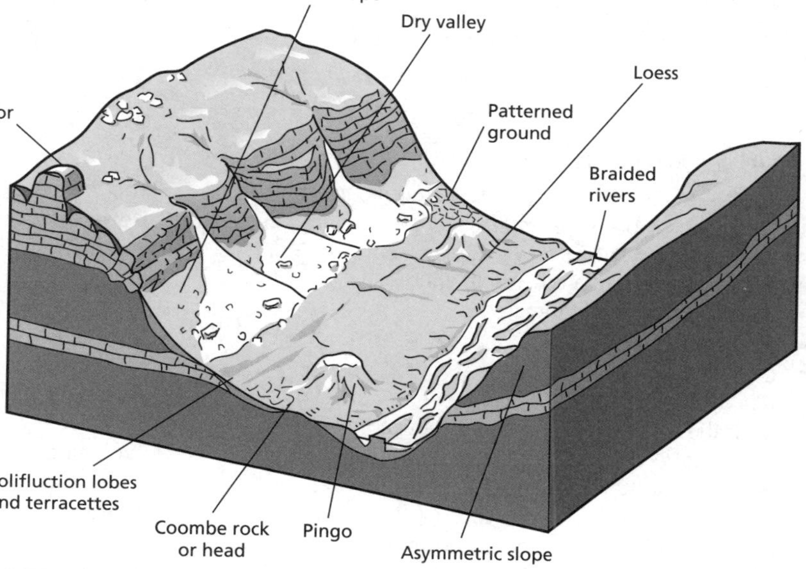

PINGOS

A **pingo** is an isolated, conical hill up to 90 m high and 800 m wide, which can only develop in periglacial areas. Pingos form as a result of the movement and freezing of water under pressure. Two types are generally identified: **open-system** and **closed-system** pingos. Where the water is from a distant elevated source, open-system pingos are formed, whereas if the supply of water is local, and the pingo is formed as a result of the expansion of permafrost, closed-system pingos are formed. Nearly 1500 pingos are found in the Mackenzie Delta of Canada. When a pingo collapses, ramparts and ponds are left.

THERMOKARST

Thermokarst refers to subsidence caused by the melting of permafrost. This may be because of broad climatic changes or local environment changes.

Local environmental changes include:

- changes in vegetation, which may affect the albedo (reflectivity of the surface)
- shifting of stream channels, which may affect the amount of heat coming in contact with permafrost
- fire, which rapidly destroys permafrost
- disruption of vegetation by human activity, which may remove surface layers and so open the permafrost to raised air temperatures in summer.

For example, the clearing of the forest for agricultural purposes near Fairbanks, Alaska, in the early 1920s, led to the development of an extensive pattern of thermokarst mounds, varying in diameter from 3 to 15 m, and in height from 0.3 to 2.4 m.

Hot, arid environments (1)

WEATHERING EROSION AND DEPOSITION

Weathering in deserts

Salt crystallization causes the decomposition of rock by solutions of salt. There are two main types of **salt crystal growth**. First, in areas where temperatures fluctuate around 26–28 °C, sodium sulphate (Na_2SO_4) and sodium carbonate (Na_2CO_3) expand by about 300%. This creates pressure on joints, forcing them to crack. Second, when water evaporates, salt crystals may be left behind. As the temperature rises, the salts expand and exert pressure on rock. Both mechanisms are frequent in hot desert regions, where low rainfall and high temperatures cause salts to accumulate just below the surface.

Disintegration is found in hot desert areas where there is a large diurnal temperature range. In many desert areas, daytime temperatures exceed 40 °C, whereas night-time ones are little above freezing. Rocks heat up by day and contract by night. As rock is a poor conductor of heat, stresses occur only in the outer layers. This causes peeling or **exfoliation** to occur. Griggs (1936) showed that moisture is essential for this to happen. In the absence of moisture, temperature change alone does not cause rocks to break down. It is possible that the expansion of many salts, such as sodium, calcium, potassium and magnesium, can be linked with the exfoliation.

Arid	Egypt	0.0001–2.0
Semi arid	Australia	0.6–1.0

Rates of weathering (mm/yr^{-1})

Weathering produces regolith, a superficial and unconsolidated layer above the solid rock. This material is easily transported and eroded, and may be used to erode other materials.

	Annual temperature (°C)	Annual rainfall (mm)	Processes
Semi arid	5–30	250–600	Strong wind action, running water
Arid	15–30	0–350	Strong wind action, slight water action

Peltier's classification of regions and their distinctive processes

WIND ACTION IN DESERTS

Many of the world's great deserts are dominated by subtropical high-pressure systems. Large areas are affected by trade winds, while local winds play a part too. Wind action is important in areas where winds:

- are strong (over 20 km/h)
- are turbulent
- come largely from a constant direction
- blow for a long period of time.
- Near the surface, wind speed is reduced by friction (the rougher the ground, the more the wind speed is reduced but the more turbulent it becomes).

Sediment is more likely to be moved if there is a lack of vegetation, and if it is dry, loose and small.

Movement of sediment is induced by **drag** and **lift** forces, but is reduced by **particle size** and **friction**. Drag results from differences in pressure on the windward and leeward sides of grains in an airflow.

There are two types of wind erosion.

- **Deflation** is the progressive removal of small material leaving behind larger materials. This forms a stony desert or reg. In some cases, deflation may remove sand to form a deflation hollow. One of the best known is the Qattara Depression in Egypt, which reaches a depth of over 130 m below sea level.
- **Abrasion** is erosion carried out by wind-borne particles. They act like sandpaper, smoothing surfaces and exploiting weaker rocks. Most abrasion occurs within a metre of the surface, since this is where the largest, heaviest, most erosive particles are carried. Examples of erosional features carved out by abrasion include yardangs, zeugens and ventifacts.
- Winds deposit the sand they carry as dunes. There are many types of dunes. Their shape and size depend on the supply of sand, the wind direction, the nature of the ground surface and the presence of vegetation.

Sand-sized particles (0.15–2.0 mm) are moved by three processes:

- **suspension** (<0.15 mm) – particles light enough to be carried substantial distances by the wind
- **saltation** (0.15–0.25 mm) – a rolling particle gains sufficient velocity for it to leave the sand surface in one or more "jumps"
- **surface creep** (0.25–2.0 mm) – larger grains are dislodged by saltating grains

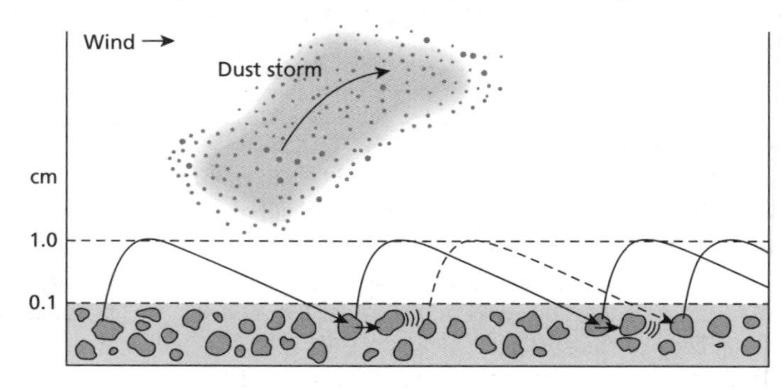

Wind transport in desert areas

Hot, arid environments (2)

THE WORK OF WATER

Water is vital for the development of many desert landforms. It is important for the operation of mechanical and chemical weathering in deserts, and it is important for erosion, too. There are a number of sources of water in deserts:

- Rainfall may be low and irregular but it does occur, mostly as low-intensity events, although there are occasional flash floods.
- Deflation may expose the water table to produce an oasis.
- Rivers flow through deserts – these can be classified as exotic (exogenous), endoreic and ephemeral.

Exotic or **exogenous** rivers are those which have their source in another, wetter environment and then flow through a desert. The Nile in Egypt is an exotic river, being fed by the White Nile, which rises in the equatorial Lake Victoria, as are the Blue Nile and Atbara, which rise in the monsoonal Ethiopian Highlands.

Endoreic rivers are those that drain into an inland lake or sea. The River Jordan, which drains into the Dead Sea, is a good example.

Ephemeral rivers are those which flow seasonally or after storms. Often they are characterized by high discharges and high sediment levels. Even on slopes as gentle as 2°, overland flow can generate considerable discharges. This is a result of many factors, including:

- an impermeable surface (in places)
- limited interception (lack of vegetation)
- rain-splash erosion displacing fine particles, which in turn seal off the surface and make it impermeable.

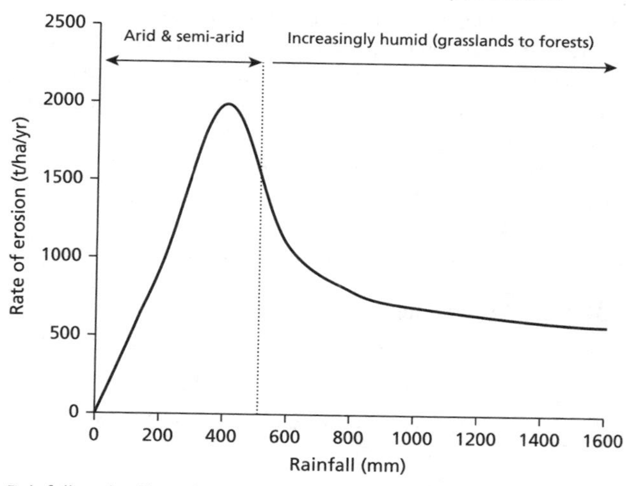

Rainfall and soil erosion

FEATURES IN THE ARID LANDSCAPE

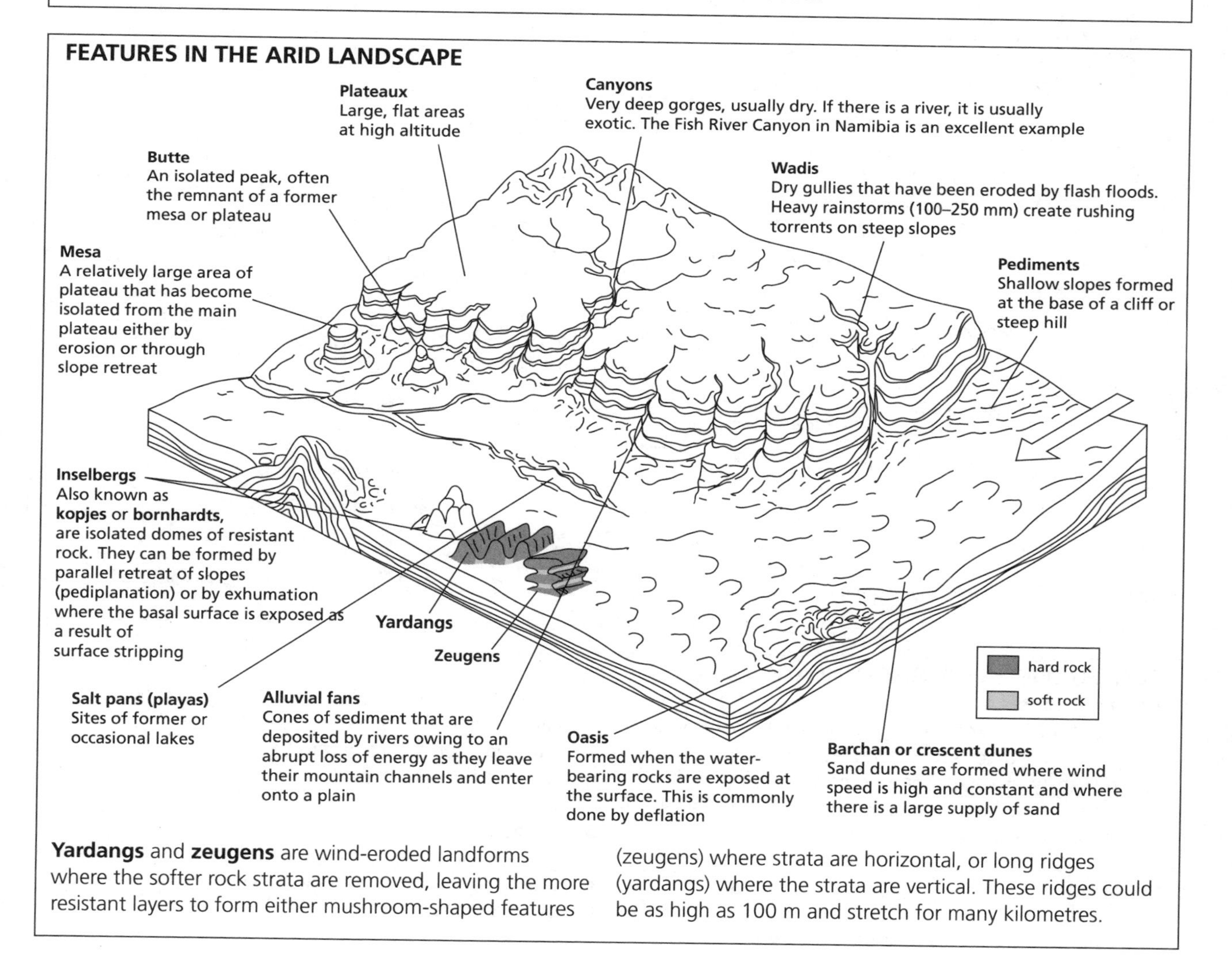

Yardangs and **zeugens** are wind-eroded landforms where the softer rock strata are removed, leaving the more resistant layers to form either mushroom-shaped features (zeugens) where strata are horizontal, or long ridges (yardangs) where the strata are vertical. These ridges could be as high as 100 m and stretch for many kilometres.

Agriculture in arid and semi-arid areas

PROBLEMS FOR FARMING

The shortage of water and the high temperatures determine many of the characteristic processes in arid and semi-arid areas, as well as many of the characteristics of their soils and ecosystems. All arid and semi-arid areas have a **negative water balance**. That means the outputs from evapotranspiration and stores of water exceed the input from precipitation (pEVT > ppt). The shortage of water can be made up by using irrigation water – i.e. by artificially increasing the amount of water that planes receive through pipes and other watering systems (central pivot irrigation, drip irrigation – see page 63).

Desert soils are arid (dry) and infertile, due to:

- a low organic content because of the low levels of biomass
- being generally very thin with few minerals
- lack of clay (the amount increases with rainfall)
- not generally being leached because of the low rainfall; hence soluble salts remain in the soil in the groundwater store and could be toxic to plants.

Salinization may occur in areas where annual precipitation is less than 250 mm. In poorly drained locations surface runoff evaporates and leaves behind large amounts of bicarbonates. The pH of soils affected by salinization is usually below 8.5. The saline soils adversely affect the growth of most crop plants by reducing the rate of water uptake by roots and germinating seeds. Plants die as a result of wilting.

DESERTIFICATION

Desertification occurs when already fragile land in arid and semi-arid areas is overexploited.

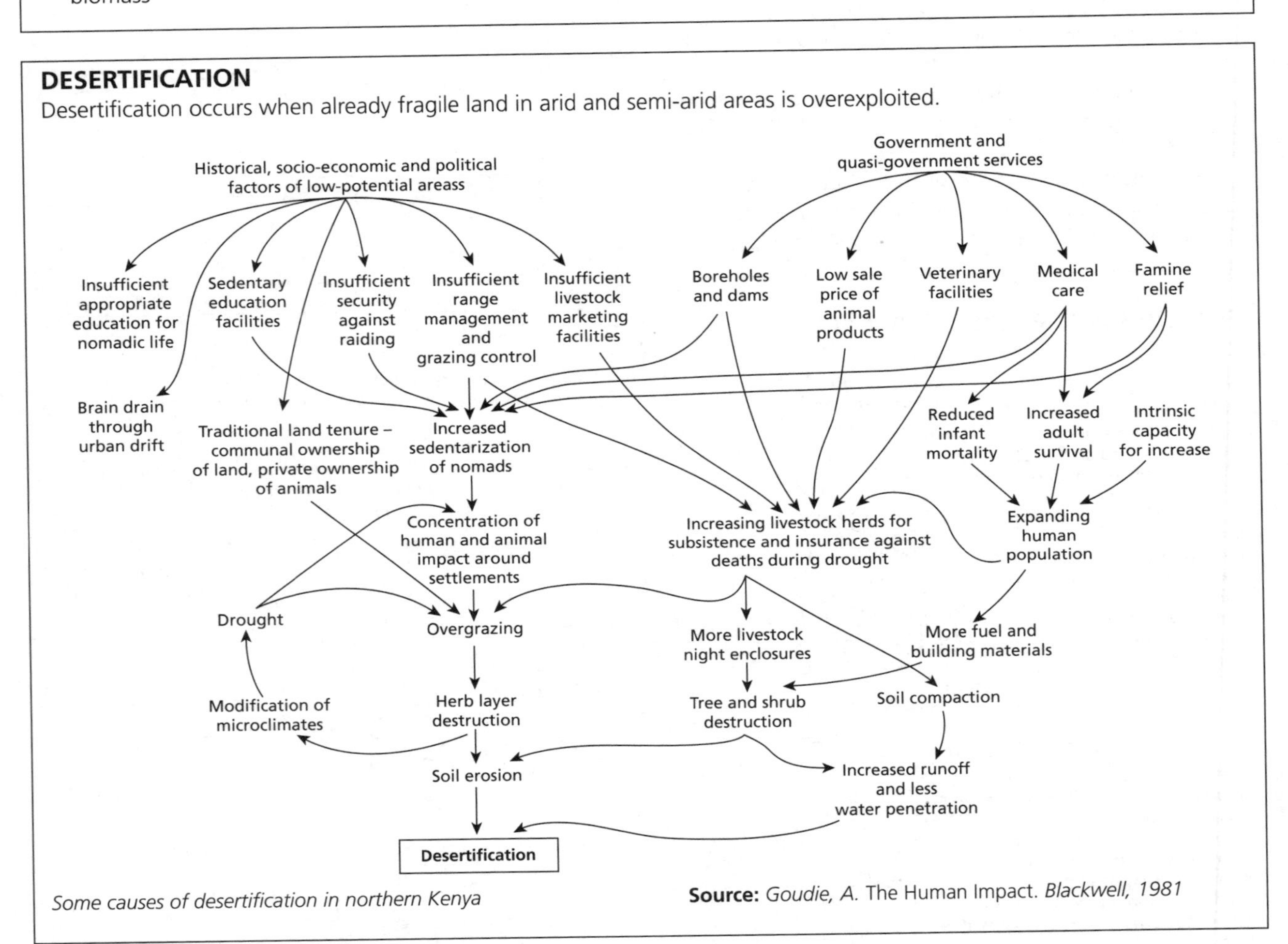

Some causes of desertification in northern Kenya

Source: *Goudie, A.* The Human Impact. *Blackwell, 1981*

CONSEQUENCES OF DESERTIFICATION

Environmental

- Loss of soil nutrients through wind and water erosion
- Changes in composition of vegetation and loss of biodiversity, as vegetation is removed
- Increased sedimentation of streams because of soil erosion, sediment accumulations in reservoirs

Economic

- Reduced income from traditional economy (pastoralism and cultivation of food crops)
- Decreased availability of fuelwood, necessitating purchase of oil/kerosene
- Increased dependence on food aid
- Increased rural poverty

Social and cultural

- Loss of traditional knowledge and skills
- Forced migration due to food scarcity
- Social tensions in reception areas for migrants

Mineral extraction in periglacial areas

FRAGILITY OF PERIGLACIAL AREAS

Periglacial areas are fragile for two reasons. First, the ecosystem is highly susceptible to interference, because of the limited number and diversity of species involved. The extremely low temperatures limit decomposition, and hence **pollution**, especially oil spills, have a very long-lasting effect on periglacial ecosystems. Second, **permafrost** is easily disrupted. The disruption of permafrost poses significant problems. Heat from buildings and pipelines, and changes in the vegetation cover, rapidly destroy it. Thawing of the permafrost increases the active layer, and subsequent settlement of the soil causes subsidence. Consequently, engineers have either built structures on a bed of gravel, up to 1 m thick for roads, or have used stilts.

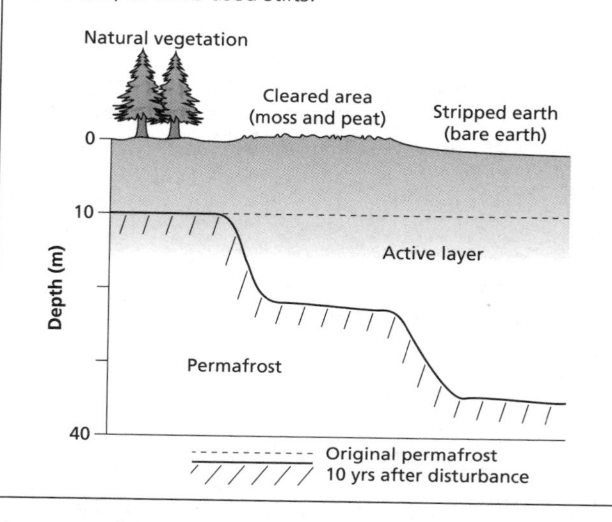

FROST HEAVE

Close to rivers, owing to an abundant supply of water, frost heave is very significant and can lift piles and structures out of the ground. Piles for carrying oil pipelines therefore need to be embedded deep in the permafrost to overcome mass movement in the active layer. In Prudhoe Bay, Alaska, they are 11 m deep. However, this is extremely expensive: each one cost more than $3000 in the early 1970s.

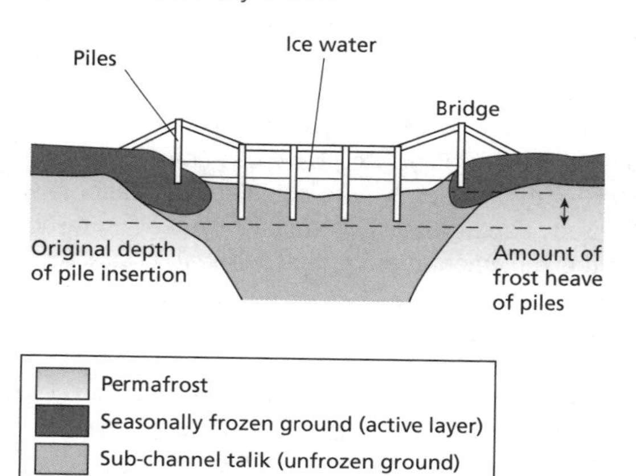

PROBLEMS IN THE USE OF PERIGLACIAL AREAS

The hazards associated with the use of periglacial areas are diverse and may be intensified by human impact. Problems include mass movements such as avalanches, solifluction, rockfalls, frost heave, icings, as well as flooding, thermokarst subsidence, low temperatures, poor soils, a short growing season and a lack of light.

For example, the Nenetsky tribe in the Yamal Peninsula of Siberia have suffered as a result of the exploitation of oil and gas. Oil leaks, subsidence of railway lines, destruction of vegetation, decreased fish stocks, pollution of breeding grounds, reduced caribou numbers, etc. have all happened directly or indirectly as a result of human attempts to exploit this remote and inhospitable environment.

Services are difficult to provide in periglacial environments. It is impossible to lay underground networks and so **utilidors**, insulated water and sewage pipes, are provided above ground. Waste disposal is also difficult because of the low temperatures.

Unstable permafrost pipeline above ground

Radiators for ammonia cooling system

Fibreglass and polyurethane insulation, to keep oil warm and pumpable in winter shutdowns

Steel pipe

Oil

Teflon-coated shoes allows pipe to slide

Active layer

Pipes for liquid ammonia cooling system – disperses summer heat, retains permafrost

Slurry backfilled in and around vertical support

Pipe anchored only every 250–550 m. Zig-zag line allows pipe to expand and contract (temperature range) and adjust to earthquakes

Unstable permafrost pipeline buried
(i.e. where above-ground pipe would block caribou migration)

Insulation

Oil

Refrigerated brine pumped through small pipes – keeps ground frozen

Earthquake and other pipe fractures

Automatic valves close, limiting spillage to an average of 15,000 barrels of oil

Problems with pipelines

ALPINE PERIGLACIAL AREAS

Alpine periglacial areas also suffer environmental pressures. Here the concerns are more than damage to the physical environment, as traditional economies have declined at the expense of electrochemical and services industries, especially tourism. An elaborate infrastructure is required to cope with the demands of an affluent tourist population, and this may undermine the natural environment and traditional societies.

Resource development in hot, arid areas

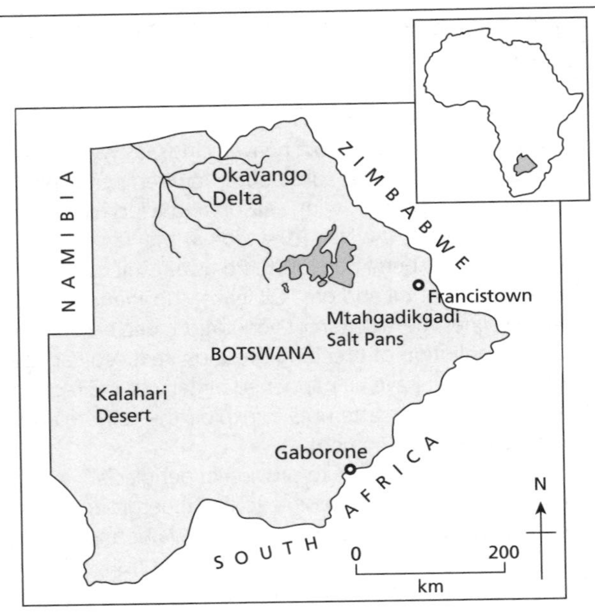

RURAL DEVELOPMENT IN BOTSWANA

Botswana is an African success story. Since its independence in 1966 it has been transformed from a largely rural society dependent on livestock to a middle-income country with a diversified rural economy. Its success is largely due to the discovery of diamonds in 1967, and the investment of its wealth into social and infrastructural projects. Mineral extraction, principally diamond mining, dominates economic activity, though tourism is a growing sector due to the country's conservation practices and extensive nature reserves. There has been some success in creating jobs in textiles and car manufacturing, but not enough to absorb all the unemployed. Unemployment officially was 23.8% in 2004, but unofficial estimates place it closer to 40%. An expected levelling off in diamond-mining production overshadows long-term prospects.

The impact of minerals, especially diamonds, has been considerable. Not only does the mining industry create jobs, it also earns foreign exchange, acts as a catalyst for industrial development and helps develop the infrastructure.

Land use	1% arable, 0% permanent crops, 99% other
Population (2008)	1.8 million
Age structure	0–14 years = 35%, 15–64 years = 61%, 65+ years = 4%
Birth rate (2008)	23‰
Death rate (2008)	14‰
Infant mortality rate	44‰
Life expectancy	50 years
Unemployment	23%
GDP by sector	Agriculture 1.6%, Mining 36%, Manufacturing 15.5%, Services 46.9%
PPP	$14,700/head

Botswana factfile

PROBLEMS AND ISSUES IN BOTSWANA

Mining and cattle ranching have led to problems:

- The diamond reserves are finite.
- Development based on diamonds has led to inequalities – 54% of the rural population live below the poverty line.
- Population growth is rapid – the total population was 1.3 million in 1991 and is likely to exceed 2 million in 2011. The population is youthful, although there has been an increase in the prevalence of AIDS. This is disproportionately affecting the working population. HIV/AIDS infection rates are the second highest in the world and threaten Botswana's impressive economic gains. Partly as a result of mining development, there is increasing rural–urban migration, especially of young adults. This is having severe implications in rural areas.

Economic development in Botswana has generated multiple environmental problems (such as rangeland degradation; loss of trees for fuelwood; depletion of groundwater resources; reductions in wildlife populations; erosion of arable land). The majority of Botswana's environmental policies are wildlife related. Since 1968, several National Parks Acts have been passed and nearly 40% of the country is a protected area. A National Conservation Strategy Co-ordinating Agency is responsible for the implementation of the conservation strategy.

DEFENCE AND WEAPONS TESTING IN AUSTRALIA

There has been a growth in defence and weapons testing in Australia's arid lands. Woomera, for example, was built as a rocket-launch town, and nuclear testing by the UK has occurred at Maralinga and Emu. Some radioactive waste has been buried in the area, which is now closed to the public. This has made the land worthless. It is highly unlikely that this sort of exercise will be repeated in the future.

Tourism in Zuni Pueblo, New Mexico

ZUNI PUEBLO

Zuni Pueblo is the largest of the 19 New Mexico Pueblos, with more than 1800 km^2 of land and a population of over 10,000. It is considered the most traditional of all the Pueblos, with a unique language, culture and history, resulting in part from its geographic isolation in a remote area of one of the most sparsely populated regions of the USA.

Zuni needs to develop tourism in ways compatible with maintaining and enhancing the lifestyle and sense of community that currently exists, and in ways that conserve its natural and cultural resources. Tourism is a double-edged sword: more often than not it destroys what it originally set out to enhance.

Many visitors are familiar with Zuni because of the reputation of Zuni jewellery, arts and crafts. Visitors are also attracted to the landscape of the Zuni River valley and the dramatic sandstone mesas.

Zuni is part of the south-west tourist itinerary as one of the stops in Indian country. According to Butler's tourism life-cycle model, Zuni is still in the exploration stage but could soon enter the consolidation stage. At present most of the population derive their income from art, though there are efforts to increase the share of income from agriculture and tourism. Tourism is an attractive option because of the relatively low capital investment and the potentially high economic returns. However, there are many long-lasting negative impacts of unregulated, hasty tourism development.

Sociocultural concerns

The need for Zuni control over the development of tourism is to safeguard against the negative consequences that could affect the social and cultural life of the community. As a result there has been very limited external involvement and influence in the demands to develop tourism. Although there have been proposals for motel complexes, casinos and golf courses, none has been implemented.

Zuni culture continues to retain its integrity and social traditions in spite of its existence within the USA. For example, photographic, audio or video recordings, drawings or other documentation of Zuni religious events are prohibited.

Environmental concerns

Water, air, soil and biodiversity are resources that can easily be affected by tourism. Water, in particular, is an issue in Zuni, as its domestic water supply is limited. Some developments are allowed. For example, big-game hunting by non-Indians is permitted, as long as they have a Zuni guide.

Overall, the impact of tourism has been limited, largely because it has been controlled. Tourism in Zuni has evolved in a way that enables culture and the environment to survive. Nevertheless, tourism will almost certainly become a part of Zuni society, and in the early stage Zuni has been careful to consider the social, economic and environmental costs.

- Check-in with the Visitor Center before starting your visit to Zuni Pueblo. Remember, you are visiting an active community of residents' daily lives and homes – not a museum or theme park.
- Consider capturing visual memories instead of photographs! Assume that ALL "cultural" activities within the Pueblo are off-limits to photograph, video or audio record or sketch unless specifically informed otherwise. Always inquire first and ask permission before photographing any activity involving people. NO photography is permitted of images inside the Old Mission.
- Observe with quiet respect any traditional dances and events that you may encounter. Applause is as inappropriate as in a church setting.
- Exercise common sense by not climbing around fragile archaeological structures or adobe walls. Removal of artifacts or objects from these areas is a Federal offence.
- Respect our community by not using alcohol or drugs and not bringing weapons.
- Hike only in designated areas (check at Visitor Center) and not around archaeological ruin sites.

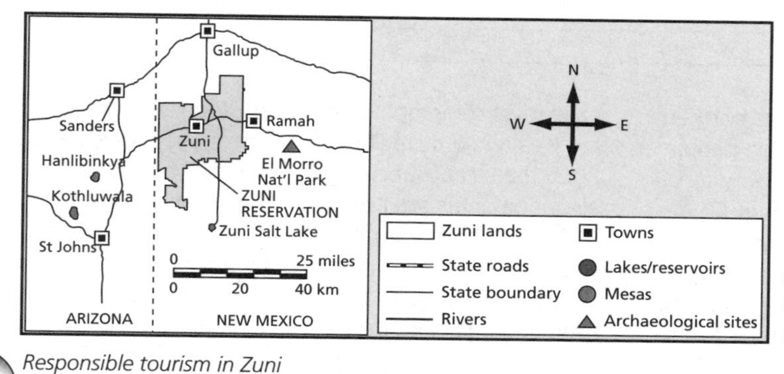

Note: *Indigenouts populations – a native population – the earliest inhabitants in a specific geographical region – who may have experienced colonization.*

Responsible tourism in Zuni

EXTENSION

Visit

www.ashiwi.org/ the official website of the Zuni tribe.

Sustainability in extreme environments

Global climate change may alter the environmental conditions in some extreme environments, and thus their location could change. There are suggestions that the world's hot deserts may become wetter. This would be great for their farming and food supply. In contrast, some predictions suggest that the Middle East will get much drier and hotter this century. By 2100, rainfall is predicted to decrease by 30% across Turkey, Lebanon, northern Syria, western Iran and Afghanistan. There are also fears that hot deserts will spread into other areas. Italy now has a programme of helping the countries of North Africa to combat desertification, partly in order to stem the increasing tide of refugees attempting to reach Europe.

DESERTIFICATION IN EUROPE

The Sahara has crossed the Mediterranean, forcing thousands to migrate as a lethal combination of soil degradation and climate change turns parts of southern Europe into desert. Up to a third of Europe's soil could eventually be affected. A fifth of Spanish land is so degraded that it is turning into desert, and in southern Italy tracts of land are abandoned and technically desert.

In areas such as drought-stricken Sardinia and Sicily, economic conditions are accelerating the problem. In many places tourism is making things far worse. Water is pumped from below ground, pulling salt water from the sea into the aquifers. Imagine how much water it takes to maintain an 18-hole golf course for tourists.

The sustainable use of soil is one of Europe's greatest environmental, social and economic challenges. In some parts of Europe, the degradation is so severe that it has reduced the soil's capacity to support human communities and ecosystems and resulted in desertification. Because it can take hundreds or thousands of years to regenerate most soils, the damage occurring today is effectively irreversible.

In Europe up to 150 million hectares are at high risk of erosion. Deterioration is at a critical point in Mediterranean countries, while the situation is no better in eastern Europe, where 41% of agricultural land in Ukraine is at risk of erosion.

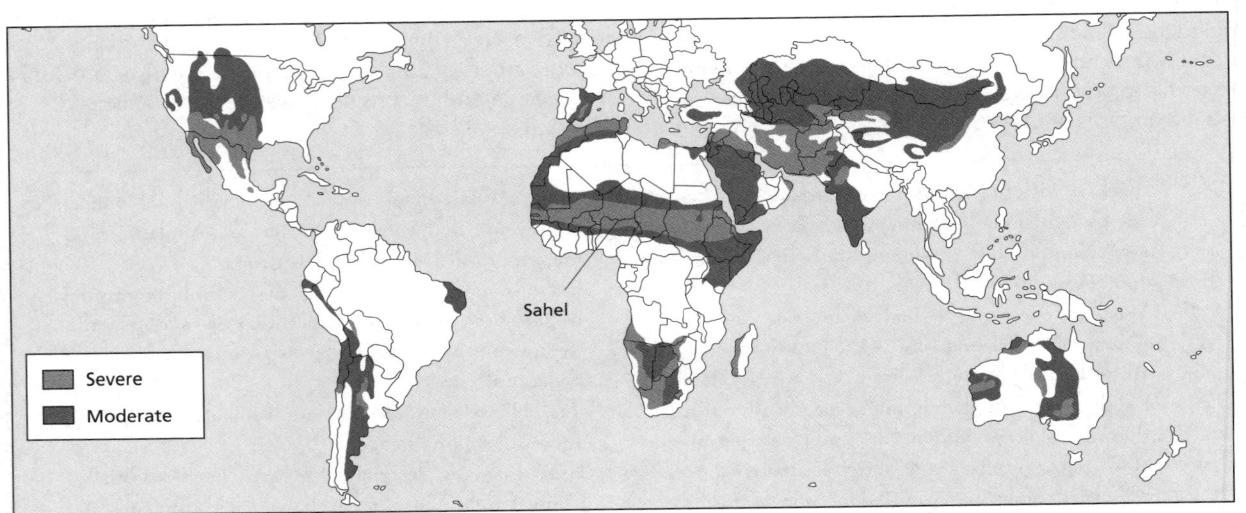

Areas at risk of desertification

CHANGES IN COLD ENVIRONMENTS

Until recently, Canada paid little attention to its northern region. Only 104,000 of the country's 33 million people live north of the 60th parallel. Two things are now forcing them to pay attention. The first is climate change. The warming climate has made minerals once locked in the ice accessible, just when their prices are high, unleashing an exploration boom. Second, people who live in the north are demanding and getting more of a say in their future.

There is no dispute that the Arctic is warming. Arctic temperatures have increased at almost twice the global average in the last 100 years; 70,000 km^2 of sea ice (an area about the size of Ireland) is disappearing annually.

A warming climate brings many problems for the Inuit. Unpredictable sea ice can be fatal. Life is becoming more expensive: snowmobiles must take longer routes, buildings are weakened by melting permafrost and, ironically, in 2006 the local council in Kuujjuaq felt obliged to buy 10 air-conditioners after temperatures reached 31°C.

The effects of climate change – more shipping, mining, and oil and gas exploration – may threaten the environment and with it the Inuit's traditional life, based on hunting and fishing. Some want development – but on their terms. In 2006 Nunavut's economy grew by 5.8%, second only to that of oil-rich Alberta. Much of the boost came from the opening of the territory's first diamond mine. Of the 130 companies exploring in Nunavut this year, 32 are looking for uranium. Others are seeking gold, diamonds, silver, zinc, nickel, copper, iron ore and sapphires.

Definitions and characteristics (1)

DEFINITIONS

- **Hazard**: a threat (whether natural or human) that has the potential to cause loss of life, injury, property damage, socio-economic disruption or environmental degradation
- **Hazard event**: the occurrence (realization) of a hazard, the effects of which change demographic, economic and/or environmental conditions
- **Disaster**: a major hazard event that causes widespread disruption to a community or region, with significant demographic, economic and/or environmental losses, and which the affected community is unable to deal with adequately without outside help
- **Vulnerability**: the geographic conditions that increase the susceptibility of a community to a hazard or to the impacts of a hazard event
- **Risk**: the probability of a hazard event causing harmful consequences (expected losses in terms of death, injuries, property damage, economy and environment)

Geophysical		Biological	
Climate and meteorological	**Geological and geomorphological**	**Floral**	**Faunal**
Snow and ice	Avalanches	Fungal diseases, e.g. athlete's foot, Dutch elm disease, wheat stem rust	Bacterial and viral diseases, e.g. influenza, malaria, smallpox, rabies
Droughts	Earthquakes	Infestations, e.g. weeds, water hyacinth	Infestations, e.g. rabbits, termites, locusts
Floods	Erosion (such as soil erosion and coastal erosion)	Hay fever	Venomous animal bites
Frosts	Landslides	Poisonous plants	
Hail	Shifting sand		
Heatwaves	Tsunami		
Tropical cyclones	Volcanic eruptions		
Lightning and fires			
Tornadoes			

Types of hazards and disasters

LOCATION OF NATURAL HAZARDS

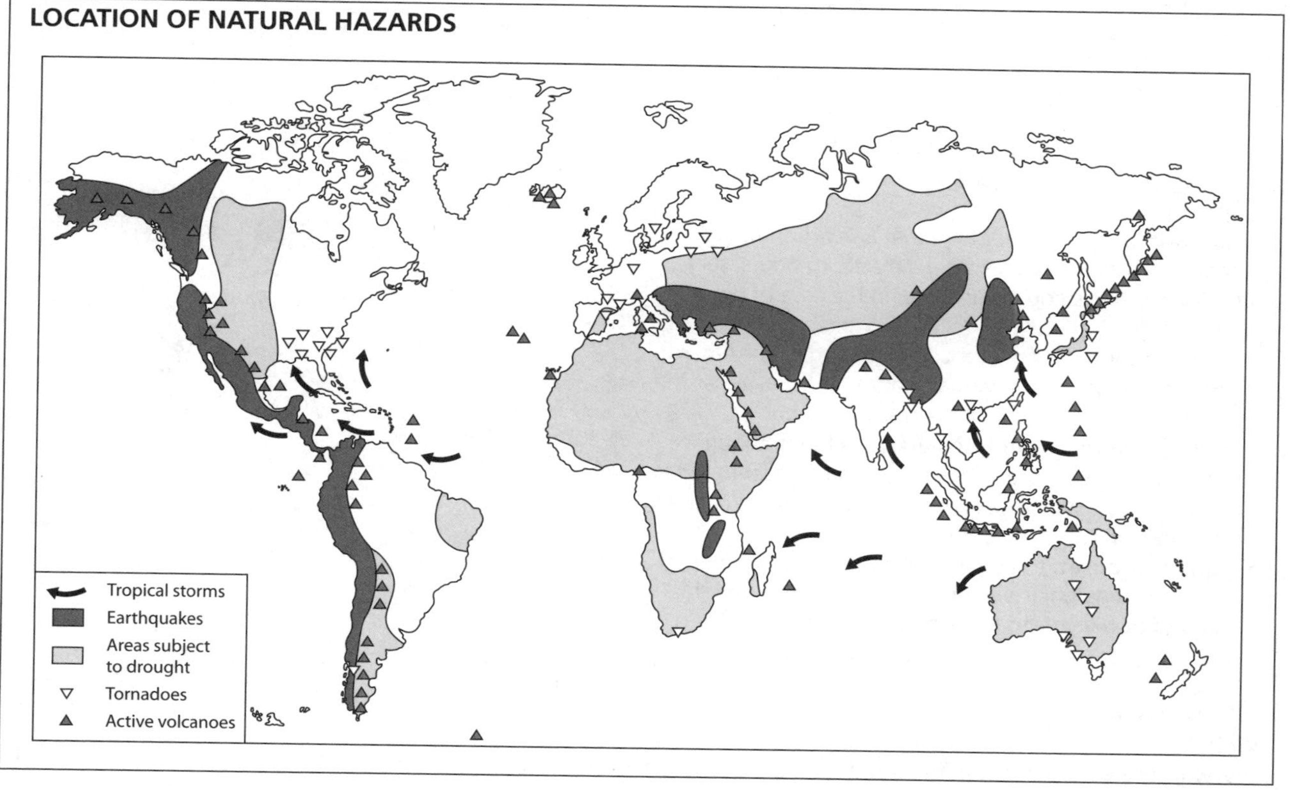

Definitions and characteristics (2)

DISTRIBUTION OF EARTHQUAKES

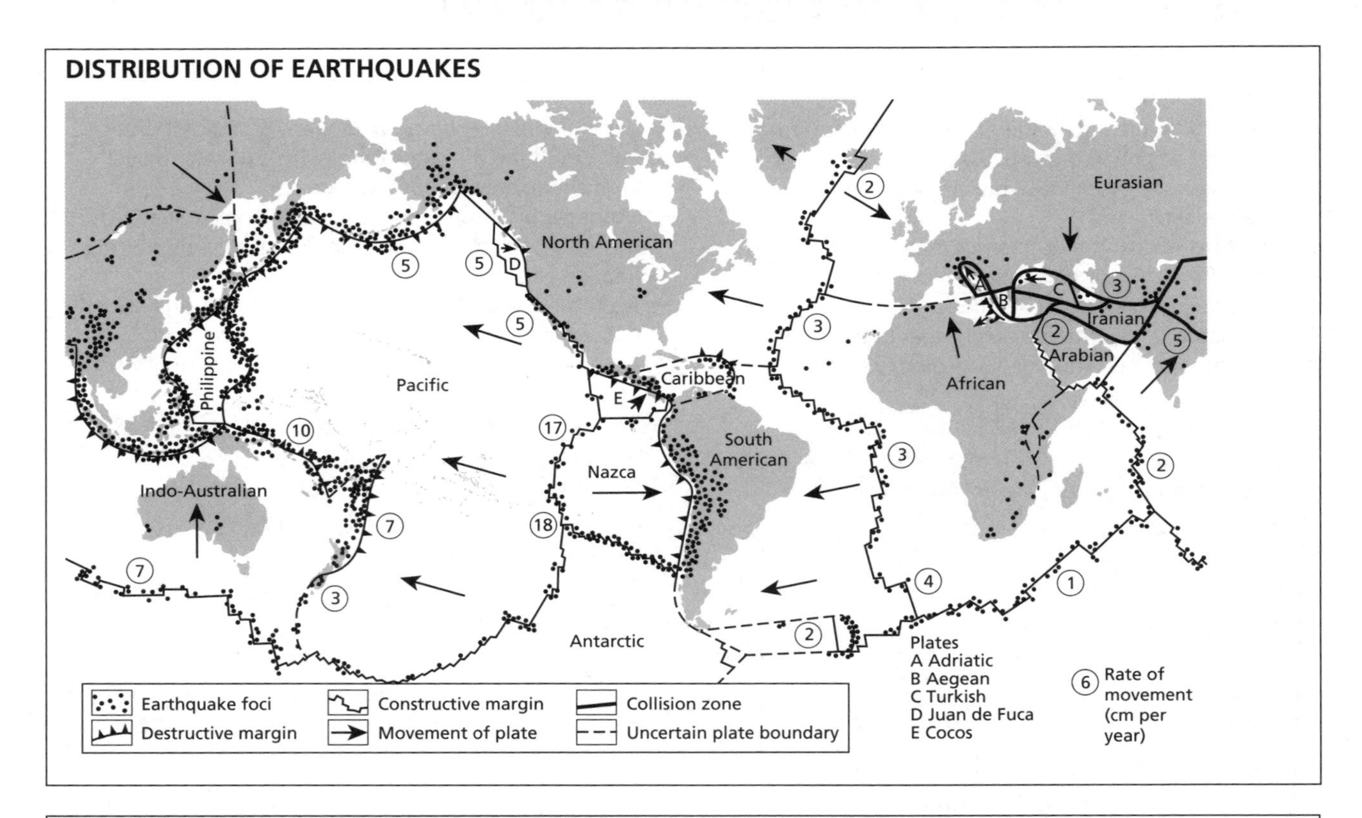

CHARACTERIZING HAZARDS AND DISASTERS

It is possible to characterize hazards and disasters in a number of ways:

- **Magnitude:** the size of the event, e.g. Force 10 on the Beaufort scale, the maximum height or discharge of a flood, or the size of an earthquake on the Richter scale.
- **Frequency:** how often an event of a certain size occurs. For example, a flood 1 m high may occur, on average, every year. By contrast, in the same stream a flood of 2 m might occur only every 10 years. The frequency is sometimes called the recurrence interval (Gumbel's laws). The larger the event, the less frequently it occurs. However, it is the very large events that do most of the damage (to the physical environment, to people, properties and livelihoods).
- **Duration:** the length of time that an environmental hazard exists. This varies from a matter of hours, such as with urban smog, to decades, in the case of drought, for example.
- **Areal extent:** the size of the area covered by the hazard. This can range from very small scale, such as an avalanche chute, to continental, as in the case of drought.
- **Spatial concentration/dispersion** is the distribution of hazards over space; whether they are concentrated in certain areas, such as tectonic plate boundaries, coastal locations, valleys and so on.
- **Speed of onset:** this is rather like the "time-lag" in a flood hydrograph. It is the time difference between the start of the event and the peak of the event. It varies from rapid events, such as the Kobe earthquake, to slow time-scale events such as drought in the Sahel of Africa.
- **Regularity** (or temporal spacing): some hazards, such as cyclones, are regular; whereas others, such as earthquakes and volcanoes, are much more random.

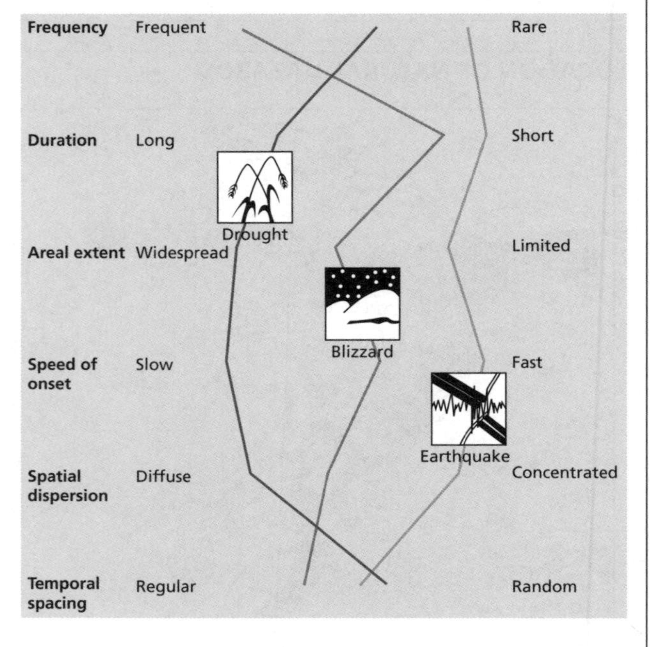

The characteristics of hazards

EXTENSION

Visit

www.intute.ac.uk/sciences/hazards/ for a general introduction to world hazards.

Earthquakes

An earthquake is a sudden, violent shaking of the earth's surface. Earthquakes occur after a build-up of pressure causes rocks and other materials to give way. Most of this pressure occurs at plate boundaries when one plate is moving against another. Earthquakes are associated with all types of plate boundaries.

The **focus** refers to the place beneath the ground where the earthquake takes place. **Deep-focus earthquakes** are associated with subduction zones. **Shallow-focus earthquakes** are generally located along constructive boundaries and along conservative boundaries. The **epicentre** is the point on the ground surface immediately above the focus.

CAUSES OF EARTHQUAKES

Some earthquakes are caused by human activity, such as:

- nuclear testing
- building large dams
- drilling for oil
- coal mining.

FACTORS AFFECTING EARTHQUAKE DAMAGE

The extent of earthquake damage is influenced by the following:

- **Strength of earthquake** and **number of aftershocks:** the stronger the earthquake, the more damage it can do, e.g. an earthquake of 6.0 on the Richter scale is 100 times more powerful than one of 4.0; the more aftershocks there are, the greater the damage that is done.
- **Population density:** an earthquake that hits an area of high population density, such as in the Tokyo region of Japan, could inflict far more damage than one which hits an area of low population and building density.
- **Type of buildings:** MEDCs generally have better quality buildings, more emergency services and the funds to cope with disasters. People in MEDCs are more likely to have insurance cover than those in LEDCs.
- **Time of day:** an earthquake during a busy time, such as rush hour, may cause more deaths than an earthquake at a quiet time. There are fewer people in industrial and commercial areas on Sundays; there are more people in homes at night.
- **Distance from the centre** of the earthquake: the closer a place is to the centre (epicentre) of the earthquake, the greater the damage that is done.
- **Type of rocks and sediments:** loose materials may act like liquid when shaken; solid rock is much safer. Buildings should be built on flat areas formed of solid rock.
- **Secondary hazards** such as mudslides and tsunami (high sea waves), fires, contaminated water, disease, hunger and hypothermia.

Country	Year	Death toll (est.)	Richter scale
Indonesia	2004	248,000	9.1
Kashmir, Pakistan	2005	86,000	7.6
Bam, Iran	2003	30,000	6.6
Chengdu, China	2008	78,000	7.9

The world's worst earthquakes by death toll in the 21st century

DEALING WITH EARTHQUAKES

People cope with earthquakes in a number of ways.

The three basic options from which they can choose are:

- to do nothing and accept the hazard
- to adjust to living in a hazardous environment – strengthen their home
- to leave the area.

The main ways of dealing with earthquakes include:

- better forecasting and warning
- building design, building location and emergency procedures.

Ways of predicting and monitoring earthquakes include:

- crustal movement – small-scale movement of plates
- changes in electrical conductivity
- strange and unusual animal behaviour, especially carp fish
- historic evidence – whether there are trends in the timing of earthquakes in a region.

Volcanoes

A volcano is an opening through the earth's crust through which hot molten magma (lava), molten rock and ash are erupted onto the land. Most volcanoes are found at plate boundaries, although there are some exceptions, such as the volcanoes of Hawaii. Some eruptions let out so much material that the world's climate is affected for a number of years.

Magma refers to molten materials inside the earth's interior. When the molten material is ejected at the earth's surface through a volcano or a crack at the surface, it is called **lava**.

The **chamber** refers to the reservoir of magma located deep inside the volcano. A **crater** is the depression at the top of a volcano following a volcanic eruption. It may contain a lake. A **vent** is the channel which allows magma within the volcano to reach the surface in a volcanic eruption.

KEY FACTS

- The greatest volcanic eruption was Tambora in Indonesia in 1815. Some 50–80 km^3 of material was blasted into the atmosphere.
- In 1883 the explosion of Krakatoa was heard as far as 4776 km away.
- The largest active volcano is Mauna Loa in Hawaii, 120 km long and over 100 km wide.

TYPES OF VOLCANO

The shape of a volcano depends on the type of lava it contains. Very hot, runny lava produces gently sloping **shield volcanoes** (Hawaiian type), while thick material produces **cone-shaped volcanoes** (Plinian type). These may be the result of many volcanic eruptions over a long period of time. Part of the volcano may be blasted away during eruption. The shape of the volcano also depends on the amount of change there has been since the volcanic eruption. Cone volcanoes are associated with destructive plate boundaries, whereas shield volcanoes are characteristic of constructive boundaries and hot spots (areas of weakness within the middle of a plate).

Hawaiian type
Runny basaltic lava which travels down sides in lava flows. Gases escape easily

Plinian type
Gas rushes up through sticky lava and blasts ash and fragments into sky in huge explosion. Gas clouds and lava can also rush down slopes. Part of volcano may be blasted away during eruption

Active volcanoes have erupted in recent times, such as Mount Pinatubo in 1991 or Montserrat 1997, and could erupt again. **Dormant volcanoes** are volcanoes that have not erupted for many centuries but may erupt again, such as Mount Rainier in the USA. **Extinct volcanoes** are not expected to erupt again. Kilamanjaro in Kenya is an example of an extinct volcano.

VOLCANIC STRENGTH

The strength of a volcano is measured by the volcanic explosive index (VEI). This is based on the amount of material ejected in the explosion, the height of the cloud it causes, and the amount of damage caused (see page 000). Any explosion above level 5 is considered to be very large and violent. So far there has never been a level 8.

VOLCANIC ERUPTIONS

Volcanic eruptions eject many different types of material. **Pyroclastic flows** are super-hot (700 °C) flows of ash and pumice (volcanic rock) at speeds of over 500 km/h. In contrast, **ash** is very fine-grained but very sharp volcanic material. **Cinders** are small-sized rocks and coarse volcanic materials. The volume of material ejected varies considerably from volcano to volcano.

Eruption	Date	Volume of material ejected
Mt St Helens, USA	1980	1 km^{-3}
Mt Vesuvius, Italy	AD79	3 km^{-3}
Mt Katmai, USA	1912	12 km^{-3}
Mt Krakatoa, Indonesia	1883	18 km^{-3}
Mt Tambora, Indonesia	1815	80 km^{-3}

The biggest volcanic eruptions

PREDICTING VOLCANOES

The main ways of predicting volcanoes include:

- seismometers to record swarms of tiny earthquakes that occur as the magma rises
- chemical sensors to measure increased sulphur levels
- lasers to detect the physical swelling of the volcano
- ultrasound to monitor low-frequency waves in the magma resulting from the surge of gas and molten rock, as happened at Pinatubo (Philippines), El Chichon (Mexico) and Mount St Helens.

LIVING WITH THE VOLCANO

People often choose to live in volcanic areas because they are useful.

- Some countries such as Iceland or the Philippines were created by volcanic activity.
- Volcanic soils are rich, deep and fertile, and allow intensive agriculture to take place.
- Volcanic areas are important for tourism.
- Some volcanic areas are very symbolic and are part of the national identity, such as Mount Fuji in Japan.

EXTENSION

Visit Volcano World at **http://volcano.oregonstate.edu/** for current volcanic activity.

Hurricanes

Hurricanes are intense hazards that bring heavy rainfall, strong winds and high waves, and cause other hazards such as flooding and mudslides. Hurricanes are characterized by enormous quantities of water. This is due to their origin over moist tropical seas. High-intensity rainfall and large totals of up to 500 mm in 24 hours invariably cause flooding. The path of a hurricane is erratic; hence it is not always possible to give more than 12 hours' notice. This is insufficient for proper evacuation measures.

Hurricanes develop as intense low-pressure systems over tropical oceans. Winds spiral rapidly around a calm central area known as the eye. The diameter of the whole hurricane may be as much as 800 km, although the very strong winds that cause most of the damage are found in a narrower belt, up to 300 km wide. In a mature hurricane, pressure may fall to as low as 880–970 millibars. This, and the strong contrast in pressure between the eye and the outer part of the hurricane, lead to strong gale-force winds.

Hurricanes move excess heat from low latitudes to higher latitudes. They normally develop in the westward flowing air just north of the equator (known as an easterly wave). They begin life as small-scale tropical depressions, localized areas of low pressure that cause warm air to rise. These trigger thunderstorms which persist for at least 24 hours and may develop into tropical storms, which have greater wind speeds of up to 117 km/h (73 mph). However, only about 10% of tropical disturbances ever become hurricanes, storms with wind speeds above 118 km/h (above 74 mph).

For hurricanes to form, a number of conditions are needed:

- Sea temperatures must be over 27 °C. (Warm water gives off large quantities of heat when it is condensed – this is the heat which drives the hurricane.)
- The low-pressure area has to be far enough away from the equator so that the Coriolis force (the force caused by the rotation of the earth) creates rotation in the rising air mass – if it is too close to the equator, there is insufficient rotation and a hurricane would not develop.

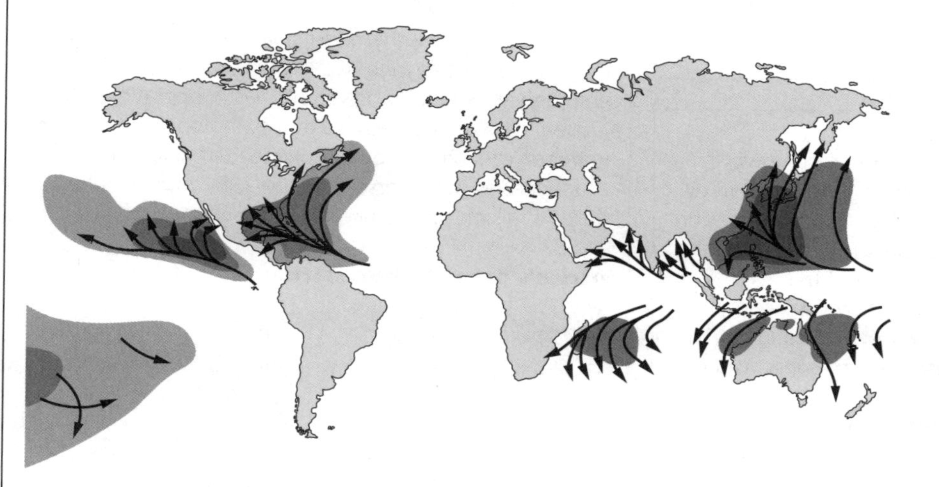

The location of the main hurricane tracks

HURRICANE KATRINA

Hurricane Katrina was the USA's worst natural disaster in living memory. The storm hit land near New Orleans on 29 August 2005 at a speed of some 230 km/h (145 mph.) Katrina was a category 4 hurricane, but what set it apart from other hurricanes was the way it lingered rather than passed through.

Over 1830 people were killed in the USA. Economists suggest Hurricane Katrina cost the US economy $80 billion. The rescue operation was criticized for not doing enough to help the poorest members of the population. Many of those left without help were from the poor neighbourhoods, many of which were the worst hit by the hurricane.

EXTENSION

Visit **http://www.nhc.noaa.gov/** for the National Hurricane Centre.

CYCLONE NARGIS

Some 134,000 people died in Cyclone Nargis, which struck in Burma in May 2008. As many as 95% of all buildings in the affected area were demolished by the cyclone. Winds exceeding 190 km/h (118 mph) and torrential rain devastated the area.

The Burmese government identified 15 townships in the Irrawaddy delta that had suffered the worst. Seven of them had lost 90–95% of their homes, with 70% of their population dead or missing. International frustration mounted as disaster management experts failed to get the necessary visas to enter the country.

The land in the Irrawaddy delta is very low-lying. It is home to an estimated 7 million of Burma's 53 million people. Nearly 2 million of the densely packed area's inhabitants live on land that is less than 5 m above sea level, leaving them extremely vulnerable.

As well as the cost in lives and homes is the agricultural loss to the fertile delta – considered Burma's rice bowl.

Droughts

VARIATIONS IN RAINFALL

A large proportion of the world's surface experiences dry conditions. Semi-arid areas are commonly defined as having a rainfall of less than 500 mm per annum, while arid areas have less than 250 mm, and extremely arid areas less than 125 mm per annum. In addition to low rainfall, dry areas have **variable rainfall**. As rainfall total decreases, variability increases. For example, areas with a rainfall of 500 mm have an annual variability of about 33%. This means that in such areas rainfall could range between 330 mm and 670 mm. This variability has important consequences for vegetation cover, farming and the risk of flooding.

DEFINING DROUGHT

Drought is an extended period of dry weather leading to conditions of extreme dryness. Absolute drought is a period of at least 15 consecutive days with less than 0.2 mm of rainfall. Partial drought is a period of at least 29 consecutive days during which the average daily rainfall does not exceed 0.2 mm.

EUROPE'S DROUGHT OF 2003

Estimates for the death toll from the French heatwave in 2003 were as high as 30,000. Harvests were down by between 30% and 50% on 2002. France's electricity grid was also affected, as demand for electricity soared as the population turned up air conditioning and fridges. At the same time, nuclear power stations, which generate around 75% of France's electricity, were operating at a much reduced capacity because there was less water available for cooling.

Portugal declared a state of emergency after the worst forest fires for 30 years. Temperatures reached 43 °C in Lisbon in August 2003 – 15 °C hotter than the average for the month. Over 1300 deaths occurred in the first half of August, and up to 35,000 ha of forest, farmland and scrub were burned. Some fires were, in fact, deliberately started by arsonists seeking insurance or compensation money. Over 70 people were arrested.

The prolonged heatwave left some countries facing their worst harvests since the end of the Second World War. Some countries that usually export food were forced to import it for the first time in decades. Across the EU, wheat production was down 10 million tonnes, about 10%.

ARID CONDITIONS

Arid conditions are caused by a number of factors.

- The main cause is the global atmospheric circulation. Dry, descending air associated with the **subtropical high-pressure belt** is the main cause of aridity around at 20–30 °N.
- In addition, distance from sea, or **continentality**, limits the amount of water carried across by winds.
- In other areas, such as the Atacama and Namib deserts, **cold offshore currents** limit the amount of condensation into the overlying air.
- Others are caused by intense **rain-shadow effects**, as air passes over mountains. This is true of the Patagonian desert.
- A final cause, or range of causes, are human activities. Many of these have given rise to the spread of desert conditions into areas previously fit for agriculture. This is known as desertification, and is an increasing problem.

DROUGHT IN AFRICA

In 2003 parts of southern Ethiopia were experiencing the longest drought anyone had known. The world's largest emergency food aid programme was in operation, but it proved inadequate. Because of a sixth poor rainy season in three years, 20 million people needed help. The situation was now worse than the 1984 famine, when only 10 million people needed food.

Africa's "at risk" population

Ethiopia	20 million
Zimbabwe	7 million
Malawi	3.2 million
Sudan	2.9 million
Zambia	2.7 million
Angola	1.9 million
Eritrea	1 million
Plus around 7.3 million across Swaziland, Congo, Uganda, Congo-Brazzaville, Lesotho and Mozambique	

People seen as under threat of famine in Africa

EXTENSION

Analyzing data/comparing two events

Look again at the two boxes above this one. Both relate to drought in 2003. The scale of the problem in Ethiopia (20m) dwarfs all other countries – so much so that they appear unaffected – yet nearly 3 million people were at risk in both Sudan and Zambia. In contrast, in Europe, wheat production was down by only 10% (down by 30–50% on 2002) and countries were able to import food. What they have in common is the shortage of food – but the scale is very different.

Technological hazards

Technological hazards are extremely wide-ranging, and include war, nuclear (radioactive) material, oil spills, industrial accidents and contamination of water and soil.

In these hazards, the misuse of technology endangers lives and property. It is generally people, not the technology, that have caused the disaster.

Class	Examples
Multiple extreme hazards	Nuclear war (radiation), recombinant DNA, pesticides
Extreme hazards	
Intentional biocides	Chainsaws, antibiotics, vaccines
Persistent teratogens	Uranium mining, rubber manufacture
Rare catastrophes	Liquified natural gas (LNG) explosions, commercial aviation (crashes)
Common killers	Auto crashes, coal mining (black lung)
Diffuse global threats	Fossil fuel (CO_2 release), sea surface temperatures (ozone depletion)
Hazards	Saccharin, aspirin, appliances, skateboards, bicycles

A classification of technological hazards

Source: *Smith, K.* Environmental hazards. *Routledge, 1992*

INDUSTRIAL POLLUTION: CASE STUDY – BHOPAL

On 2 December 1984, toxic gas settled over the sleeping population of Bhopal, the capital of Madhya Pradesh state, central India. The leak came from a pesticide plant owned by the American transnational company (TNC) Union Carbide. The gas, methyl isocyanate (MIC), attacks the internal organs, especially the lungs, preventing oxygen entering the bloodstream.

The impacts

When the MIC escaped, over half a million people were exposed to the fumes. Within hours, thousands had died and tens of thousands were suffering from blindness, skin complaints and breathing difficulties. The disaster continues to affect the local population today, and is the world's worst industrial disaster.

The effects of the gas leak may be worse than at first thought. A survey in 2003 showed that children exposed to the gas (and some born to parents exposed to the gas) show signs of growth retardation – they are shorter, with smaller heads. Although the local administration claims that 3000 people died on the night, other estimates are as high as 20,000. The official death total by 2004 was 12,000. Between 150,000 and 600,000 have been injured or suffer ill health as a result of the leak.

Enduring impacts

Bhopal's miscarriage rate is seven times the average for India, and there are large numbers of cases of respiratory illnesses and cancers. Bhopal has one of the highest rates of lung cancer in India, and there are very high rates of breast cancer, too. Cancer rates are significantly higher among the section of the population affected by the leak.

Whose responsibility?

In 1992 Union Carbide made a one-off payment of $470 million to the Indian government. However, many survivors still await adequate compensation. When Union Carbide left the site in 1999, thousands of tonnes of toxins and chemicals remained. Some of these have seeped into the soil and water, some washed there by monsoon rains. Recent tests in the water suggest levels of contamination 500 times higher than recommended World Health Organization (WHO) amounts. Union Carbide is now part of Dow Chemicals, the world's largest chemicals firm. Dow Chemicals' annual sales are worth $32.6 billion. Cleaning the plant would cost an estimated $23 million.

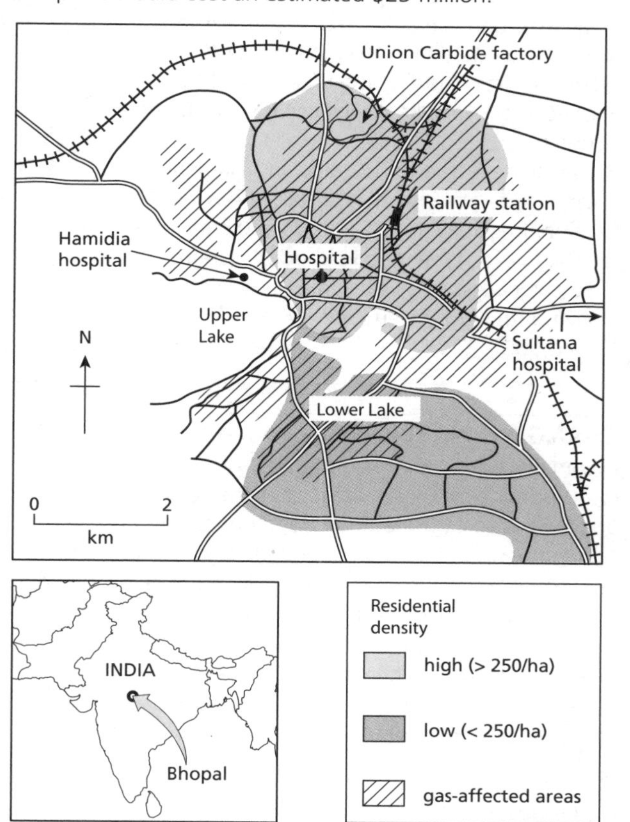

Areas in Bhopal affected by gas from the Union Carbide factory

Wells

Why people live in hazardous environments

A DISTINCTION

A hazard is a perceived natural event which threatens both life and property; a disaster is the realization of this hazard. A distinction can therefore be made between extreme events in nature, which are not environmental hazards (because people and/or property are not at risk) and environmental hazards in which people and/or property are at risk.

Environmental hazards are caused by people's use of dangerous environments. A large part of environmental hazards is caused by human behaviour, namely the failure to recognize the potential hazard and act accordingly. Hence the term "natural hazard" is not a precise description, as natural hazards are not just the result of "natural" events.

WHY DO THE POOR OFTEN LIVE IN HAZARDOUS ENVIRONMENTS?

Environmental hazards occur only when people and property are at risk. Although the cause of the hazard may be geophysical or biological, this is only part of the explanation. It is because people live in hazardous areas that hazards occur. So why do people live in such places? The **behavioural** school of thought considers that environmental hazards are the result of natural events. People put themselves at risk by, for example, living in floodplains. By contrast, the **structuralist** school of thought stresses the constraints placed on the (poor) people by the prevailing social and political system of the country. Hence, poor people live in unsafe areas – such as steep slopes or floodplains – because they are prevented from living in better areas. This school of thought provides a link between environmental hazards and the underdevelopment and economic dependency of many developing countries.

RESOURCE OR HAZARD?

People choose to live in certain environments because of the resources they bring. Deltas provide water, silt, fertile soils and the potential for trade and communications. They are also subject to floods, as shown by the 2008 floods in the Irrawaddy delta (Burma) and those caused by Hurricane Katrina in New Orleans (2005). Such events are rare. Most of the time water levels operate at a level where they can be considered a resource.

The same is true for volcanic environments. These may provide rich fertile soils and minerals to mine; they may attract tourists and create new land. However, when the volcano is erupting it may be necessary to evacuate, as in the case of Plymouth in Montserrat (1997) and Chaiten, Chile (2008).

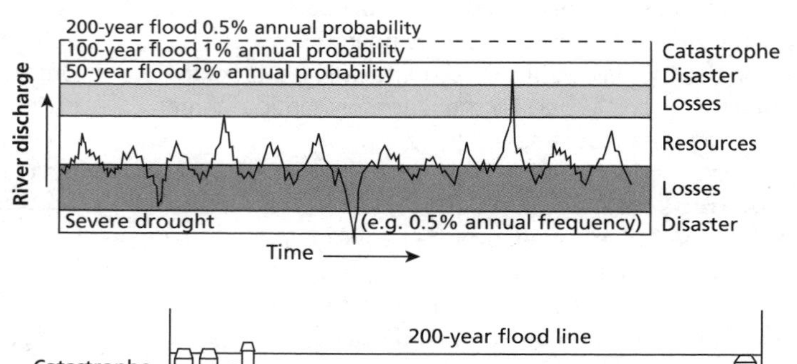

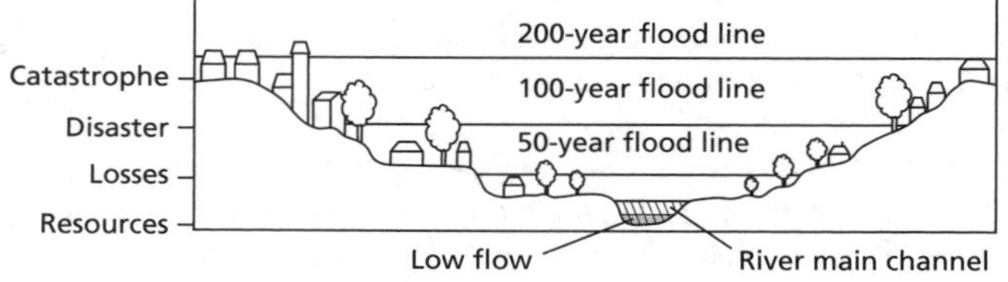

Flood recurrence intervals

CHANGING PATTERNS?

In some locations, the sheer number of people triggers hazards. For example, in megacities the volume of vehicles on roads almost inevitably causes air quality to decline. The concentration of manufacturing industry in certain regions (e.g. south-east China and south-east India) is also linked with a decline in air quality, increasing water pollution and acidification. As more people move into urban areas – whether into slums or formal housing – the risk of hazards increases, since there are more people living in the area and there is more alteration of the natural habitat.

In some areas, changing climate patterns are putting people at risk. For example, in southern Spain and Portugal, increasingly dry years are turning large areas into desert. This natural process is compounded by overuse of water for golf courses and recreational facilities. Consequently, groundwater levels are declining, soils are drying, vegetation is dying and the land is becoming desertified. This leads to increased risk of wind and water erosion, and further declines in productivity.

EXTENSION

Visit **http://www.fhrc.mdx.ac.uk/resources/publications.html** for free downloads on flooding and impacts/responses.

Vulnerability

DEFINITION

- **Vulnerability:** the geographic conditions that increase the susceptibility of a community to a hazard or to the impacts of a hazard event

THE PROGRESSION OF VULNERABILITY

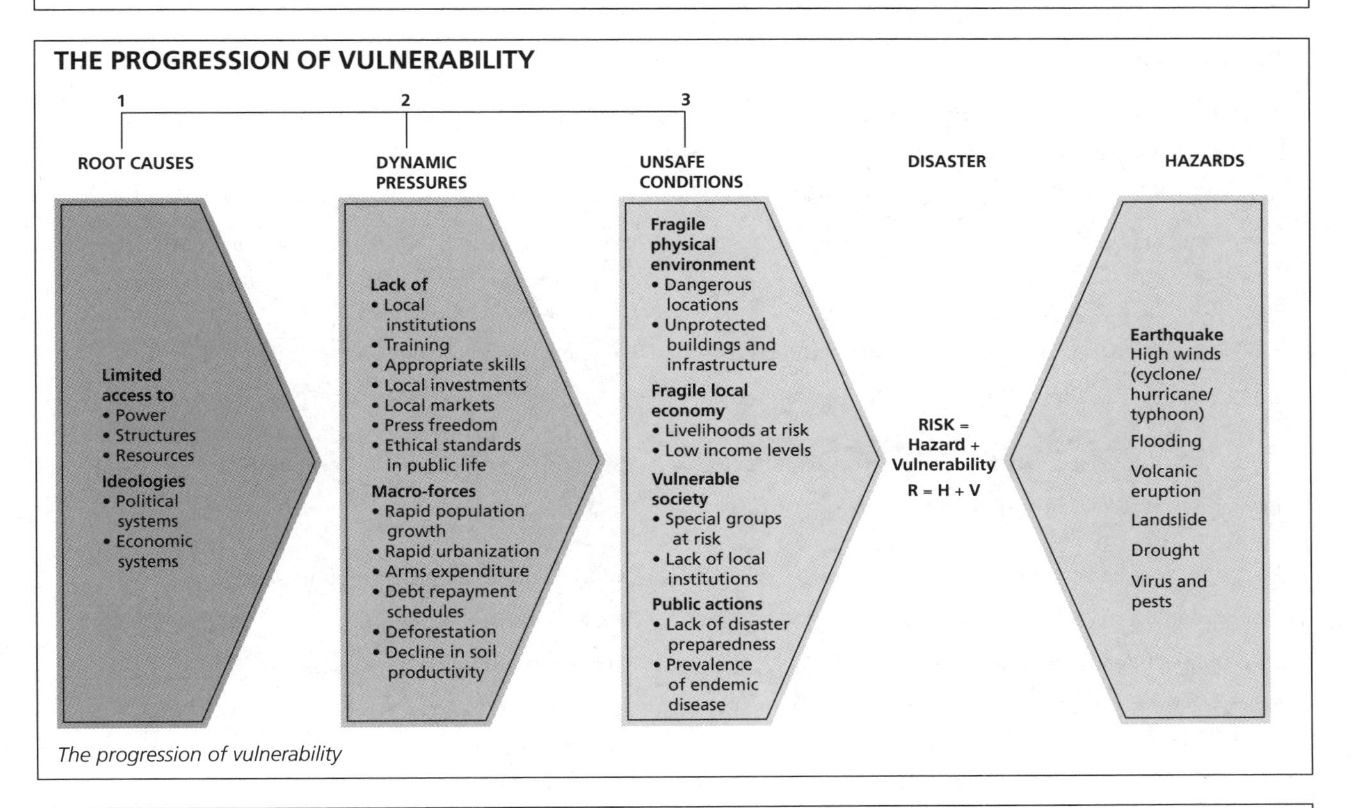

The progression of vulnerability

FACTORS AFFECTING VULNERABILITY

The concept of vulnerability encompasses not only the physical effects of a natural hazard but also the status of people and property in the area. Many factors can increase one's vulnerability to natural hazards, especially catastrophic events. Aside from the simple fact of living in a hazardous area, vulnerability depends on:

- population density – a large number of rapidly growing cities occur in hazardous areas; large urban areas such as New Orleans in the US are especially vulnerable to natural hazards
- understanding of the area – recent migrants into shanty towns may be unaware of some of the natural hazards posed by that environment
- public education – educational programmes in Japan have helped reduce the number of deaths in earthquakes
- awareness of hazards – the 2004 tsunami in south Asia alerted many people to the dangers that tsunamis cause
- the existence of an early warning system – the number of deaths from hurricanes in the USA is usually low partly because of an effective early warning system
- effective lines of communication – the earthquake in Sichuan (China) in 2008 brought a swift response from the government, who mobilized 100,000 troops and allowed overseas aid into the country
- availability and readiness of emergency personnel – there were many deaths following Cyclone Nargis in Burma due to a shortage of trained personnel)
- insurance cover – generally it is the poor who have no insurance cover and they are most likely to be affected in a natural hazard as their housing quality is poor
- construction styles and building codes – there was criticism during the Sichuan earthquake that many schools were destroyed (by implication, poorly built), whereas government buildings remained standing
- the nature of society – the failure of the Burmese government to allow aid to the victims of Cyclone Nargis in 2008 increased the death rate due to disease and malnutrition
- cultural factors that influence public response to warnings.
- Many of these factors help explain why poor countries are much more vulnerable to natural hazards than are industrialized countries.

EXTENSION

http://webra.cas.sc.edu/hvri/ is the home page for the Hazards and Vulnerability Research Institute from the University of South Carolina.

Risk and risk relationships

DEFINITION

- **Risk:** the probability of a hazard event causing harmful consequences (expected losses in terms of death, injuries, property damage, economy and environment)

RISK PERCEPTION

Factors tending to increase risk perception	Factors tending to reduce risk perception
Involuntary hazard (radioactive fallout e.g. Chernobyl, 1986)	Voluntary hazard (professional mountaineers)
Immediate impact (e.g. Cyclone Nargis, Burma, 2008)	Delayed impact (e.g. drought in Ethiopia, 2003, 2008)
Direct impact (e.g. Sichuan earthquake, 2008)	Indirect impact (e.g. drought in Spain and Portugal and the effect on tourism)
Dreaded impact (e.g. cancer, AIDS)	Common accident (car crash)
Many fatalities per disaster (e.g. Hurricane Katrina, 2005)	Few fatalities per disaster (e.g. UK floods, 2007)
Deaths grouped in space or time (e.g. Bhuj earthquake, India, 2000)	Deaths random in space and time (stomach cancer)
Identifiable victims (e.g. chemical plant workers, Bhopal)	Statistical victims (cigarette smokers)
Processes not well understood (nuclear accident e.g. Sellafield, UK)	Processes well understood (flooding)
Uncontrollable hazard (e.g. Hurricane Katrina)	Controllable hazard (ice on motorway)
Unfamiliar hazard (tsunami, e.g. Indonesia, 2004)	Familiar hazard (river flood)
Lack of belief in authority (young population)	Belief in authority (university scientist)
Much media attention (nuclear hazards e.g. Chernobyl; Mozambique floods, 2000)	Little media attention (factory discharge in water or atmosphere)

Factors influencing public risk perception, with examples of relative safety judgments

Factors affecting the perception of risk

At an individual level, there are three important influences on an individual's response:

- experience – the more experience of environmental hazards, the greater the adjustment to the hazard
- material well-being – those who are better off have more choice
- personality – is the person a leader or a follower, a risk-taker or a risk-minimizer?
- Ultimately, in terms of response, there are just the three options: do nothing and accept the hazard; adjust to the situation of living in a hazardous environment; leave the area. It is the adjustment to the hazard that we are interested in.

PREDICTING VOLCANOES

It is virtually impossible to monitor all active volcanoes. Satellites offer the prospect of global coverage from space and being developed for remote warning systems.

In the 1991 Mount Pinatubo eruption, over 320 people died, mostly due to collapse of ash-covered roofs. Many more lives were saved because early warnings were issued and at least 58,000 people were evacuated from the high-risk areas.

Management of the 1991 eruption seems to have been well-coordinated and effective:

- State-of-the-art volcano monitoring techniques and instruments were applied.
- The eruption was accurately predicted.
- Hazard-zonation maps were prepared and circulated a month before the violent explosions.
- An alert and warning system was designed and implemented.
- The disaster response machinery was mobilized on time.

EARTHQUAKE PREDICTION

The most reliable predictions focus on:

- measurement of small-scale ground surface changes
- small-scale uplift or subsidence
- ground tilt
- changes in rock stress
- anomalies in the earth's magnetic field
- changes in radon gas concentration
- changes in electrical resistivity of rocks.

One intensively studied site is Parkfield, California, on the San Andreas fault. Parkfield is heavily instrumented: strain meters measure deformation at a single point; two-colour laser geodimeters measure the slightest movement between tectonic plates; and magnetometers detect alterations in the earth's magnetic field, caused by stress changes in the crust. Nevertheless, the 1994 Northridge earthquake was not predicted and it occurred on a fault that scientists did not know existed. Technology helps, but not all the time.

Disasters

DEFINITIONS

- **Hazard:** a threat (whether natural or human) that can cause loss of life, injury, property damage, socio-economic disruption or environmental degradation
- **Disaster:** a major hazard that causes widespread disruption with significant demographic, economic and/or environmental losses, and which the affected community is unable to deal with adequately without outside help

The distinction between the two is not always clear cut.

STAGES IN A DISASTER

I Preconditions	
Phase I	*Everyday life* (years, decades, centuries)
	"Lifestyle" risks, routine safety measures, social construction of vulnerability, planned developments and emergency preparedness.
Phase 2	*Premonitory developments* (weeks, months, years)
	"Incubation period" – erosion of safety measures, heightened vulnerability, signs and problems misread or ignored.
II The disaster	
Phase 3	*Triggering event or threshold* (seconds, hours, days)
	Beginning of crisis; "threat" period: impending or arriving flood, fire, explosion; danger seen clearly; may allow warnings, flight or evacuation and other pre-impact measures. May not, but merging with:
Phase 4	*Impact and collapse* (instant, seconds, days, months)
	The disaster proper. Concentrated death, injury, devastation. Impaired or destroyed security arrangements. Individual and small group coping by isolated survivors. Followed by or merging with:
Phase 5	*Secondary and tertiary damages* (days, weeks)
	Exposure of survivors, post-impact hazards, delayed deaths.
Phase 6	*Outside emergency aid* (weeks, months)
	Rescue, relief, evacuation, shelter provision, clearing dangerous wreckage, "organized response". National and international humanitarian efforts.
III Recovery and reconstruction	
Phase 7	*Clean-up and "emergency communities"* (weeks, years)
	Relief camps, emergency housing. Residents and outsiders clear wreckage, salvage items. Blame and reconstruction debates begin. Disaster reports, evaluations, commissions of inquiry.
Phase 8	*Reconstruction and restoration* (months, years)
	Reintegration of damaged community with larger society. Re-establishment of "everyday life", possibly similar to, possibly different from pre-disaster. Continuing private and recurring communal grief. Disaster-related development and hazard-reducing measures.

Temporal sequences or phases that may be involved in disasters, with reported durations and selected features of each phase

TWO DISASTERS OF 2008

The Sichuan earthquake, China

In May 2008, an earthquake registering 7.9 on the Richter scale devastated the Chinese province of Sichuan. Over 69,000 people were killed and nearly 18,000 people were missing as a result of the earthquake. A further 4.8 million people were made homeless. Many rivers were blocked by landslides and formed 34 "quake lakes". The risk of landslides was increased by the arrival of the summer rains. The Chinese government received praise for its swift rescue attempts and its willingness and openness to receive foreign aid.

Cyclone Nargis, Burma

In contrast, the Burmese government received considerable criticism for the way it dealt with Cyclone Nargis. Over 134,000 people were killed and a further 56,000 people were missing. The disaster cost an estimated $10 billion damage. However, the event is also a man-made disaster. The Burmese military rulers refused international aid at first.

Adjustment and response

COPING WITH HAZARDS

How people adjust to hazards depends on:

- the type of hazard
- the risk (probability) of the hazard – several factors influence how people view risk
- the likely cost (loss) caused by the hazard.

Ways of managing the consequences of a hazard include:

- modifying the hazard event, through building design, building location and emergency procedures
- improved forecasting and warning
- sharing the cost of loss, through insurance or disaster relief.

BUILDING DESIGN

A single-storey building has a quick response to earthquake forces. A high-rise building responds slowly, and shockwaves are increased as they move up the building. If the buildings are too close together, vibrations may be amplified between buildings and increase damage. The weakest part of a building is where different elements meet. Elevated motorways are therefore vulnerable in earthquakes because they have many connecting parts. Certain areas are very much at risk from earthquake damage – areas with weak rocks, faulted (broken) rocks, and on soft soils. Many oil pipelines and water pipelines in tectonically active areas are built on rollers, so that they can move with an earthquake rather than fracture.

AN EARLY VERSION OF ALTERNATIVE ADJUSTMENTS TO NATURAL HAZARDS

Class of adjustments	Earthquakes	Floods	Snow
Affect the cause	No known way of altering the earthquake mechanism	Reduce flood flows by: land-use treatment; cloud seeding	Change geographical distribution by cloud seeding
Modify the hazard	Stable site selection: soil and slope stabilization; sea wave barriers; fire protection	Control flood flows by: reservoir storage; levées; channel improvement; flood fighting	Reduce impact by snow fences; snow removal; salting and sanding of highways
Modify loss potential	Warning systems; emergency evacuation and preparation; building design; land-use change; permanent evacuation	Warning systems; emergency evacuation and preparation; building design; land-use change: permanent evacuation	Forecasting; rescheduling; inventory control; building design; seasonal adjustments (snow tyres, chains); seasonal migration; designation of snow emergency routes
Adjust to losses			
Spread the losses	Public relief; subsidized insurance	Public relief; subsidized insurance	Public relief; subsidized insurance
Plan for losses	Insurance and reserve funds	Insurance and reserve funds	Insurance and reserve funds
Bear the losses	Individual loss-bearing	Individual loss-bearing	Individual loss-bearing

ADJUSTMENTS TO DROUGHT SUGGESTED BY PEASANT FARMERS IN NIGERIA AND TANZANIA

	Northern Nigeria	Tanzania
Change location	Nothing permanent	Nothing permanent
Change use	Nothing	Drought-resistant crops, irrigation
Prevent effects	Store food for next year; seek work elsewhere temporarily; seek income by selling firewood, crafts, or grass; expand fishing activity; plant late cassava; plant additional crop	More thorough weeding; Cultivate larger areas; work elsewhere; tie ridging; planting on wet places; sending cattle to other areas; sell cattle to buy food; staggered planting
Modify events	Consult medicine men; pray for end of drought	Employ rainmakers; pray
Share	Turn to relatives; possible government relief	Send children to kinsmen; government relief; store crops; move to relative's farm; use savings
Bear	Suffer and starve; pray for support	Do nothing

Short-term, mid-term and long-term responses after an event

CHANGING PRIORITIES

In the immediate aftermath of a disaster the main priority is to rescue people. This may involve the use of search and rescue teams and sniffer dogs. Thermal sensors may be used to find people alive among the wreckage.

The number of survivors decreases very quickly. Few survive after 72 hours, although there were reports from Sichuan of people surviving for nearly 20 days – the number is extremely low, however.

Rehabilitation refers to people being able to make safe their homes and be able to live in them again. Following the UK floods of 2007, some people were unable to return to their homes for over a year. For some residents in New Orleans, rehabilitation was not possible, so reconstruction (rebuilding) was necessary. This can be a very long, drawn-out process, taking up to a decade for major construction projects. The time-scales involved are shown in the model of disaster recovery below.

As well as dealing with the aftermath of a disaster, governments try to plan to reduce impacts of future events. This is sometimes called **hazard mitigation**. This was seen after the south Asian tsunami of 2004. Before the event, a tsunami early warning system was not in place in the Indian Ocean. Following the event, as well as emergency rescue, rehabilitation and reconstruction, governments and aid agencies in the region developed a system to reduce the impacts of future tsunamis. It is just part of the progress needed to reduce the impact of hazards and to improve safety in the region.

Periods	Emergency	Restoration	Reconstruction I	Reconstruction II
Capital stock	Damaged or destroyed	Patched	Rebuilt (replacement)	Major construction (commemoration, betterment, development)
Normal activities	Ceased or changed	Return and function	Return at pre-disaster levels or greater	Improved and developed

Maximal
Coping activity
Minimal
0.5 1 2 3 4 5 10 20 30 40 50 100 200 300 400 500
Disaster event
Time in weeks following disaster

Sample indicators

Completion of search and rescue
End of emergency shelter or feeding
Clearing rubble from main arteries

Restoration of major urban services
Return of refugees
Rubble cleared

Attain pre-disaster levels of capital stock and activities

Completion of major construction projects

A model of disaster recovery for urban areas

Disaster impact pyramid

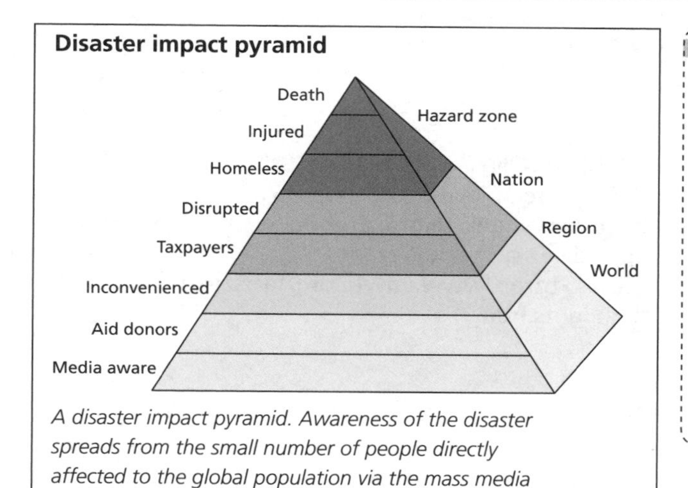

A disaster impact pyramid. Awareness of the disaster spreads from the small number of people directly affected to the global population via the mass media

EXTENSION

Scales in geography

The diagrams on this page illustrate two different scales in geography. One is spatial scale – ranging from the local (hazard zone) to the global. Some of these may in fact overlap – the national/regional, for example. The other scale is the temporal scale – in this example days, weeks, years and decades. Geological timescales go up to hundreds of millions of years, and in terms of climate and population change, we think in terms of change over the next century. Good geographers consider a range of different spatial and temporal scales – both past and future.

Tourism, sport, leisure and recreation

DEFINITIONS

- **Leisure:** any freely chosen activity or experience that takes place in non-work time
- **Recreation:** a leisure-time activity undertaken voluntarily and for enjoyment. It includes individual pursuits, organized outings and events, and non-paid (non-professional) sports
- **Sport:** a physical activity involving events and competitions at the national and international scale with professional participants
- **Tourism:** travel away from home for at least one night for the purpose of leisure. This definition excludes day trips, some of which may be international trips. There are many possible subdivisions of tourism. Subgroups include **ecotourism** – tourism focusing on the natural environment and local communities; **heritage tourism** – tourism based on a historic legacy (landscape feature, historic building or event) as its major attraction; **sustainable tourism** – tourism that conserves primary tourist resources and supports the livelihoods and culture of local people
- **Mass tourism:** an organized form of large-scale tourism, in which travel, accommodation and meals are booked and paid for in advance.
- **Leakage:** refers to the money that "escapes" from a tourist destination and makes its way to other countries via airline companies, hotel companies, MNCs, food importers etc.

There are many difficulties in applying these definitions. For example, definitions of sport, leisure and recreation overlap, and participation in them may be simultaneous. Someone may play golf or go swimming or skiing while they are on holiday.

THE GROWTH OF TOURISM

For much of history, tourism and travel were difficult, expensive, uncomfortable and dangerous, so the desire to travel had to be very strong. Nowadays visiting other places is considered to be a natural part of life and most people expect to travel at least on an annual basis, if not more often. As it becomes less difficult and more affordable, more and more people travel and for a greater variety of motives.

- The social and economic emancipation of the urban middle class, and especially the working class, was very important for the growth of tourism.
- The emergence of paid holiday, which provides sufficient lengths of time for people to plan trips, as well as the ability to afford trips, are equally important.
- Mass tourism is possible only with the development of efficient and affordable systems of transport, and these need to be sufficiently large-scale to take large numbers of people.
- Modern tourism requires an organizational backup system and provision of infrastructure and personnel able to run the tourism business. Such facilities include accommodation, transport, entertainment and retailing, as well as travel.

CHANGES IN TRANSPORT AND COMMUNICATIONS

The advent of jet airliners, in particular the wide-bodied jets with increased passenger capacity and extended range, halved both journey times and the real cost of air travel. Tourism to distant destinations would not have grown to the extent that it has if passengers were still being offered the fares, travel times and comfort of the 1950s.

ECONOMIC AND POLITICAL STABILITY

Across large areas of the world, general levels of prosperity have been rising since the 1950s. Political stability is important too. In western Europe, from the Second World War until the late 1990s, there was almost a complete absence of major political and military conflict; this is not the case in eastern Europe, however, where tourism is less important.

FACTORS AFFECTING GROWTH

Tourism has increased because tourists are more competent at travel, are more relaxed about travelling and wish to travel more. A number of factors are behind this:

- increased education levels and better training of personnel within the tourist industry
- increased acquisition of foreign languages
- travel procedures such as customs and airport check-in counters becoming less of a constraint
- the use of IT (computers) to provide details on availability of flights, accommodation, etc.
- globalization of credit cards, facilitating financial transactions and purchases
- improved telecommunications, making it easier to keep in touch with developments at home
- standardized forms of accommodation and other services in international hotels, restaurant chains and car-hire offices, reducing the sense of dislocation that foreign travel might create.

In recent years, the growth of tourism has slowed. This is due to a combination of factors, including:

- the tightening up of airport security
- rises in the price of oil
- decreased disposable incomes
- increased awareness of individual carbon expenditure.

EXTENSION

Visit the World Tourism Organization at **http://www.unwto.org**. Click on tourism highlights for up to date statistics and graphs.
See also **http://www.unwto.org/facts/eng/highlights.htm**.

Changing patterns of international tourism

CHANGING PATTERNS

Traditionally, international tourism has been dominated by western Europe, as both a receiving and a generating region. This is due to a number of factors:

- an established tradition in domestic tourism that converts easily into international tourism
- a mature and developed pattern of infrastructure, such as transport, hotels and travel companies
- a large variety of natural and man-made attractions
- a large population that is affluent and mobile
- a range of climatic zones, which facilitates summer and winter tourism.

GLOBAL PATTERNS

- There has been a reduction in the share of tourists attracted to regions in Europe and the Americas.
- There are relatively static positions in areas of chronic underdevelopment in developing countries in parts of Africa and south Asia (including India) and the politically unstable Middle East.
- There has been a huge expansion in tourism into east Asia and the Pacific, centred on Thailand, Singapore, Indonesia, Hong Kong, Japan and Australia.

For an area to grow there must be primary and secondary resources. **Primary tourist/recreational resources** are the pre-existing attractions for tourism or recreation (those not built specifically for the purpose), including climate, scenery, wildlife, indigenous people, cultural and heritage sites. **Secondary tourist/ recreational resources** include accommodation, catering, entertainment and shopping.

FACTORS AFFECTING TOURISM

A whole range of physical, social, economic and political factors affect tourism. Some of these change over time.

Factors	Examples
Natural landscape	Mountains, Nepal; biodiversity, Monteverde cloud forest, Costa Rica; coasts, Mediterranean; forests, Amazon rainforest; deserts, Tunisia; polar areas, Iceland; rivers, Grand Canyon
Climate	Hot, dry areas are attractive to most tourists; seasonality of climate leads to seasonality of tourism
Cultural	Language, customs, clothing, food, architecture and theme parks. Examples include: recreation, Paris; religion, Mecca; education, Oxford
Social	Increasing affluence, leisure time, longer holidays, paid holidays, better mobility, better transport, more working women, age of tourists and stage in life cycle
Economic	Exchange rates, foreign exchange, employment, multiplier effects, infrastructure, leakages
Political	E.g. the 2001 terrorist attacks on the USA, resulted in fewer overseas visits taken by US civilians
Sporting events	Events such as the World Cup (Korea–Japan 2002), the Olympic Games (Beijing 2008) lead to a small boom in tourism

BUTLER'S MODEL OF EVOLUTION OF TOURIST AREAS

1. **Exploration:** A small number of tourists, new location, exotic adventurous travel, minimal impact.
2. **Involvement:** If tourists are accepted and if tourism is acceptable, the destinations become better known. There are improvements in the tourist infrastructure. Some local involvement in tourism may begin.
3. **Development:** Inward investment takes place. Tourism becomes a big business. Firms from MEDCs control, manage and organize tourism, leading to more package tours, more holidays and less local involvement.
4. **Consolidation:** Tourism becomes an important industry in an area or region. It involves not just the provision of facilities but also marketing and advertising. Former agricultural land is used for hotels. Facilities such as beaches and hotel swimming pools may become reserved for tourists. Resentment begins and there is a decelerating growth rate.
5. **Stagnation:** There is increased local opposition to tourism and an awareness of the problems it creates. Fewer new tourists arrive.
6. **Decline**: The area decreases in popularity. International operators move out and local involvement may resume. Local operators may be underfunded; hence there is a decline in tourism. It is possible for the industry to be rejuvenated, as in UK coastal resorts in the 1990s.

Leisure at the international scale: sport

There are significant spatial variations in the participation of sport, and in international sporting success.
The **participation rate** refers to the proportion of a population that takes part in a specific sporting activity.

GLOBAL PARTICIPATION IN TWO SPORTS

1 Cross-country

Rank	Country	Number of participants in global top 200	Per capita GNI (2007) US$
1	Kenya	38	1,600
2	Ethiopia	24	700
3	USA	13	46,000
4	Japan	12	33,800
5	Morocco	10	3,800
6	France	10	33,800
7	Spain	9	33,700
8	Portugal	7	21,800
9	UK	6	35,300
10	Russia	6	14,600

2 Golf

Rank	Country	Number of participants in global top 200	Per capita GNI (2007) US$
1	USA	96	46,000
2	UK	27	35,300
3	Japan	16	33,800
4	Australia	16	37,500
5	Sweden	13	36,900
6	South Africa	5	10,600
7	Argentina	4	13,000
8	Spain	3	33,700
9	Ireland	2	45,600
10	Germany	2	34,400

FACTORS AFFECTING PARTICIPATION IN SPORT

There are many variations in sporting activity by nations. The USA, for example, is strong on baseball, American football and basketball. In contrast, cricket is largely played in parts of the former British Empire. Gaelic football and hurling are played almost exclusively in Ireland and Australian Rules is played mostly in Australia. Global sports include football and athletics.

Physical factors

A number of physical factors have an impact on sporting participation and success. Examples include the following:

- Skiing and winter sports are associated with areas, such as the Alps, that have regular and reliable snow in winter.
- Coastal areas with large plunging breakers produce ideal conditions for surfing, such as in Hawaii and California.
- Hilly areas can promote mountain biking, as in the case of Wales.
- Rivers and lakes promote fishing.
- It has been suggested that the increase in red blood cell concentration at high altitude favours long-distance runners in the high-altitude regions of Kenya and Ethiopia.

Human factors

Most sports take place in sporting venues, such as tennis courts, football pitches and swimming pools. Thus, physical geographical factors might not be as important as human factors. There is a strong correlation between **economic wealth** and provision of sporting facilities. Most golf courses are found in MEDCs and NICs. An obvious exception to this is the large number of golf courses that may be located in LEDCs for the benefit of tourists.

There is also evidence that **political factors** influence the provision of sporting infrastructure. Geographers have identified different "models" of national sports systems:

- The Eastern Bloc model uses sports to show how successful the communist system is.
- The Emergent Nation model for South-East Asia and Africa often uses organizations such as the police force and the army to develop its sporting talent.
- The American model is based on competition and rewarding success.

In the UK, the government has attempted to use sport to develop underprivileged areas and rural areas.

In some LEDCs, a lack of funding means that sports resources are limited. This is especially important in explaining variations in success in sport. Within the education system, boarding schools are often seen as having superior facilities to day schools (non-boarding).

Social factors are also important. Some people cannot afford the membership fees associated with certain sports. Golf clubs are generally expensive. Boxing is a sport generally associated with a working-class population (although Oxford University and Cambridge University both have boxing clubs). Polo is another sport that is largely the preserve of the wealthy.

Finally, there are the **cultural factors** that influence participation in sport. A good example is the low participation of Muslim women in athletics and swimming. The convention for Muslim women to remain robed means that successful Muslim athletes, such as the Moroccan middle-distance runner Hasna Benhassi, receive much criticism at home.

Case study of an international sports event

ATLANTA, 1996

The 1996 Olympic Games were the first to be used for economic development and, in economic terms, were a huge success. The Games were funded solely with private money, largely due to the ability of the Atlanta business community to see the economic implications of attracting the Games. It should be noted, however, that despite economic success, there were criticisms about the way the Games were run, mostly concerning the widely distributed Olympic sites and transport problems.

Unlike previous Games, the Atlanta Olympiad had a clear purpose beyond that of sport. Its organizers intended to use the games to:

- attract new business
- help economic regeneration
- act as a catalyst for inner-city regeneration.

The organizers held to the ideal of economic development. By the end of the Games, 18 new companies had been formed in the city, creating 3100 jobs.

One of the Games' great achievements was to stimulate investment in infrastructure which would otherwise have been deferred. For example, the airport was redeveloped comprehensively. Extra facilities were gained by the Georgia State University and the Georgia Institute of Technology, home of the Olympic Village. Since the Games, new housing has been built close to the former Olympic Village and old industrial buildings have been converted or gentrified into "loft" spaces, resembling some of the trendiest areas of New York but on sale at about one-third of the cost.

Thus the downtown area has been partly redeveloped by the construction of sports facilities. One of the most successful attractions in the city centre is the Centennial Olympic Park, the largest new urban park to be opened in the USA since 1945. The 10 ha park, which cost $57 million, replaced a scattering of unsightly parking lots and mostly run-down buildings in the heart of downtown Atlanta.

ATLANTA AND REGIONAL ECONOMIC DEVELOPMENT

Atlanta used its status as an Olympic city to attract new business set-ups and relocations. For example, in 1994, 223 companies moved to the Atlanta metropolitan area, with a further 260 relocating in 1995. The same pace continued during 1996 and then slowed slightly.

The city was a popular choice for relocation and expansion even before it was awarded the Games. Most investors in the south of the USA go there because Atlanta is the predominant place in the region. The city's location – within two hours' flying time of 80% of the population – and its excellent transport infrastructure are its two main attractions. Also, Atlanta's telecommunications capacity, especially in the wake of the Olympic-related digital and fibre-optics developments, exceeds that of any other US city. Finally, land prices in Atlanta are lower than in cities such as New York and Los Angeles.

THE ADVANTAGES AND DISADVANTAGES OF HOSTING THE OLYMPIC GAMES

Advantages	Disadvantages
Prestige – it is considered an honour to host the event and if the Games are a success the host city gains in reputation	There may be financial problems – Montreal made a loss of over $1 billion in 1976 and the debt took years to pay off
Economic spin-offs – trade and tourism in particular	Some events attract terrorists – the shooting of Israeli athletes at the 1972 Munich Olympics is an example
It unites the country and gives a sense of pride	A large number of visitors puts a strain on hotels, transport, water supplies, etc.
It gives a boost to sports facilities – and other facilities. Cities build or improve their facilities to host events	Large events are security risks – due to the international television coverage they are now prime terrorist targets
The event may generate a profit through sales of radio and TV rights, tickets and merchandise, as well as spending in hotels, restaurants, etc.	If an event does not do well, the host country's image suffers. The host will have difficulty attracting other events – if, indeed, it wants to

Case study of a national tourist industry: Spain

GROWTH AND DEVELOPMENT

Spain is a classic example of post-1945 growth in tourism, with over 34 million tourists annually. Spain illustrates many of the problems that resort areas encounter as they reach capacity, and the tendency for tourist places to drift downmarket, setting in motion a downward spiral. The key factors that led to the rise of mass tourism to Spain include:

- its attractive climate
- its long coastline
- the accessibility of Spain to countries in north-west Europe
- the competitive price of Spanish tourism, especially accommodation and dining
- the distinctive Spanish culture.

Over half the foreign visitors to Spain come from France, Germany and the UK. Most of the travellers head for the south coast, for holidays based on sun, sea and sand. Over 70% of tourists are concentrated into just six regions, namely the coastal areas and the Mediterranean islands. The rapid growth of tourism has led to many unforeseen developments. For example, Torremolinos has changed dramatically – before 1960 it was a small fishing village and a tourist resort for only select tourists. However, the town became popular as a centre for package tours and rapid, uncontrolled developments led to the area being swamped by characterless buildings, a lack of open space, limited car parking and inaccessible sea frontage. Overcommercialization, crowding of facilities such as bars, beaches and streets, and pollution of the sea and beach also occurred.

CHANGES IN TOURISM ON THE SPANISH COSTAS

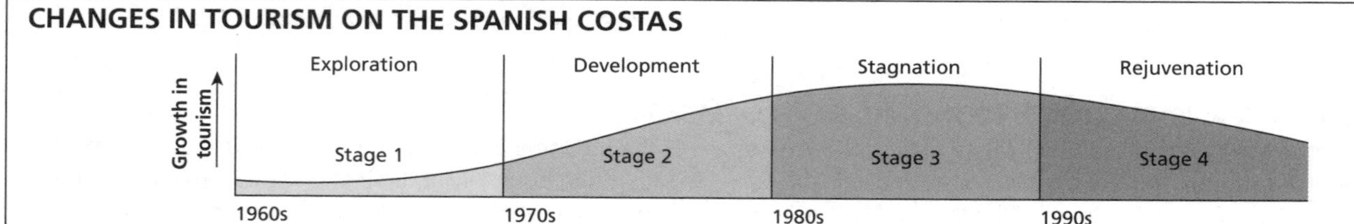

Tourism life cycle for the Costa del Sol

	1960s	1970s	1980s	1990s	2005
State of, and changes in, tourism	Very few tourists	Rapid increase in tourism. Government encouragement	Carrying capacity reached – tourists outstrip resources	Decline – world recession – cheaper, up-market hotels elsewhere	Attracting more affluent visitors
Local employment	Mainly in farming and fishing	Construction work. Jobs in cafés, hotels, shops. Decline in farming	Mainly in tourism – up to 70% in some areas	Unemployment increases as tourism decreases (20%). Farmers use irrigation	Decrease in unemployment
Holiday accommodation	Limited accommodation, very few hotels and apartments, some holiday cottages	Large blocks built (using breeze block and concrete), more apartment blocks and villas	More large hotels built, also apartments and timeshare, luxury villas	Older hotels looking dirty and run-down. Fall in house prices. Only high-class hotels allowed to be built	Development of up-market quality accommodation
Infrastructure (amenities and services)	Limited access and few amenities. Poor roads. Limited street lighting and electricity	Some road improvements but congestion in towns. Bars, discos, restaurants and shops added	E340 opened – "highway of death". More congestion in towns. Marinas and golf courses built	Bars/cafés closing. Malaga bypass and new air terminal opened	Upgrading of infrastructure that has deteriorated
Landscape and environment	Clean, unspoilt beaches. Warm sea with relatively little pollution. Pleasant villages. Quiet. Little visual pollution	Farmland built on. Wildlife frightened away. Beaches and seas less clean	Mountains hidden behind hotels. Litter on beaches. Polluted seas (sewerage). Crime (drugs, vandalism and mugging). Noise pollution	Attempts to clean up beaches and seas (EU Blue Flag beaches). New public parks and gardens opened. Nature reserves	20% of Spanish golf courses are found in Andalucia – 8% annual growth

Source: Adapted from *Baker, S. et al.* Pathways in Senior Geography. *Nelson, 1995*

Ecotourism

Ecotourism is a "green" or "alternative" form of sustainable tourism. It generally occurs in remote areas, with a low density of tourists. It operates at a fairly basic level. Ecotourism includes tourism that is related to ecology and ecosystems. These include game parks, nature reserves, coral reefs and forest parks. Ecotourism aims to give people a first-hand experience of natural environments and to show them the importance of conservation. Its characteristics include:

- planning and control of tourist developments so that they fit in with local conditions
- increasing involvement and control by local or regional communities
- appropriateness to the local area
- a balance between conservation and development, between environment and economics.

However, in areas where ecotourism occurs, there is often a conflict between allowing total access to visitors and providing them with all the facilities they desire, and with conserving the landscape, plants and animals of the area. Another conflict arises when local people wish to use the resource for their own benefit rather than for the benefit of animals or conservation.

THE MONTEVERDE CLOUD FOREST, COSTA RICA

Costa Rica attracts about a million visitors each year. Well-organized government promotions and a reputation as the safest country in Central America attract a large number of North American and European visitors. Costa Rica's tourism is unusual in that a large part of it relates to special-interest groups, such as birdwatchers, and its dispersed small-scale nature is a form of sustainable ecotourism.

Monteverde's cloud forest is situated at a height of around 1700 m. There are over 100 species of mammals, 400 species of birds, 120 species of reptiles and amphibians, and several thousand species of insects.

Early tourists were mostly scientists and conservationists from the USA studying the area's rich biodiversity. In 1974, there were just 471 visitors. Since the early 1990s, the number of tourists has stabilized at about 50,000. The nature of the tourist has changed. At the beginning they were mainly specialists. Now most of the tourists have a more general interest in the forest, and seek a balance between entertainment, adventure and knowledge. Monteverde now accounts for about 18% of Costa Rica's total tourist revenue.

THE IMPACTS OF TOURISM IN COSTA RICA

The growth and development of tourism came at a time when there was a long-term decline in agriculture in Costa Rica. Ecotourism was able to absorb some of the displaced agricultural workers in their own villages. Much of this development was small-scale. For example, 70% of the hotels in Costa Rica have fewer than 20 rooms.

New businesses have been created in Monteverde, including hotels, bed and breakfasts, restaurants, craft stalls, supermarkets, bars, riding stables, and a butterfly and botanical garden. Many of these are locally owned. Over 400 full-time and 140 part-time jobs have been created. In addition, there are indirect employment and multiplier effects. (Multiplier effects are knock-on benefits; for example, an increased number of tourists leads to more spending in shops, more jobs, increased tax revenue and more investment in infrastructure.)

Unlike many rural areas in developing countries, Monteverde is not experiencing out-migration. Indeed, the situation has been quite the opposite. Because of the developments in tourism there have been increased employment opportunities in accommodation, transport, food and communications, and this has attracted many young people. However, the growth in resident and tourist population has placed a great strain on the existing infrastructure such as water supplies, refuse collection, electricity and telecommunications. In addition, the price of land has soared.

There have been benefits other than employment. Controlled access to the cloud forest, with the use of local people employed as guides, has meant a reduction of visitor impact on parts of the forest. Local arts and crafts have been rejuvenated. The increase in the number of small businesses means that income should be more evenly distributed. Formal and informal education programmes have been strengthened, and the local community is even more aware of the value of the natural resources than when they were farmers. Furthermore, the education is two-way: the villagers learn from the tourists just as the tourists learn from the villagers.

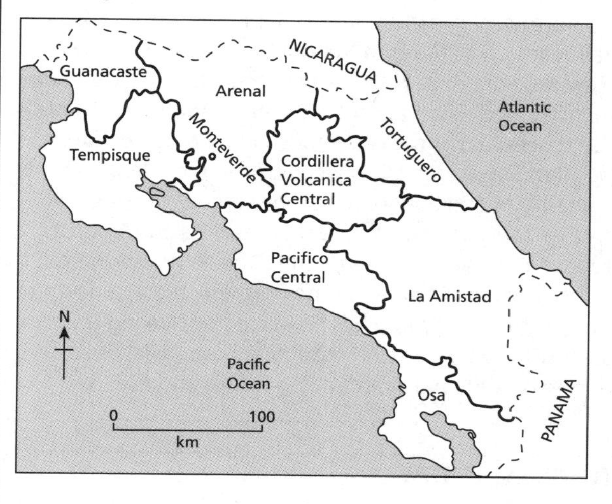

Tourism as a development strategy

TOURISM DEVELOPMENT IN TUNISIA

Tunisia has an established tourist industry benefiting from its Mediterranean location and its tradition of low-cost package holidays from Western Europe. The Tunisian government has actively promoted tourism; between 1970 and 1992 the number of hotels rose from 212 to over 550, and the number of bed spaces increased from 34,000 to about 135,000. Nearly 2 million European tourists entered Tunisia in 1992, in addition to over 1.5 million visitors from North Africa. However, tourist numbers have fluctuated and Tunisia has been unable to attract the high-spending US visitors, largely due to the rise of Islamic fundamentalism and political instability in the region.

In Tunisia, the tourist industry earns over US$900 million and employs over 50,000 people. Given the state involvement in tourism in Tunisia, a high proportion of its earnings remain in the country. However, most of the employment is low-paid and unskilled, such as waiters, kitchen staff and cleaners, while many of the managers are foreign workers. To combat this, the Tunisian government has established a number of training schools. The country also needs to develop its agricultural sector in order to provide food for the tourist market and to reduce expensive imports.

TOURISM PLANNING IN TUNISIA

The Tunisian government has developed a series of five-year National Development Plans. The development of tourism has been an increasingly important element of these plans. The Seventh National Development Plan (1986–89) set up the following targets:

- bed spaces to increase by 19% to 118,000
- bed occupancy to increase by 42% to 18 million bed nights
- direct employment to increase by 13% to 46,000
- total investment to increase by 72% to approximately $1435 million
- annual tourism receipts to increase from 4% to $932 million.

Much of the investment was carried out by private companies, although the government also contributed. The range of investments included:

- infrastructural investments, especially transport routes
- promotion and marketing, especially since the recession of the 1980s and the Gulf War of the early 1990s
- training programmes
- regional initiatives aimed at diversification of attractions and the development of new tourist areas.

Until recently, most tourism in Tunisia has been on the north-east coastline around Tunis and the Bay of Hammanet. The Seventh National Development Plan, however, announced several new tourist areas, including an integrated resort at Port el Kantaoui with over 13,000 bed spaces, a marina, restaurants and a range of sports facilities. Smaller schemes were planned for Hergla and Gamarth. In addition, projects on the undeveloped northern Tunisian coasts and proposals for new tourist access to the Arridge Interior in southern Tunisia were announced. At Tabarka, on the northern coast, a new integrated tourist route has been created, linking the coast with the desert and mountain oasis at Tamerza. This follows the old Arab trading routes and uses accommodation in modern versions of the traditional caravaneserai (hotels or staging posts).

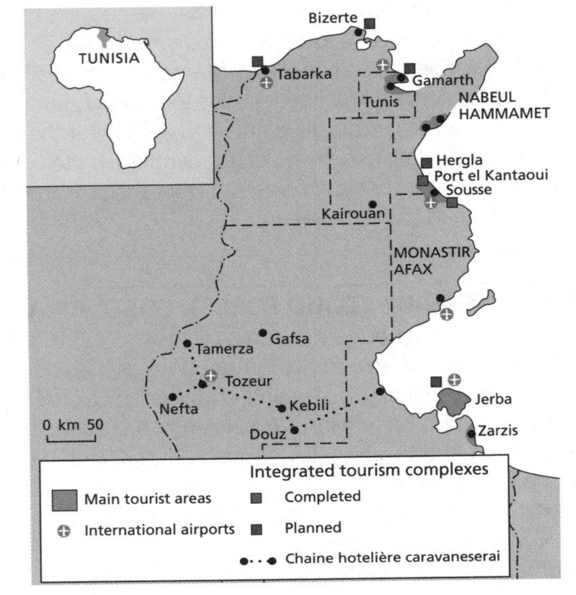

Tourist developments in Tunisia

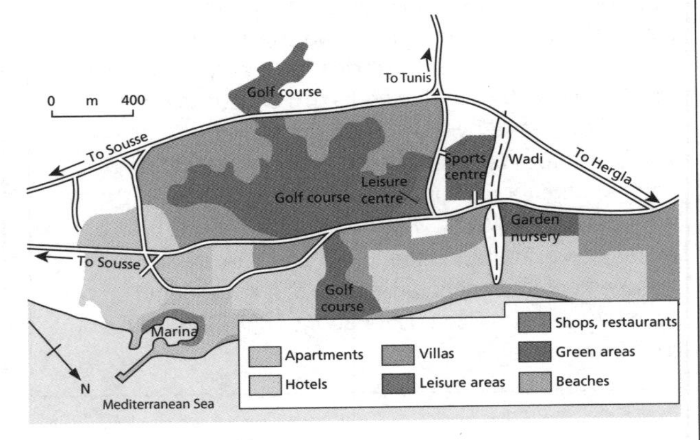

Port el Kantaoui, Tunisia

TOURISM TODAY

France, Germany, Italy and the UK are the four traditional tourist markets, though Tunisia lost roughly 500,000 tourists from Germany after the events of 9/11. From 2003–04, it regained tourists, and 2007 saw arrivals increasing by 3% on 2006.

A national sports league: rugby in South Africa

Rugby is one of South Africa's big three sports, alongside soccer and cricket. The country has fared extremely well on the world stage.

For the disadvantaged people of the old apartheid South Africa, rugby was the white person's game and, even more so, the game of the Afrikaner. Traditionally, most communities of colour played soccer, while for white communities rugby was the winter sport of choice.

SUPER 14

The Super 14 competition features 14 regional teams from South Africa, New Zealand and Australia, with South Africa providing five teams, New Zealand five and Australia four.

The South African teams in the competition are:

- the Sharks – made up of players from the Natal Sharks (based in Durban, Natal and KwaZulu)
- the Stormers – made up of players from provincial teams the Western Province and Boland Cavaliers (Cape Winelands and the west coast of the Western Province)
- the Central Cheetahs – made up of players from provincial teams the Cheetahs, Griquas and Griffons (largely northern Free State)
- the Cats – made up of players from provincial teams the Lions, Pumas and Leopards (Johannesburg, North West and Mpumalanga)
- the Bulls – made up of players from provincial teams the Blue Bulls and Falcons (Northern Transvaal, East Rand and Pretoria).

Back in 1993, the South African team Transvaal beat the New Zealand team Auckland to win the first Super 10 competition. A South African team next won in 2007.

VODACOM CUP

The Vodacom Cup has become an important competition on the South African rugby calendar. It takes place at the same time as the Super 14 competition – starting in late February and finishing in mid-May – and thus creates a platform for talented young players who might otherwise not get a chance to make their mark.

It has also been a fertile breeding ground for strong players from previously disadvantaged backgrounds, thanks to the enforcement of quotas. Quotas, successfully implemented lower down, now extend through the higher levels of South African rugby, including the Super 14.

The Vodacom Cup is divided into two sections – North and South – with the top two teams advancing to the semi-finals and playing a cross-section matches of one-versus-two for a place in the final.

The North is made up of the Golden Lions, Griffons, Leopards, Pumas, Falcons, Blue Bulls and Griquas. The South's teams are the Mighty Elephants, Boland Cavaliers, Border Bulldogs, Free State Cheetahs, Eagles, Western Province and KZN Wildebeests (KwaZulu Natal).

SUPPORTERS

Most of the supporters are very localized or regionally based. This is perhaps inevitable, given the regional/local structure of rugby teams in South Africa. Supporters have traditionally been mainly white people but, increasingly, more black and coloured people follow the sport. Former President Nelson Mandela was a keen rugby fan and "claimed" the shirt of the South African captain, following the country's first Rugby World Cup triumph.

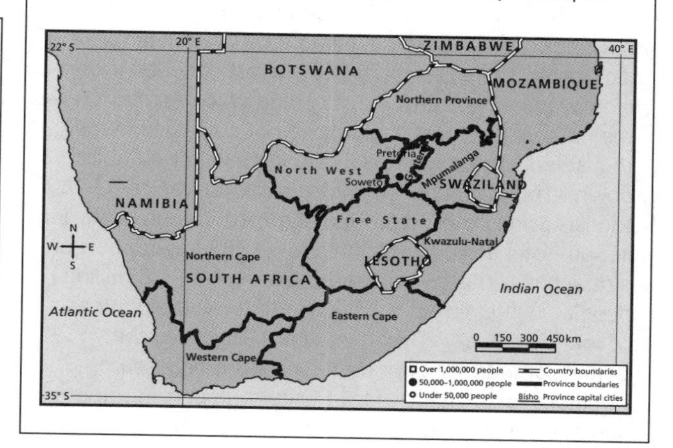

CURRIE CUP

The Currie Cup is the premier provincial rugby competition in South Africa, and was first contested in 1892. The format of the Currie Cup varied from year to year, and finals were held intermittently until 1968, after which the final became an annual event.

Up to and including 2007, the most successful province in the history of the Currie Cup is Western Province (Western Cape) with 32 titles (4 shared), followed by the Blue Bulls (Northern Transvaal) with 22 (4 shared), the Lions (Transvaal) with 9 (1 shared), the Natal Sharks (Natal) with 4, and the Cheetahs (Free State) with 4 (1 shared). Other teams that have lifted the trophy include Griquas (Northern Cape) (three times) and Border (Eastern Cape) (twice, both shared).

For many years the biggest rivalry in South African rugby was between Western Province and the Blue Bulls. During the early to mid-1990s, this was superseded by a three-way rivalry between Natal, the Lions and Western Province.

The Blue Bulls have returned to Currie Cup prominence, however, while the Free State Cheetahs won three titles in succession, from 2005 to 2007, including sharing the Currie Cup with the Blue Bulls in 2006.

The Currie Cup takes place roughly between July and October. The format divides 14 teams into 8 Premier Division and 6 First Division teams.

The teams, in alphabetical order, are: Blue Bulls, Boland Cavaliers, Border Bulldogs (East London), Eagles, Falcons (East Rand and Gauteng), Free State Cheetahs, Golden Lions (Johannesburg), Griffons (Welkom), Griquas (Northern Cape), Leopards (Mpumalanga), Mighty Elephants (Port Elizabeth), Natal Sharks, Pumas (North West) and Western Province (Western Cape).

Leisure at the local scale

CARRYING CAPACITY

Carrying capacity refers to the maximum number of visitors or participants that a site or an event can satisfy at one time. It is customary to distinguish between **environmental carrying capacity**, the maximum number before the local environment becomes damaged, and **perceptual carrying capacity**, the maximum number before a specific group of visitors considers the levels of impact, such as noise, to be excessive – for example, young mountain-bikers may be more crowd-tolerant than elderly walkers. (See also page 117.)

TOURISM IN VENICE

The historic centre of Venice comprises 700 ha, with buildings protected from alteration by government legislation. There is a conflict of interest between those employed in the tourist industry (and who seek to increase the number of tourists) and those not employed in the tourist industry (and who wish to keep visitor numbers down). The optimum carrying capacity for Venice is 9780 tourists using hotel accommodation, 1460 tourists staying in non-hotel accommodation and 10,857 day-trippers on a daily basis. This gives an annual total of over 8 million people – a figure that is 25% greater than the number of tourists actually arriving in Venice. However, the pattern of tourism is not even. There are clear seasonal variations, with an increase in visitor numbers in summer and at weekends. Research has estimated that an average of 37,500 day-trippers a day visit Venice in August. A ceiling of 25,000 visitors a day has been suggested as the maximum carrying capacity for the city.

Exceeding the carrying capacity has important implications for the environment and its long-term preservation. The environmental carrying capacity (concerned with preservation) and the economic carrying capacity (concerned with economic gain) have different values, but the 25,000 figure is a useful benchmark.

In 2000, the carrying capacity of 25,000 visitors was exceeded on over 200 days, and on 7 days the visitor numbers exceeded 100,000.

The large volume of visitors travelling to Venice creates a range of social and economic problems for planners. The negative externalities of overpopulation stagnate the centre's economy and society through congestion and competition for scarce resources. This in turn has resulted in a vicious circle of decline, as day-trippers, who contribute less to the local economy than resident visitors, replace the resident visitors as it becomes less attractive to stay in the city.

A number of measures have been made to control the huge number day-trippers. These include:

- denying access to the city by unauthorized tour coaches via the main coach terminal
- withdrawing Venice and Veneto region's bid for EXPO 2000.

Nevertheless, the city continues to market the destination, thereby alienating the local population.

The excessive numbers of day-trippers have also led to a deterioration in the quality of the tourist experience. This is significant in that it highlights problems affecting many historic cities around the world, especially those in Europe.

TOURISM IN THE BRECON BEACONS, WALES

The Brecon Beacons National Park is located in the south of Wales and is one of the closest national parks to people living in cities such as London, Birmingham and Bristol.

The Llanthony Valley is a microcosm of all that is bad about tourism. Tourists bring little or no benefit to the area, but cause disruptions, irritations and problems. Farmers have at times experienced difficulty moving animals and large machinery, found their gates blocked and been disrupted by pony-trekkers and sightseers driving slowly. For the tourists, the trip is merely a pleasurable drive and they gain little or no understanding about the community, the landscape or the heritage that they have passed through.

Nevertheless, it is possible to integrate local communities into tourism. One attempt to involve the local community in tourism in Wales is the South Pembrokeshire Partnership for Action with Rural Communities (SPARC).

One SPARC action plan improved infrastructure, footpaths and routes linking tourist sites. Residents become involved in tourism developments in many ways:

- Local produce is used wherever possible.
- The majority of visitors stay in locally owned and managed accommodation.
- The service sector is locally owned.
- Local manufacturers are encouraged to tap the tourist market for gifts, souvenirs, crafts and other projects.

EXTENSION

http://www.biodiversity.ru/coastlearn/tourism-eng/con_capacity.html is a useful website with good case studies of sustainable tourism at a local scale, concepts, a glossary and a section on the impacts of unsustainable tourism.

Leisure in urban areas

THE LEISURE HIERARCHY

A simple hierarchy can be established, depending on population size and number of people needed to support a leisure or sporting activity. In most small settlements there are few facilities available. However, as settlement size increases (and the *threshold population* increases), settlements are able to offer a greater variety of leisure and recreation facilities with increasingly specialist functions. The area the settlement serves (the *sphere of influence*) increases in size. Higher order functions are more centralized. A simple hierarchy is outlined below.

Community size	Recommended facilities	Activities offered
Village (pop. 500–1500)	Community hall; community open space	Badminton; keep fit; yoga; football; cricket
Small country town (pop. 2500–6000)	As above, plus: tennis courts; sports hall; swimming pool	As above, plus: tennis; netball; gym; hockey
Capital city	National sports centre for selected sports	As above, plus: bowling; golf; skateboarding; judo; karate; home grounds of sports clubs (football, rugby, hockey); athletics ground; grounds/ stadia for international fixtures

INTRA-URBAN SPATIAL PATTERNS

The diagram shows the distribution of leisure facilities around a typical small or medium-sized town. In most small and medium-sized cities there is a concentration of leisure facilities and tourist attractions in the central area of the city, while on the periphery there are increasing numbers of sports and leisure centres, garden centres and country parks. The central area contains the main concentration of restaurants, cinemas, theatres and other facilities that do not require much space. Finally, there may be some leisure facilities dispersed into neighbourhoods, such as parks and recreation grounds, and community centres.

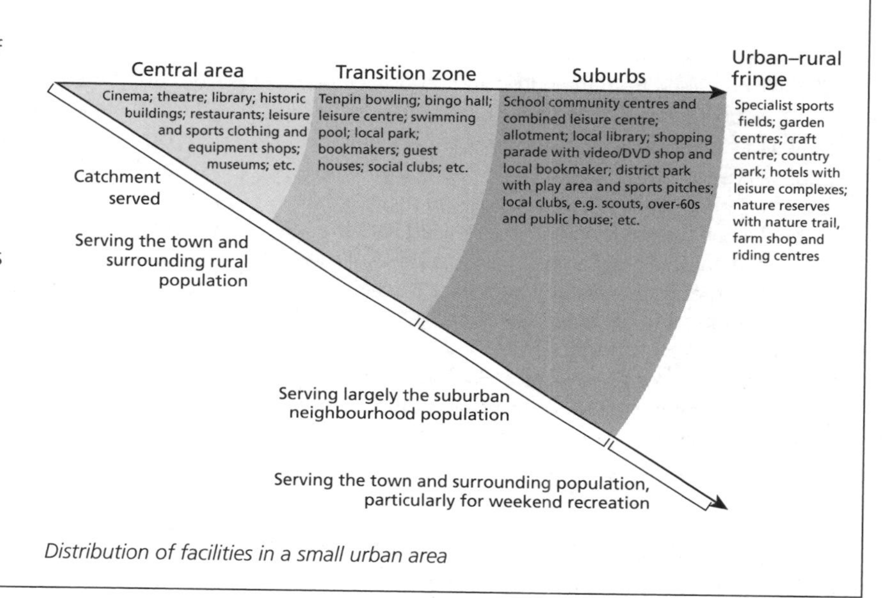

Distribution of facilities in a small urban area

TOURIST FACILITIES IN URBAN AREAS

Urban areas are important for tourism because they are:

- destinations in their own right
- gateways for tourist entry
- centres of accommodation
- bases for excursions.

The tourism business district

In most urban areas there is a distinct pattern in the distribution of tourist activities and facilities. The tourist centre of the city is often referred to as the RBD (recreational business district) or the TBD (tourist business district). In many cities, the tourist business district and the central business district (CBD) coincide.

Tourist facilities in urban areas include accommodation, catering and shopping. Most tourist-related accommodation is found in urban areas, and urban infrastructure and accessibility is vital in the location of hotels and guesthouses.

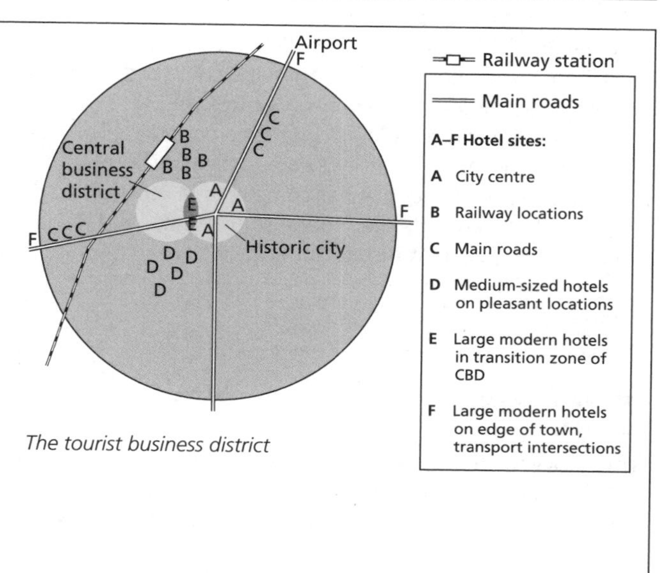

The tourist business district

Sport and urban regeneration

Sport has great potential for tourism and economic development. For example, the 1991 World Student Games were held in Sheffield, UK, and attracted over 5500 competitors from 110 countries. Four new sports arenas were built specially for the Games, costing almost £150 million, including the 25,000-capacity Don Valley Stadium. These facilities remain important not only for Sheffield but for a large region surrounding the area.

THE OLYMPIC GAMES AND REGENERATION

Barcelona 1992

An example of a city that has benefited from hosting the Olympic Games is Barcelona. The 1992 Olympics marked a watershed in the city's economic development, acting as a springboard in its drive to become one of Europe's leading cities.

The Olympics were more than just an expensive marketing campaign. They brought big investments, not just in sports facilities but in major infrastructure projects such as telecommunications and roads. These in turn attracted investment badly needed by a city which had seen its long-established and old-fashioned industrial base decline in the 1970s and 1980s.

Transport improvements were crucial. Over £6.5 billion, raised from the public and private sectors, was spent on enlarging the port, building an additional runway, establishing a high-speed rail link and improving public transport. This infrastructure, coupled with no less than five university campuses, has helped bring high-technology industries. Sony, Sharp, Hewlett Packard, Pioneer, Panasonic and Samsung are just some of the electronics groups represented in or near Barcelona.

Tourism has also benefited. Barcelona attracts an increasing number of visitors. More and more cruise ships either start or end their voyages at Barcelona's port. Many of their passengers choose to stay a night or two in town, either at the beginning or the end of the cruise.

London 2012

One important element in bringing the Games to London was the prospect of regeneration in the East End of London. In addition to this, the Games should bring an economic boom to London and the UK as a whole.

The main effect of staging the Games in 2012 will be the complete transformation of the Lower Lea Valley, the largest remaining regeneration opportunity in inner London. The area is home to one of the most deprived communities in the UK, with high unemployment. Key Olympic venues will be built there, including an 80,000-seat stadium, a world-class aquatic centre, a velodrome and BMX track and a three-arena sports complex. There will also be an Olympic Village to accommodate up to 17,800 athletes and officials.

Up to 12,000 permanent jobs will be created in the area of the Olympic Park alone, as well as thousands of temporary ones. At least 7000 jobs will be created in the construction sector.

The Games should also provide a boost for the tourism industry as more than half a million visitors head to the UK. Analysts believe Olympics-generated income from tourism that year could reach £2 billion.

THE COMMONWEALTH GAMES – MANCHESTER 2002

In 1995, Manchester was awarded the 2002 Commonwealth Games. It also made an unsuccessful bid for the 2000 Olympic Games. Despite losing the Olympic bid, Manchester has benefited from the construction of new facilities and an improved infrastructure.

The city's infrastructure has improved considerably. Greater Manchester's motorway box – the equivalent of the M25 around London – was completed in 1997. Its tram network has been extended to the stadium site, and road approaches spiral into the stadium parking area, making parking relatively easy.

The central theme of Manchester's approach has been urban regeneration. The east side of Manchester has not enjoyed the same economic buoyancy of the city centre or south Manchester; nor has it benefited from major projects such as the redevelopment of Trafford Park and Salford Quays, the old Ship Canal docklands, in west Manchester.

The stadium is located within the city's inner ring road and is on a derelict site once contaminated by heavy engineering waste and gasworks. Since the area was decontaminated and cleaned up, there have a number of new developments such as the national velodrome, a world-class cycling stadium, and a national indoor tennis centre.

EXTENSION

Visit

http://www.london-2012.co.uk/Urban-regeneration/ for a case study of urban regeneration in the lower Lea Valley and a history of the lower Lea Valley. For a different angle on sports and urban regeneration visit **http://www.independent.co.uk/sport/olympics/after-the-party-what-happens-when-the-olympics-leave-town-901629.html**

Principles of sustainable tourism

SUSTAINABLE TOURISM

Sustainable development has been defined as development that meets the needs of the present without compromising the ability of future generations to meet their own needs. Sustainable tourism therefore needs to:

- ensure that renewable sources are not consumed at a rate that is faster than the rate of natural replacement
- maintain biological diversity (biodiversity)
- recognize and value the aesthetic appeal of environments
- respect local cultures, livelihoods and customs
- involve local people in development processes
- promote equity in the distribution of the costs and benefits of tourism.

The concept of sustainable tourism has often used the idea of carrying capacity. Carrying capacity can be thought of in three main ways:

- **physical carrying capacity**: the measure of absolute space, e.g. the number of spaces within a car park
- **ecological carrying capacity**: the level of use that an environment can sustain before environmental damage occurs
- **perceptual carrying capacity**: the level of crowding that a tourist will tolerate before deciding that a location is too full.

PRINCIPLES OF SUSTAINABLE TOURISM

Sustainable tourism is that which:

- operates within natural capacities for the regeneration and future productivity of natural resources
- recognizes the contribution of people in the communities, with their customs and lifestyles linked to the tourism experience
- accepts that people must have an equitable share in the economic benefits of tourism.

This entails:

- **using resources sustainably:** the sustainable use of natural, social and cultural resources is crucial and makes long-term business sense
- **reducing overconsumption and waste:** this avoids the cost of restoring long-term environmental damage and contributes to the quality of tourism
- **maintaining biodiversity:** maintaining and promoting natural, social and cultural diversities is essential for long-term sustainable tourism and creates a resilient base for industry
- **supporting local economies:** tourism which supports a wide range of local economic activities and which takes environmental costs and values into account both protects these economies and avoids environmental damage
- **involving local communities:** the full involvement of local communities in the tourism sector not only benefits them and the environment in general but also improves the quality of the tourism experience
- **training staff:** staff training which integrates sustainable tourism into work practices, along with recruitment of local personnel at all levels, improves the quality of the tourism product
- **marketing tourism responsibly:** such as encouraging tourists to visit sites during off-peak periods to reduce visitor numbers and when ecosystems are most robust; marketing provides tourists with full and responsible information, increases respect for the natural, social and cultural environments of destination areas and enhances customer satisfaction
- **undertaking research:** ongoing monitoring by the industry, using effective data collection and analysis, is essential to help solve problems and to bring benefits to destinations, the industry, tourists and the local community
- **integrating tourism into planning:** this entails tourism and development which are integrated into national and local planning policies, and management plans which undertake environmental impact assessments and increase the long-term viability of tourism
- **better information provision:** providing tourists with information in advance and *in situ* (e.g. through visitor centres) about tourist destinations.

The key objectives for sustainable tourism are:

- quality of the environment
- maximizing the economic benefit.

MANAGING TOURISTS

The usual ways of controlling tourists are to use:

- spatial zoning
- spatial concentration or dispersal of tourists
- restrictive entry or pricing.

Spatial zoning defines areas of land that have different suitabilities or capacities for tourists. Honeypot sites are commonly provided. These are locations that attract tourists by virtue of their promotion and provision of information, refreshment and parking, and then prevent further penetration of tourists into more fragile environments. The Grand Canyon is a good example. Elsewhere, restrictions on tourists may be achieved through pricing. In the USA, national parks charge an entry fee; whereas, in the UK, entry to national parks is free.

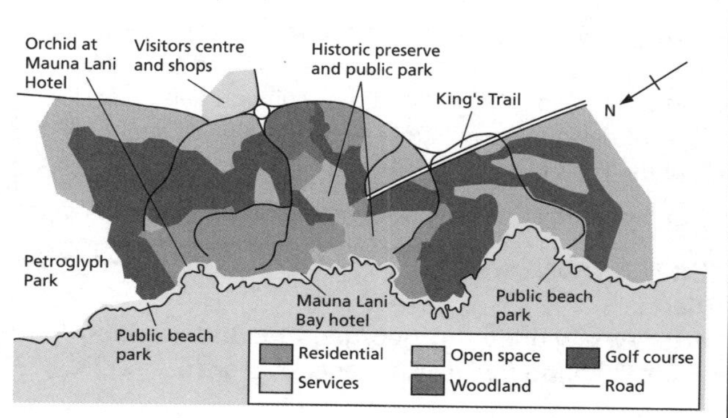

Land-use zoning in honeypots

Measuring health

DEFINITIONS

- **Infant mortality rate** (IMR) $= \frac{\text{total no. of deaths of children} < 1 \text{ year old}}{\text{total no. of live births}} \times 1000$ per year
- **Life expectancy** (E_o)**:** the average number of years that a person can be expected to live, usually from birth, given that demographic factors remain unchanged
- **Disability-adjusted life years (DALYs):** a health measure based on years of "healthy" life lost by being in poor health or in a state of disability
- **Calorie intake:** the amount of food (measured in calories) per person per day
- **Access to safe water:** access to water that is affordable, at sufficient quantity and available without excessive effort and time
- **Access to health services:** usually measured in the number of people per doctor or per hospital

INFANT MORTALITY RATE (IMR)

Infant mortality rates vary from a low of 2‰ in Iceland to over 150‰ in Angola, Afghanistan and Sierra Leone. There is a very strong correlation between types of country and IMR. Countries with a high human development index (HDI) have a low IMR, and those with a low HDI have a high IMR. The region with the highest IMRs is sub-Saharan Africa, with an average IMR of 102‰ in 2005.

LIFE EXPECTANCY

Life expectancy varies from over 80 years in a number of rich countries, such as Sweden, Japan and France, to under 40 years in Zambia, Zimbabwe, Angola and Swaziland. Swaziland has the lowest at under 32 years. The reason for the low and declining life expectancy in many sub-Saharan countries is HIV/AIDS.

Most countries would expect to see life expectancy rise over time. As a country develops, it should have better food supply, clean water and adequate housing. However, a number of countries saw their life expectancy fall between 1970 and 2000–05. These include Zambia (from 50 to 39 years) and Zimbabwe (from 55 to 40 years). In contrast, in Burma (Myanmar), one of the poorest nations in Asia, life expectancy rose from 53 to nearly 60 years.

DISABILITY-ADJUSTED LIFE YEARS

The disability-adjusted life years (DALYs) can show interesting patterns. The DALYs for outdoor air pollution (OAP) show high levels in countries such as Angola, Turkey, Libya and Romania. In contrast, there are low levels in South Africa, Australia, Canada and the USA. Part of the explanation may be population density and level of development. Part may be the type of economy and the fuels burnt.

EXTENSION

Visit
http://www.unep.org/geo/geo4/media/graphics/Zoom/1.05.jpg for life expectancy by region.

CALORIE INTAKE

The intake of food varies from a low of just over 1500 calories per person per day in Afghanistan and Eritrea to highs of over twice that amount in the developed world. The largest intakes are seen in countries such as the USA (3774 calories), Portugal (3740) and Greece (3721). Newly industrializing countries such as China and India are associated with rising food intakes – 2951 and 2459 calories, respectively.

ACCESS TO SAFE WATER

Access to safe water varies from 100% in countries with a high HDI, such as Iceland and Norway, to a low of 22% in Ethiopia, a country with a low HDI. Approximately 83% of the world's population have access to safe water. However, in sub-Saharan Africa the proportion is just 55%. Parts of east Asia also experience a lack of safe water.

ACCESS TO HEALTH SERVICES

Access to health services varies from one doctor per 100,000 people in Burundi and one doctor per 50,000 people in Mozambique, to one doctor per 280 people in Hungary and Iceland. In China there are 610 people per doctor and in India there are 1960 people per doctor. However, inequalities in health services are not just a question of the number of people per doctor or hospital bed; they are also to do with the facilities available in hospitals and clinics, a feature which reinforces the inequalities. It would be wrong to consider merely the quantity of resources per capita, but it is usually impossible to assess their quality.

EXTENSION

Visit
www.who.int/quantifying_ehimpacts/national/countryprofile/oapdalyshighres.jpg
for the DALYs associated with outdoor air pollution (OAP)
www.unep.org/geo/geo4/media/graphics/Zoom/2.12.jpg
for indoor pollution and DALYs

Health-adjusted life expectancy (HALE)

HEALTH-ADJUSTED LIFE EXPECTANCY

HALE is an indicator of the overall health of a population. It combines measures of both age- and sex-specific health data and mortality data into a single statistic. HALE indicates the number of expected years of life equivalent to years lived in full health, based on the average experience in a population. Thus, HALE is not only a measure of quantity of life but also quality of life.

Life expectancy and HALE generally increase with educational attainment. However, the difference between these measures diminishes as education level rises. Therefore, less highly educated people are doubly worse off. Not only do they have shorter life expectancies, but they also shoulder a higher burden of ill health during their shorter lifetimes than their more highly educated counterparts.

Compared with conventional life expectancy, which considers all years as equal, to calculate HALE, years of life are weighted by health status. In a survey in Canada, a health utility index obtained from 1994–95 National Population Health Survey data was used to measure health status. Traditional life expectancy and HALE figures were compared to estimate the burden of ill health. The results showed:

- The social burden of ill health is higher for women than for men.
- It is highest among those in "early" old age, not among the very elderly.
- Sensory problems and pain comprise the largest components of the burden of ill health.
- Higher socio-economic status confers a dual advantage – longer life expectancy and a lower burden of ill health.

CALCULATING HALE

The World Health Organization (WHO) uses life expectancy tables and Sullivan's method (the number of remaining years, at a particular age, which an individual can expect to live in a healthy state) to compute the HALE for countries. The calculation method also includes a weight assigned to each type of disability, adjusted for the severity of the disability.

Mortality data for the calculation of life tables are obtained from death registration data reported annually to WHO. For countries without such data, available survey and census sources containing information on child and adult mortality are analysed and used to estimate life expectancy tables.

Lack of comparable and reliable data on mortality and disease prevalence

A major challenge with the HALE indicator is the lack of reliable data on mortality and morbidity, especially from low-income countries. Other problems with the indicator include the lack of comparability of self-reported data from health interviews.

	Male	Female	Total
Austria	69.3	73.5	71.4
Belgium	68.9	73.3	71.1
Bulgaria	62.5	66.8	64.6
Cyprus	66.7	68.5	67.6
Czech Republic	65.9	70.9	68.4
Denmark	68.6	71.1	69.8
Estonia	59.2	69.0	64.1
Finland	68.7	73.5	71.1
France	69.3	74.7	72.0
Germany	69.6	74.0	71.8
Greece	69.1	72.9	71.0
Hungary	61.5	68.2	64.9
Iceland	72.1	73.6	72.8
Ireland	68.1	71.5	69.8
Italy	70.7	74.7	72.7
Latvia	58.0	67.5	62.8
Lithuania	58.9	67.7	63.3
Luxembourg	69.3	73.7	71.5
Malta	69.7	72.3	71.0
Netherlands	69.7	72.6	71.2
Norway	70.4	73.6	72.0
Poland	63.1	68.5	65.8
Portugal	66.7	71.7	69.2
Romania	61.0	65.2	63.1
Slovakia	63.0	69.4	66.2
Slovenia	66.6	72.3	69.5
Spain	69.9	75.3	72.6
Sweden	71.9	74.8	73.3
Switzerland	71.1	75.3	73.2
UK	69.1	72.1	70.6
EU-27	–	–	70.3
EU-15	–	–	71.7

Health-adjusted life expectancy at birth for men, women and the total population, 2002

Note: *– no data available*

Source: *WHO-HFA, 2007,* **www.euphix.org/object_document/o4992n27073.html**

Note: IHD = heart disease; COPD = chronic disruptive pulmonary disorder (lungs)

Variations in health

CHANGES IN LIFE EXPECTANCY1950–2000

Area	Years					
	1950–55	1960–65	1970–75	1980–85	1990–95	1995–2000
World	46	52	58	61	64	66
Developed countries	67	70	71	73	74	74
Less developed countries	41	48	55	59	62	64
Africa	38	42	46	49	53	54
Asia	41	48	56	60	65	66
Latin America (and Caribbean)	51	57	61	65	69	70
Europe	66	70	71	72	73	73
North America (USA and Canada)	69	70	72	75	76	77

Life expectancy at birth by world region, 1950–2000

Source: *Yaukey, David and Douglas L. Anderton.* Demography: The Study of Human Population. *Prospect Heights, IL: Waveland, 2001*

Global life expectancy rose from 46 years in 1950–55 to 66 years in 1995–2000. During this period, life expectancy rose in all regions. However, in many sub-Saharan African countries, largely due to the impact of HIV/AIDS, life expectancy is now falling and the gains made by the 1990s have fallen into reverse.

Reasons for the rise in life expectancy include greater food production, greater availability of clean water, better living conditions, and better healthcare, especially for the young and the old. Nevertheless, national statistics can hide many features, such as large-scale geographic (spatial) variations within countries. In Brazil, for example, death rates are much higher in the shanty towns than in the better-quality environments. Regional differences occur, too. People in the richer south-east region of Brazil live longer than those living in the poorer north-east. Sometimes there are racial differences. The IMR among black people in South Africa is higher than the IMR among the white population, although the differences are falling.

EPIDEMIOLOGICAL MODEL

One of the main changes in a country's health profile is the shift from infectious or contagious diseases (**epidemics**) to diseases causing a gradual worsening in health (**degenerative diseases**). This is known as the epidemiological transition. For example, a country in an early stage of development would be expected to have a large number of deaths and illnesses from infectious diseases such as respiratory diseases, measles and gastro-enteritis (diarrhoea and vomiting). By contrast, we would expect an MEDC to have more deaths and illnesses due to heart attack, stroke and cancers, diseases which are not infectious or communicable. The exception to this is the rise in AIDS, and with it TB, in MEDCs in the first decade of the 21st century.

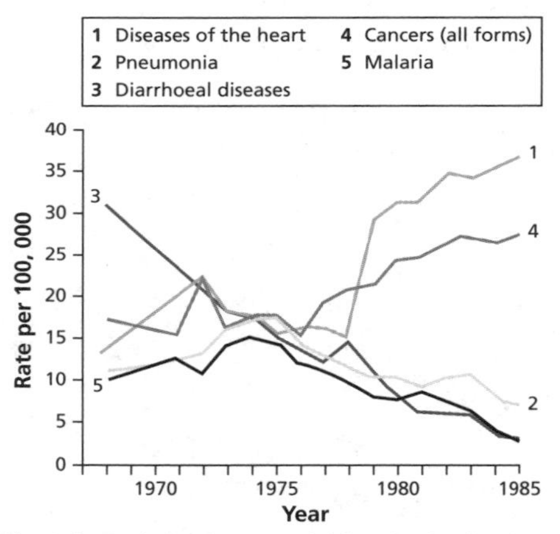

Epidemiological transition model for Thailand, 1968–85

PREVENTION RELATIVE TO TREATMENT

The type of healthcare that is available also varies. Some healthcare is preventive – trying to prevent illnesses from developing. Most healthcare, however, is **curative** – curing symptoms after they have developed.

In South Africa, 8% or more of GNI is spent on the national health system, including both the public and private health sectors. On average, 60% of this is spent in the private sector, which provides care to 20% of the population. The majority of the population (80%) relies on the public health system for healthcare. This sector receives 40% of total expenditure on health.

Prevention is a good way to reduce the burden of disease and improve the quality of life

Individuals living in poverty are more likely to experience delays in receiving appropriate treatment, or to lack access to water and sanitation within their dwelling.

Global availability of food

GLOBAL PATTERNS OF FOOD INTAKE

The map of global patterns of food intake below shows that the largest calorie intakes (over 3500 calories per person per day) are in the USA and parts of western Europe (e.g. Portugal, Greece, France and Austria). In contrast, the lowest levels of calorie intake (under 2000 calories per person per day) occur in parts of sub-Saharan Africa, Mongolia and Afghanistan. There are generally high levels (over 3000 calories) in Europe, North America, Australia, parts of northern Africa and China. Relatively low levels (2000–2500 calories) are found across the Sahel, south Asia and parts of southern Africa. Most of South America, eastern Europe and South-East Asia have an intake of between 2500 and 3000 calories.

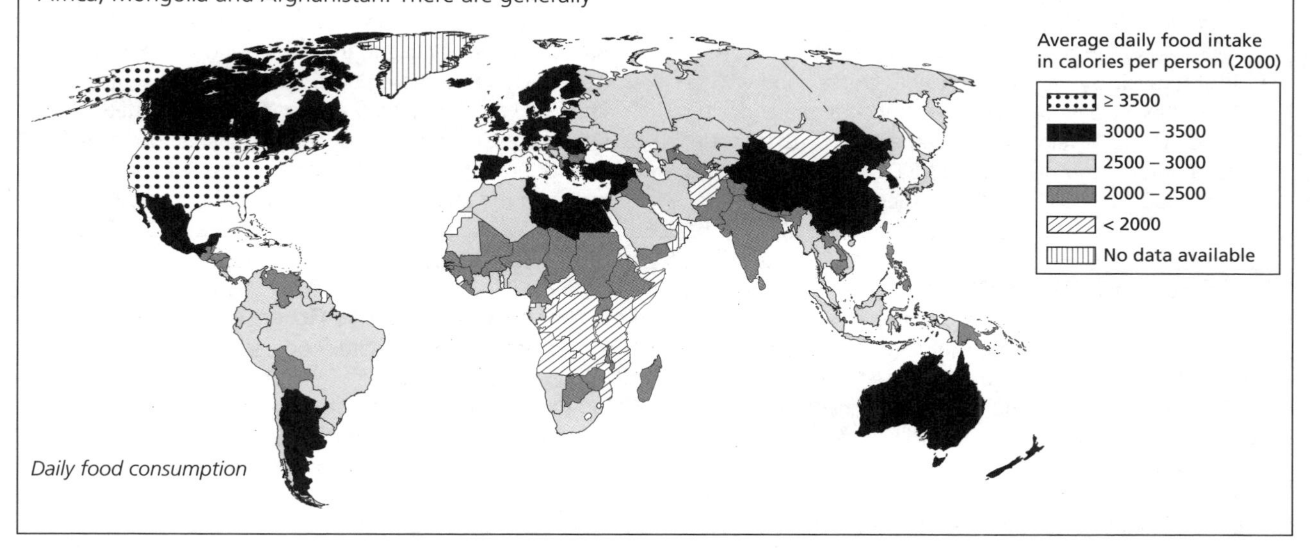

Daily food consumption

TYPES OF FOOD SHORTAGE

- Malnutrition – a diet that is lacking (or has too much) in quantity or quality of foods
- Deficiency diseases – lack of specific vitamins or minerals
- Kwashiorkor – lack of protein in the diet
- Marasmus – lack of calories/energy in the diet
- Obesity – too much energy/protein foods
- Starvation – limited/non-existent intake of food
- Temporary hunger – a short-term decline in the availability of food to a population in an area
- Famine – a long-term decline in the availability of food in a region

FOOD SECURITY

Two commonly used definitions of food security come from the United Nation's Food and Agriculture Organization (FAO) and the United States Department of Agriculture (USDA):

- "Food security exists when all people, at all times, have access to sufficient, safe and nutritious food to meet their dietary needs and food preferences for an active and healthy life." (FAO)
- "Food security for a household means access by all members at all times to enough food for an active, healthy life. Food security includes at a minimum (i) the ready availability of nutritionally adequate and safe foods, and (ii) an assured ability to acquire acceptable foods in socially acceptable ways (that is, without resorting to emergency food supplies, scavenging, stealing, or other coping strategies)." (USDA)

FAD and FED

Much of the early literature on hunger, famine and malnutrition were reports on climate and its effect on food supplies, and on the problems of transport, storage and relief organizations. Such studies were often grouped under the umbrella term of **food availability deficit** (FAD), which implied that food deficiencies were caused by local shortages due to physical factors.

More recently, the literature has been heavily influenced by political and economic factors. Sen (1981) observed that not all food shortages caused hunger, and increased hunger could be observed in areas where food production was, in fact, increasing. This has been the case in India, Ethiopia and Sudan. FAD could not therefore be seen as a complete explanation of the causes of malnutrition, nor did it link hunger with the distribution of resources and poverty. In the analysis of the population "at risk" of malnutrition, it became clear that it was important to look also at the political and economic system in which food is produced, distributed and consumed. This included not just the physical factors which affected yield, but also people's access to food and the conditions which cause that access to alter, i.e. **food entitlement deficit** (FED). Sen's work has generally been accepted, although it is important to consider physical factors such as precipitation and environmental degradation as potential triggers of famines.

Areas of food sufficiency

INCREASING FOOD OUTPUT

Ways of improving food production are well known:

- Genetically engineered **high-yielding varieties** (HYVs): India feeds twice as many people as Africa on just 13% of the land area.
- **Fertilizers, pesticides and herbicides:** fertilizer use in Africa is less than 10% of Chinese levels.
- **Irrigation:** the North Sinai Development Canal running from the River Nile Delta to the Sinai peninsula, for example, will irrigate 62,000 km^2 of desert.
- **Biotechnology** has the capacity to create another Green Revolution (see below). However, much of the agricultural research and development is carried out by large-scale companies (agribusinesses) in MEDCs and is concerned with food for MEDC markets rather than LEDCs.
- Two mechanisms which have a powerful influence on farming are **markets** and **human productivity**. Farmers will increase output in response to guaranteed prices and guaranteed markets. In part, this was the cause of the food mountains and lakes in Europe in the 1980s (see page 124). In order to increase production, it is necessary to pay farmers properly. Nowhere is this more needed than in LEDCs, where agriculture has stagnated relative to industrialization. To keep the better educated, more skilled labour in rural areas, improved pay and working conditions are needed. Otherwise the migration of highly qualified personnel will continue to have the same effect as soil erosion – reducing the ability of the land to feed the population.

THE GREEN REVOLUTION

The Green Revolution is the application of science and technology to increase food productivity. It includes a variety of techniques, such as genetic engineering to produce HYVs of crops and animals, mechanization, pesticides, herbicides, chemical fertilizers and irrigation water.

HYVs are the flagship of the Green Revolution. During 1967–68 India adopted Mexican Rice IR8, which had a short stalk and a larger head than traditional varieties, and yielded twice as much grain. However, it required considerable amounts of water and nitrogen. Up to 55% of crops in India are HYVs and 85% in the Philippines. By contrast, only 13% of Thailand's crops are HYVs.

However, population growth is more rapid than the increase in food production. In India, for example, by 2010 the population will reach 1.15 billion people and food production will need to increase by 50% to match demand. But much of India's land is of limited potential.

The consequences of the Green Revolution

The main benefit is that more food can be produced:

- Yields are higher.
- Up to three crops can be grown each year.
- More food should lead to less hunger.
- More exports create more foreign currency.

However, there are many problems:

- Not all farmers adopt HYVs – some cannot afford it.
- As the cost has risen, indebtedness has increased.
- Rural unemployment has increased due to mechanization.
- Irrigation has led to salinization – 20% of Pakistan's and 25% of central Asia's irrigated land is affected.
- Soil fertility is declining as HYVs use up nutrients. They can be replenished by fertilizers but these are expensive.
- Developing countries are dependent on many rich countries for the inputs.

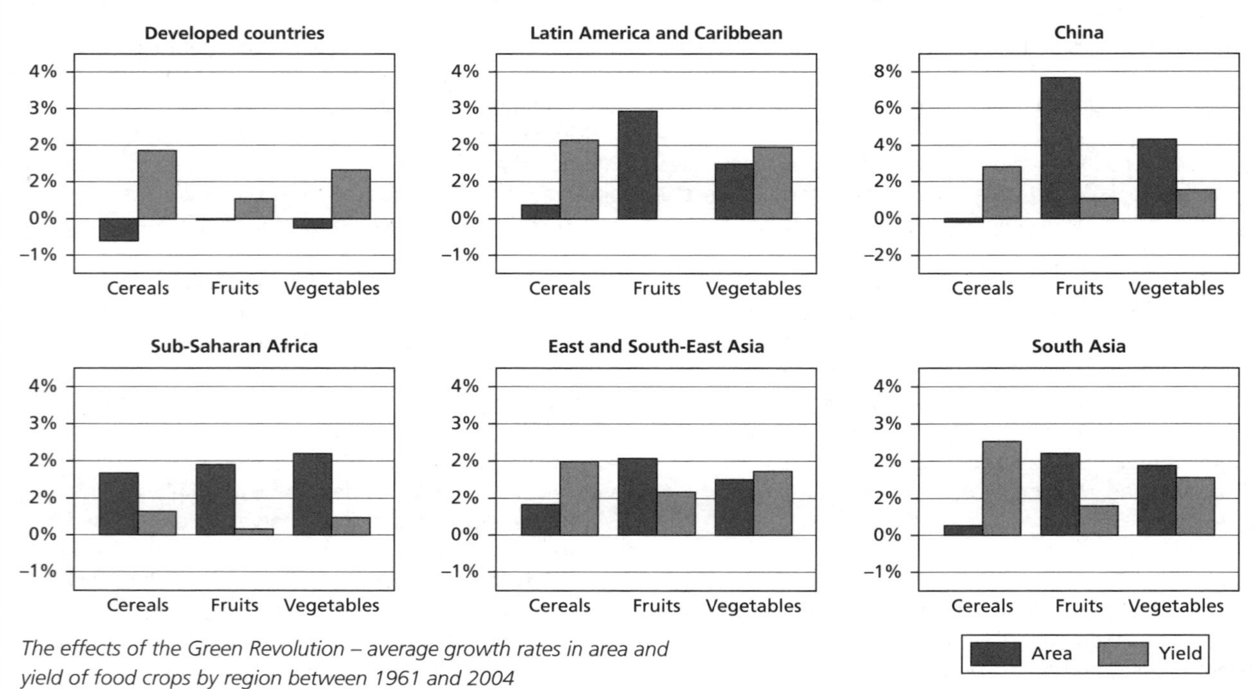

The effects of the Green Revolution – average growth rates in area and yield of food crops by region between 1961 and 2004

Areas of food deficiency

WHY ARE THERE FOOD SHORTAGES?

There are a number of environmental, demographic, political, social and economic factors that cause food shortages around the world:

- Soaring oil and energy prices in 2007 have pushed up the cost of food production dramatically: fertilizer is up more than 70%; fuel for tractors and farm machinery is up 30%; the price of pesticides, which depend on oil, has increased; as have labour costs, the cost of producing rice and the cost of transporting it. This oil-driven inflation seems to be the underlying factor in the current rice crisis.
- Demand is rising as the global population grows and as people in emerging economies such as China and India use increasing affluence to buy more meat, eggs and dairy products. Over 30% of the world's grain now goes to feeding animals rather than people directly. Farming one hectare of decent land can produce 169 kg of protein from grain, but one hectare given over to beef farming will produce only 24 kg of protein.
- Droughts in grain-producing areas of the world have hit harvests in the last few years. Grain stocks are at a historic low.
- Cyclone Nargis in Burma (2008) and cyclones in Bangladesh have reduced the supply of grain on the world market.
- Biofuels are competing with food for arable land, with both the USA and the EU mandating their use. About 30% of the US corn crop is expected to be diverted to biofuels in 2008.
- For years, food experts warned that chronic underinvestment in agriculture in developing countries, by governments and donors alike, would one day spell disaster. In 1986, 20% of foreign aid spent by rich countries was devoted to agriculture in the developing world. By 2006, that share had shrunk to less than 3%. African governments, more wary of the political clout of their urban citizens, now spend less than 5% of their budgets on supporting farming and the rural communities, which are home to the poorest two-thirds of their populations.
- Dumping of excess crops by the West throughout the cheap-food era, combined with the market-orientated World Bank structural adjustment policies of the 1980s and 1990s that closed down government marketing boards designed to guarantee price stability, served to squeeze much of the remaining life out of African farming.
- Farm subsidies in the West make farm products more competitive than those produced in developing countries. Farm support schemes in the West cost poor families in developing countries $100 billion a year in lost income.
- Speculative trading in agricultural commodities has grown dramatically. Several big investment banks have launched agricultural commodity index funds, as they look for new areas to make profits in following the 2007/8 "credit crunch". The result has been enormous fluctuations in market prices that do not appear to relate to changes in supply and demand. In 2004, $10–15 billion was invested in agricultural commodities funds; that figure is now more than $150 billion. Wall Street investment funds own 40% of US wheat futures and more than 20% of US corn futures.
- There is also the creeping disaster of climate change. Some areas will become drier and face water shortages; others may experience more extreme weather conditions. One estimate is that by 2050, half of arable land in the world might no longer be suitable for production because of water shortages and climate change. By then the global population is expected to have grown from today's 6.3 billion to 9 billion.

Different experts give different weight to each of these factors, but all agree that their coincidence has led to the current turbulence.

Food prices

Export prices for 60 internationally traded food commodities have soared, especially oils and fats.

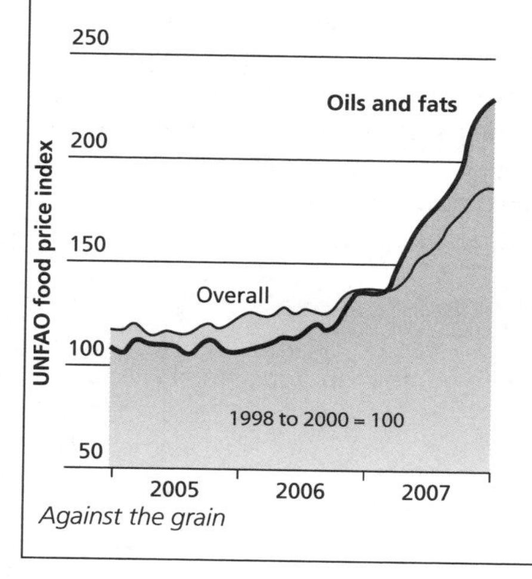

Against the grain

Oil consumption

Most vegetable oil is used for food purposes. But turning it into biofuels represents the fastest-growing demand for oils.

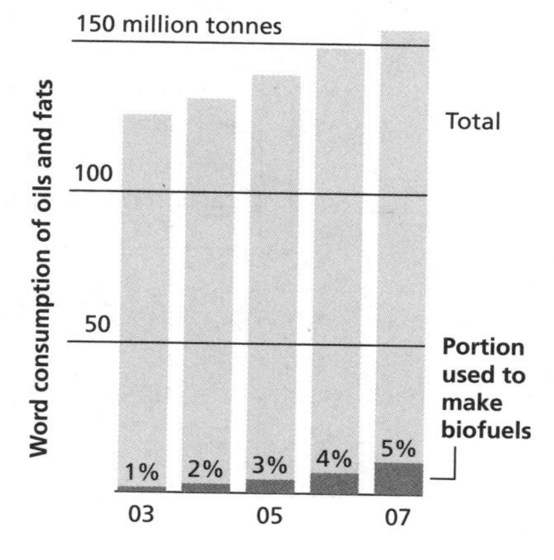

Oil prices

Farmers in the USA and China have been planting less soy, causing a shortage that has helped raise prices for palm oil, the main alternative to soybean oil.

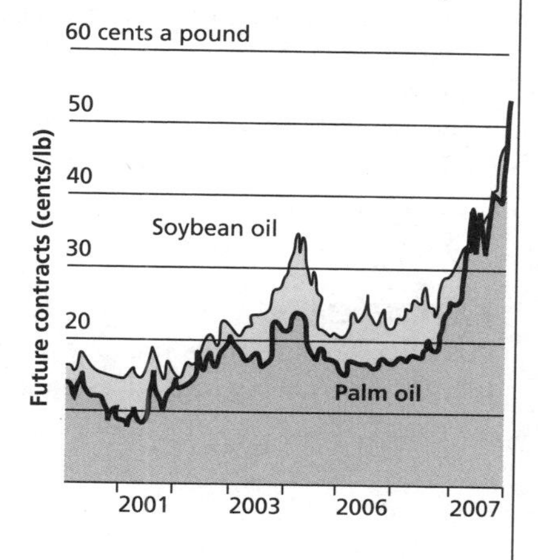

Food production and markets

There are many factors that affect the production and availability of food. Among the political factors are trade barriers, agricultural subsidies, and bilateral and multilateral agreements. Large farming companies – or agribusinesses – are often part of transnational companies (TNCs) and they have a major impact on trade and trading arrangements.

TRADING BLOCS AND FARM SUBSIDIES

A trading bloc is an arrangement among a number of countries to allow free trade among member countries but to impose tariffs (charges) on countries that may wish to trade with them. The European Union (EU) is an example of a trading bloc and it has a major impact on global food production and trade.

To increase farm productivity the EU introduced the Common Agricultural Policy (CAP). At the centre of the CAP was the system of **guaranteed prices** for unlimited production. This encouraged farmers to maximize their production, as it provided a **guaranteed market**. Imports were subjected to import duties or levies and export subsidies were introduced to make EU products more competitive on the world market.

FREE TRADE

Free trade allows a country to trade **competitively** with another country. There are no restrictions regarding what can be exported or imported. By contrast, protectionism creates **restrictions** to trade. It creates barriers to imports as well as to exports.

The advantages of free trade

- It allows countries to specialize and concentrate on their **comparative advantages**, that is, the things they do better than other countries.
- It allows countries to obtain goods and services more cheaply than if they had to produce them themselves.
- It allows countries to obtain goods year-round.
- It increases competition, promotes efficiency and reduces waste.

TRADE BARRIERS

Protection can be justified for a number of reasons. These include the following aims:

- To improve the trade balance
- To protect new home industries against old, established foreign industries
- To reduce imports
- To increase self-sufficiency
- To maintain employment levels

MULTILATERAL ARRANGEMENTS

Multilateral arrangements occur when a number of countries (such as those in the EU) agree to import goods from a number of other countries, in this example the ACP (African, Caribbean and Pacific nations).

The 1975 Lomé (Trade) Convention was important because it gave preferential access to ACP farmers to the entire EU market. Nevertheless, Caribbean bananas accounted for only 7–9% of EU banana imports.

Some producers, notably the US-owned multinational companies, were unhappy about the trade links between the EU and the Caribbean. Instead, the USA favoured free trade for bananas, so that US companies could gain access to the EU market themselves.

Smallholder growers in Dominica, Grenada, St Lucia and St Vincent supply up to 66% of bananas consumed in the UK, although their supplies to Europe as a whole are low. Their governments claim that they cannot compete against cheaper TNC-produced Central American bananas. However, there is a knock-on or **multiplier effect**. If banana-traders' boats no longer visited the main Caribbean trading ports, trade in avocados and citrus fruits could be damaged, as these are not large enough on their own to attract boats, but are a useful supplement for boats dealing with bananas.

US TNCs expanded their banana production in the mid-1990s, as they expected the European market to be opened. However, trading conditions continue to favour the ACP countries and there has been a slump in the TNCs' profits. As a result of European protectionism, the USA imposed tariffs and sanctions on 17 items of goods imported from Europe. It is ironic that the USA and Europe fell out over bananas, since neither exports bananas and bananas are a relatively insignificant item in world trade.

BILATERAL ARRANGEMENTS

A bilateral arrangement is when one consumer enters an agreement with one producer. For example, in 2007 the Caribbean island of St Lucia had cause for celebration when the British supermarket chain Sainsbury announced that all the bananas it would sell in future would be fairly traded bananas, and that nearly 100 million of these would come from St Lucia.

EXTENSION

Visit the World Food Programme at:
http://farmsubsidy.org/ to find out how much each country or company gets from the Common Agricultural Policy.

Alleviating food shortages

THE SHORT TERM

Increase production – reduction in set-aside

High prices have encouraged more food production. The EU has abandoned its compulsory programme to set aside land. Russia, Ukraine and Kazakhstan have increased production, and Australia has recovered well. Record wheat production in 2008 brought prices down. However, long-term trends are not so good.

Food aid

The World Food Programme (WFP) managed to raise all its $755 million 2008 appeal to maintain its emergency feeding programmes, largely thanks to a surprise Saudi donation of $500 million. It was able to send food aid to Burma (Myanmar) for victims of Cyclone Nargis, and to Sudan and Georgia for victims of civil conflict.

However, the WFP reaches only about 80 million of the most desperate, mostly refugees from conflicts and natural disasters. There are 700 million more chronically hungry people scattered around the world.

Seeds and fertilizer

As well as needing food to survive, the rural poor urgently need help planting next season's crops if there is to be an end to the food crisis. Millions have been forced to eat next season's seeds to survive, and the price of fertilizer (largely dependent on oil) has risen sixfold in some regions over the course of a year.

Export bans

Export bans drive prices even higher and increase market variability.

THE MEDIUM TERM

Free trade

Trade liberalization, lowering farm subsidies in the USA and undoing some of the protectionism of the EU's Common Agricultural Policy should help poor farmers in the future, but the direct impact could be to raise food prices in the developing world, as producers focus on western markets.

Biofuels

The food crisis has triggered a backlash against plant-derived fuels, which were originally hailed as an answer to global warming. With over 40% of American maize being used to make ethanol, there is clearly a clash of interests.

EXTENSION

Visit the World Food Programme at: **http://www.wfp.org/english/** for an update on current food shortages and projects to alleviate food shortages.

THE LONG TERM

Agricultural investment

Experts believe yields in Africa can be increased up to fourfold with the right help – 40% of Asian agriculture is irrigated, compared to 4% in Africa. The average Asian farmer uses 110 kg of fertilizer a year. The average African uses just 4 kg. At least a third of the crops in an average African season are lost after the harvest, largely because farmers cannot get them to markets on time.

GM crops

Agriculture experts at the UN and in developing countries do not expect GM crops on their own to radically improve yields. The main trouble, they argue, is that almost all the research has been devoted to developing crops for rich countries in the northern hemisphere.

Sustainability

Campaigners argue that the world cannot feed its population if China, India and other emerging economies want to eat like people in the West. The only long-term solution, they argue, is rethinking western lifestyles and expectations.

FAIR TRADE

Fair or ethical trade can be defined as trade that attempts to be socially, economically and environmentally responsible. It is trade in which companies take responsibility for the wider impact of their business. Ethical trading is an attempt to address failings of the global trading system. The difficulty in achieving fair trade is illustrated by the following example.

Nutmeg

Indonesia is the world's largest producer of nutmeg, accounting for about 75% of production. Its exports go mainly to the USA. Grenada in the Caribbean produces most of the rest. Over 7000 farmers produce nutmegs. The trees take about 70 years to mature, and then they produce nutmegs for up to a century.

Nutmegs used to provide a steady income. Most were exported to Europe. In 1987 Indonesia and Grenada agreed to fix export volumes and prices. Nutmegs were traded at about $7000 a tonne. However, in 1990 Indonesia restructured its economy, thereby allowing free trade in farming. The agreement with Grenada was abandoned, and Indonesia flooded the market. Prices dropped to $2000, and US importers were able to play off the two countries against each other, thereby keeping prices low.

To raise the price again, much of the surplus stock was burnt so as to reduce supply. In Grenada there was thought about uprooting nutmeg trees and replacing them with banana trees. However, bananas are not very profitable either. Many of Grenada's rural areas are now very depressed and out-migration of younger people is further increasing the country's problems.

Sustainable agriculture

PHOTOSYNTHETIC EFFICIENCY

Agriculture seeks to improve the productivity of ecosystems by applying energy subsidies (to remove competitors, apply nutrients, add or take away water, and so on). However, these produce sustainable systems only when they are sympathetic to the local ecology. There is evidence that only in exceptional cases do crop efficiencies exceed 2% – comparable to temperate forests.

Crop or ecosystem	Location	Growth period (days)	Photosynthetic efficiency (%)
Natural ecosystem			
Tropical rainforest	Ivory Coast	365	0.32
Deciduous forest	UK	180	1.07
Crops			
Sugar cane	Hawaii	365	1.95
Maize (two crops)	Uganda	135 + 135	2.35
Soybeans (two crops)	Uganda	135 + 135	0.95
Rice	Japan	180	1.93

A comparison of photosynthetic efficiency for types of vegetation and selected crops

ENERGY EFFICIENCY RATIOS

The energy efficiency ratio (EER) is a measure of the amount of energy inputs into a system compared with the outputs. In a traditional agroforestry system the inputs are very low. However, the outputs from hunting and gathering may be quite high. In contrast, the inputs into intensive pastoral farming or greenhouse cultivation may be very great but the returns may be quite low.

$$EER = \frac{\text{energy outputs}}{\text{energy inputs}}$$

Agroforestry	65
Hunter-gatherers	7.8
UK cereal farm	1.9
UK allotment	1.3
UK dairy farm	0.38
Broiler hens	0.1
Greenhouse lettuces	0.002

Agricultural system	Total energy input (10^6 kJ/ha)	Protein output (kg/ha)
Hill farming (sheep)	0.6	1–1.5
Mixed farming	12–15	500
Intensive crop production	15–20	2000
Intensive animal production	40	300

Energy input and protein yields of four major agricultural systems

SUSTAINABLE YIELD

The **sustainable yield** is the amount of food (yield) that can be taken from the land without reducing the ability of the land to produce the same amount of goods in the future, without any additional inputs. If the production of palm oil, for example, reduces the nutrient availability in the soil or the moisture in the soil, it is not sustainable. Equally, if a particular type of farming leads to the build-up of salt in the soil (salinization) or nitrates in streams (eutrophication), the type of farming is not sustainable.

FOOD MILES

Food miles refer to the distance that food travels from where it is produced to where it is consumed. It is a way of indicating the environmental impact of the food we eat. The global food industry has a massive impact on transport. Food distribution now accounts for between a third and 40% of all UK road freight. The food system has become almost completely dependent on crude oil. This means food supplies are vulnerable, inefficient and unsustainable.

The wastefulness of a Christmas dinner

The ingredients of a traditional Christmas meal may have cumulatively travelled 24,000 miles, according to a report, *Eating Oil*. Buying the ingredients in a London supermarket, the report found that poultry could have been imported from Thailand (nearly 17,000 km) runner beans came from Zambia (nearly 8000 km); carrots from Spain (1600 km); mangetout from Zimbabwe (over 8000 km); potatoes from Italy (2400 km) and sprouts from Britain, where they were transported around the country before reaching the shop (200 km). By the time trucking to and from warehouses to stores was added, the total distance the food had moved was over 38,000 km, or the equivalent of travelling around the world once.

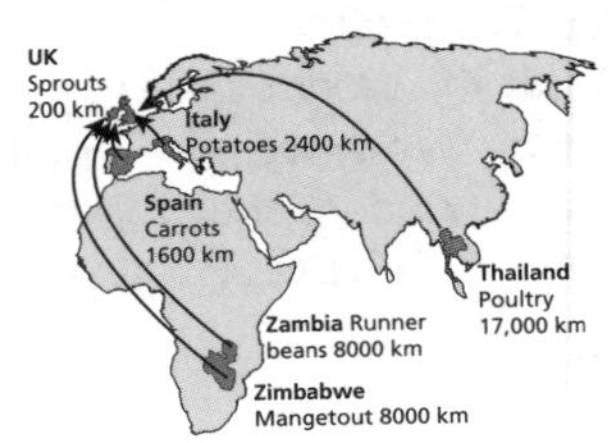

Global patterns of disease

DISEASE, POVERTY AND WEALTH

According to the World Health Organization (WHO) one-fifth of the 5.6 billion people in the world today live in extreme poverty. Almost one-third of the world's children are undernourished and half the global population does not have access to essential drugs.

Global poverty is by far the most important cause of disease and death. Of the 51 million deaths each year in the world in the 1990s, 32% were due to **infectious** and parasitic diseases – diarrhoea and dysentery, pneumonia and respiratory infections, tuberculosis, malaria and measles – from which richer countries are almost immune. By contrast, MEDCs experience more **degenerative** diseases, such as strokes, cancers and heart disease. This change in disease pattern, from mostly infectious to mostly degenerative, is known as the **epidemiological transition**. Increasingly, degenerative diseases are becoming more common even in poor countries. For example, on current trends, tobacco deaths will rise from 4 million a year in 1999 to 10 million by 2030, 70% of them in developing countries. Half of these deaths will affect people in middle age, depriving them of 20–25 years of life. Mental illness, though rarely a cause of death, is one of the biggest causes of disability worldwide. Depression is the fourth largest cause of disability worldwide, affecting rich and poor nations alike, while alcohol abuse has become a big problem among adult men in rich countries. Injuries, both intentional and accidental, are also a high and rising cause of death and disability.

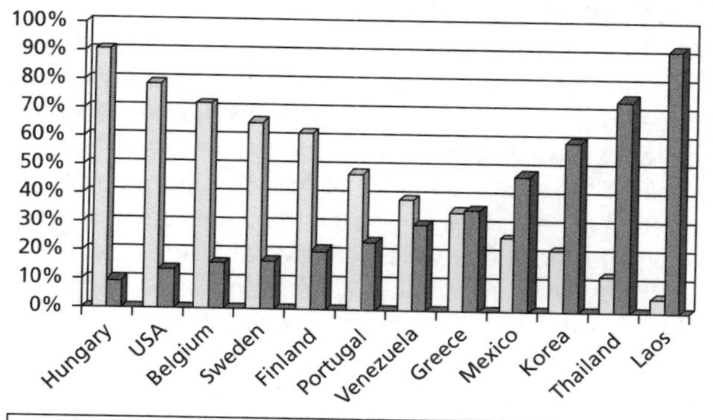

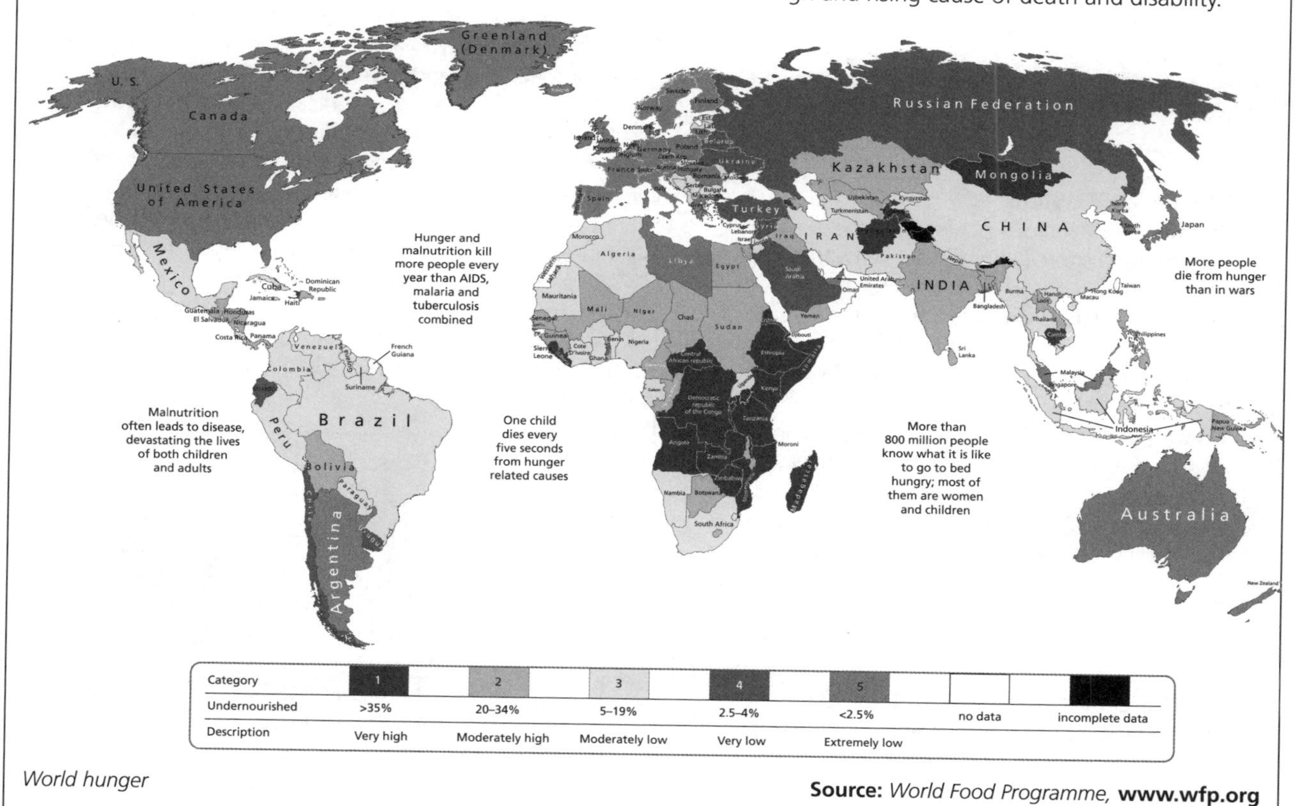

World hunger

Source: *World Food Programme*, **www.wfp.org**

EXTENSION

Visit

www.mapsofworld.com/world-top-ten/maps/countries-by-highest-death-rate-from-lung-cancer.jpg

for the top 10 countries with lung cancer. Describe the distribution of countries with high rates of the disease. Suggest reasons for the pattern you have described.

The spread of disease

DISEASE DIFFUSION

Disease diffusion refers to the spread of a disease into new locations. It occurs when incidences of a disease spread out from an initial source. The **frictional effect of distance** or **distance decay** suggests that areas that are closer to the source are more likely to be affected by it, whereas areas further away from the source are less likely to be affected and/or will be affected at a later date.

The Swedish geographer Hagerstrand is known for his pioneering work on "waves of innovation". This has formed the basis for many medical geographers who attempted to map the spatial diffusion of disease. Four main patterns of disease diffusion have been identified: expansion diffusion, contagious diffusion, hierarchal diffusion and relocation diffusion. There is also network diffusion and mixed diffusion. The diffusion of infectious disease, for example, tends to occur in a "wave" fashion, spreading from a central source.

Some physical features act as a barrier to diffusion, including mountains and water bodies, while political boundaries and economic boundaries may also limit the spread of disease.

The diffusion of disease can be identified as an S-shaped curve to show four phases: infusion (25th percentile), inflection (50th percentile), saturation (75th percentile) and waning to the upper limits.

TYPES OF DIFFUSION

- **Expansion diffusion** occurs when the expanding disease has a source and diffuses outwards into new areas.
- **Relocation diffusion** occurs when the spreading disease moves into new areas, leaving behind its origin or source, e.g. a person infected with HIV moving into a new location.
- **Contagious diffusion** is the spread of an infectious disease through the direct contact of individuals with those infected.
- **Hierarchal diffusion** occurs when a phenomenon spreads through an ordered sequence of classes or places, e.g. from cities to large urban areas to small urban areas.
- **Network diffusion** occurs when a disease spreads via transportation and social networks, e.g. the spread of HIV in southern Africa along transport routes.
- **Mixed diffusion** is a combination of contagious diffusion and hierarchal diffusion.

EXTENSION

Describing a pattern

Diffusion suggests a drop in intensity with distance from the origin. Here, the earliest impacts are along the NE coast, and in general the pattern decreases away from here. The pattern is uneven – disease spreads more easily along lines of communications (e.g. roads) and where there are more people.

WEST NILE VIRUS

West Nile Virus was first discovered in the blood of a feverish woman in Uganda's West Nile district in 1937. It is a mosquito-borne disease that first spread to the USA from Asia and Africa in 1999. Within three years it had caused eight deaths and 135 non-fatal cases; it has since spread throughout much of the country and is now considered to be endemic. West Nile Virus causes vomiting and diarrhoea, progressing to fever, confusion, muscle weakness, paralysis – and, sometimes, death.

The strain of the virus which reached New York in 1999 is thought to have come from the Middle East or Africa, possibly from an insect or an infected human travelling by air. Once there, the virus was transmitted by a local mosquito, the northern house mosquito. Infected wild birds provided a reservoir for the virus – in 1999, half the wild birds in the north-east of the New York borough of Queens were infected.

The New York outbreak caused a media frenzy, although only 62 people out of a population of 10 million were hospitalized, and only 7 died. It is estimated that even in the affected areas, less than 1% of mosquitoes are potentially dangerous, and nearly 90% of people who actually get the virus will suffer no symptoms at all and never know about it. The other 10% – mainly the old, chronically ill or HIV positive – get flu-like symptoms, and perhaps one in 200 develops the potentially fatal complication of encephalitis, which causes swelling of the brain.

Americans are now being simultaneously advised not to go out near dawn or dusk because of West Nile Virus, not to go out in the heat of the day because of the dangers of sunstroke and pollution, and not to stay inside watching TV and eating junk food because of the dangers of obesity. They are beginning to run out of options!

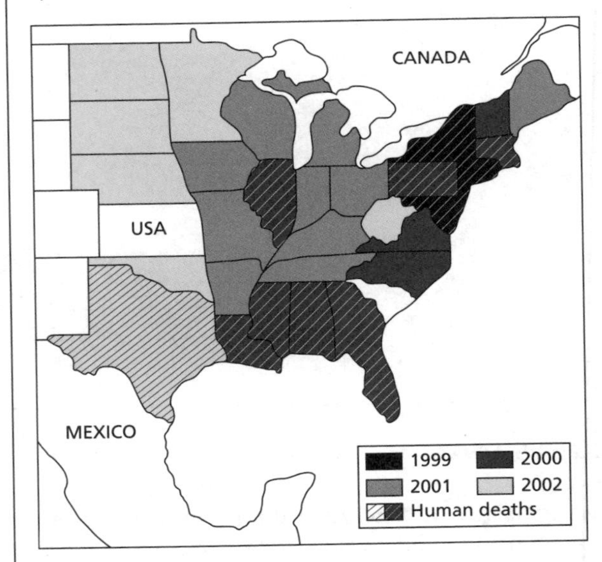

Diffusion of West Nile Virus across the USA

Geographic factors and impacts: malaria

BACKGROUND

Malaria kills up to 3 million people annually, mostly in sub-Saharan Africa, and about 500 million more people suffer from the disease. Malaria is widespread in many tropical countries; mosquito-borne diseases such as malaria and yellow fever still infect around 270 million people each year. The cost of malaria is estimated at over £1.1 billion annually.

As increasing numbers of people travel, they move into areas where malaria is endemic. The disease is affecting new victims because:

- many people are not immune to the disease
- mosquitoes are becoming more drug resistant
- mosquitoes are spreading into areas previously free of the insect
- agricultural schemes are expanding
- there is an increase in irrigation schemes
- international travel and trade are increasing as yet there is no accepted vaccine.

In southern Tanzania, up to 80% of the children are infected with the disease by the age of 6 months. There, 4% of children under the age of 5 die as a result of malaria. Pregnant women, travellers and refugees are also especially vulnerable to the disease. Deaths from malaria – concentrated among African children – could be halved to 500,000 by spending another £600 million a year on known prevention and treatment measures.

Conditions for malaria include:

- stagnant water for the mosquitoes to lay their eggs
- temperatures of >16°C for the parasite to develop within the mosquito
- temperatures below 32°C; above this large numbers of the parasites die.

Malaria can cause fever, sweating, anaemia and spleen enlargement, and it can be fatal.

EXTENSION

Summarizing data

Descriptive statistics include the mean, mode, median and range.

The *mean*, or average, is found by summing the values for all obvservations and dividing by the total number of observations.

The *mode* refers to the group that occurs most often. If there are two that occur most often, the distribution is called bimodal.

The *median* is the middle value when the data are placed in either ascending or descending order.

The *range* is the difference between the largest (maximum) and smallest (minimum) value. The *interquartile range* gives the range of the midle half of the values – it is useful because the extremes are not included.

MALARIA AND TREATMENT

The chemical DDT has been used to combat mosquitoes in countries such as Belize, Brazil, Ecuador, Ethiopia, India, Kenya and Thailand. However, DDT is believed to cause cancer, although the issue remains controversial. In the 1950s, DDT was one of the chemicals that revolutionized agriculture. The subsequent discovery that it builds up in the environment, is highly toxic to wildlife – especially invertebrates – and passes up the food chain resulted in many countries prohibiting its use. However, abandoning DDT can result in significant increases in malaria cases.

In Belize, for example, where the disease had been virtually eliminated, the malaria problem then spiralled out of control after the country stopped using DDT in 1992. Belize has a population of around 200,000, and in 1994 it reported over 10,000 malaria cases, more than at any other time in the country's history. Other factors, such as human migrations due to political unrest, might have contributed to the epidemic, but in Brazil and Ecuador a similar pattern has been found.

Using DDT for mosquito control is very different from using it for agriculture. Treating homes across the whole of Guyana, for example, which covers 215,000 km^2, uses no more DDT than would be used to spray 4 km^2 of cotton during a single growing season. The pesticide is also confined indoors.

Nonetheless, environmentalists and northern governments say that a complete ban on DDT would force health workers to explore other, less risky ways to control mosquito-borne diseases. Some countries have tried replacing DDT with organophosphates. Although these substances break down more quickly and do not linger in the environment, they are more acutely toxic and can injure health workers, who rarely wear full body protection in the hot, humid tropics. They may also be less effective as an insecticide than DDT and, since they must be frequently reapplied, are more expensive. Experience in several countries suggests that mosquitoes can rapidly develop multiple resistance to the seven or eight pesticides that are available as an alternative to DDT.

Environmentalists would like health officials to use more non-chemical methods to prevent the spread of malaria, such as bed nets, managing waterways to eliminate pools of still water where mosquitoes breed, and using predatory fish, bacteria and other biological controls to kill larvae. Health workers claim that many of these methods, while laudable, are simply not practical in the field.

EXTENSION

Visit

http://nobelprize.org/educational_games/medicine/malaria/readmore/global.html

for 'Malaria: Past and Present'.

Geographic factors and impacts: AIDS

BACKGROUND

The impact of the AIDS epidemic is increasingly being felt in many countries across the world. It has profound effects on growth, income and poverty. The annual per capita growth in half the countries of sub-Saharan Africa is falling by 0.5–1.2% as a direct result of AIDS. By 2010, per capita GDP in some of the countries hardest hit may have dropped by 8%. Heavily affected countries could lose more than 20% of GDP by 2020. Companies of all types face higher costs in training, insurance, benefits, absenteeism and illness.

One-quarter of households in Botswana (see page 13 for the population with AIDS in Botswana), where adult HIV prevalence is over 35%, can expect to lose an income earner within the next 10 years. A rapid increase in the number of very poor and destitute families is anticipated.

In sub-Saharan Africa, three-quarters of the continent's people are surviving on less than US$2 a day. The epidemic is deepening their plight. Typically, this impoverished majority has limited access to social and health services.

Households with an HIV/AIDS patient spend, on average, 20 times more on healthcare annually than households without an AIDS sufferer.

According to the FAO, 7 million farm workers have died from AIDS-related causes since 1985 and 16 million more are expected to die by 2020. Agricultural output cannot be sustained in such circumstances. The prospect of widespread food shortages and hunger is real. Some 20% of rural families in Burkina Faso are estimated to have reduced their agricultural work or even abandoned their farms because of AIDS. In 15% of these instances, children are removed from school to take care of ill family members and to regain lost income.

Families often remove girls from school to care for sick relatives or assume other family responsibilities, jeopardizing the girls' education and future prospects. In Swaziland, school enrolment is reported to have fallen by 36% due to AIDS, with girls being most affected.

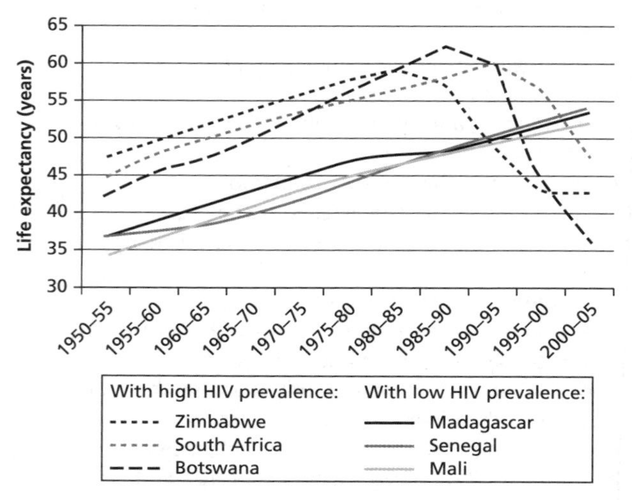

Changes in life expectancy in selected African countries with high and low HIV prevalence, 1950–2005

DEVELOPMENT AND STABILITY THREATENED

Meanwhile, the epidemic is claiming huge numbers of teachers, doctors, extension workers and other human resources. Teachers and students are dying or leaving school, reducing both the quality and efficiency of education systems. In one year alone, an estimated 860,000 children lost their teachers to AIDS in sub-Saharan Africa.

COPING WITH CRISIS

In the worst-affected countries, steep drops in life expectancies are beginning to occur, most drastically in sub-Saharan Africa, where four countries (Botswana, Malawi, Mozambique and Swaziland) now have a life expectancy of less than 40 years.

As more infants are born HIV-positive in badly affected countries, child mortality rates are rising. In the Bahamas, it is estimated that some 60% of deaths among children under the age of 5 are due to AIDS, while in Zimbabwe the figure is 70%.

HOW TO FIGHT THE VIRUS

Two success stories show that the hurdles to prevention are not impossibly high.

Senegal

Senegal is an illustration of how to stop AIDS from taking off in the first place. In its brothels, which had been regulated since the early 1970s, condom use was firmly encouraged. The country's blood supply was screened early and effectively. Vigorous education resulted in 95% of Senegalese adults knowing how to avoid the virus.

Uganda

Uganda shows that there is hope even for countries that are poor and barely literate. President Yoweri Museveni recognized the threat shortly after becoming president in 1986, and deluged the country with anti-AIDS warnings. The key to Uganda's success is twofold. Every government department took the problem seriously, and implemented its own plan to fight the virus. Second, the government recognized they could do only a limited amount, so they gave free rein to scores of non-governmental organizations (NGOs) to do whatever it took to educate people about risky sex. The climate of free debate has led Ugandans to delay their sexual activity, to have fewer partners and to use more condoms.

Urbanization

DEFINITIONS

- **Counter-urbanization:** a process involving the movement of population away from inner urban areas to new towns, new estates, commuter towns or villages on the edge or just beyond the city limits/ rural–urban fringe
- **Re-urbanization:** the development of activities to increase residential population densities within the existing built-up area of a city. This may include the redevelopment of vacant land, the refurbishment of housing and the development of new business enterprises
- **Suburb:** a residential area within or just outside the boundaries of a city
- **Suburbanization:** the outward growth of towns and cities to engulf surrounding villages and rural areas. This may result from the out-migration of population from the inner urban areas to the suburbs or from inward rural–urban movement
- **Urbanization:** the process by which an increasing percentage of a country's population comes to live in towns and cities. It may involve both rural–urban migration and natural increase
- **Urban sprawl:** the unplanned and uncontrolled physical expansion of an urban area into the surrounding countryside. It is closely linked to the process of suburbanization

URBAN POPULATIONS

Urbanization

Urbanization, defined as an increase in the percentage of a population living in urban areas, is one of the most significant geographical phenomena of the 20th century. Urbanization takes place when the urban population is growing more rapidly than the population as a whole. It is caused by a number of interrelated factors, including:

- migration to urban areas
- higher birth rates in urban areas due to the youthful age structure
- higher death rates in rural areas due to diseases, unreliable food supply, famine, decreased standard of living in rural areas, poor water, hygiene and medication.

Urban classifications

Urban populations are those living in areas with a census definition as urban. The criteria used to specify what an urban area is vary widely and it is not possible to give a single definition. However, there are a number of underlying principles:

- population size
- specific urban characteristics, such as a CBD and residential zones
- predominant economic activities, such as manufacturing and services
- an administrative function.

THE PROCESS OF URBANIZATION

Stages in urbanization in MEDCs

In many rich countries the process of urbanization is almost at an end, and the proportion of urban dwellers is beginning to fall. The progress has followed an S-shaped curve and it seems to have tailed off at 80% of the total population. For many MEDCs, there appears to be a cycle of urbanization, suburbanization, counter-urbanization and re-urbanization.

The process of urbanization

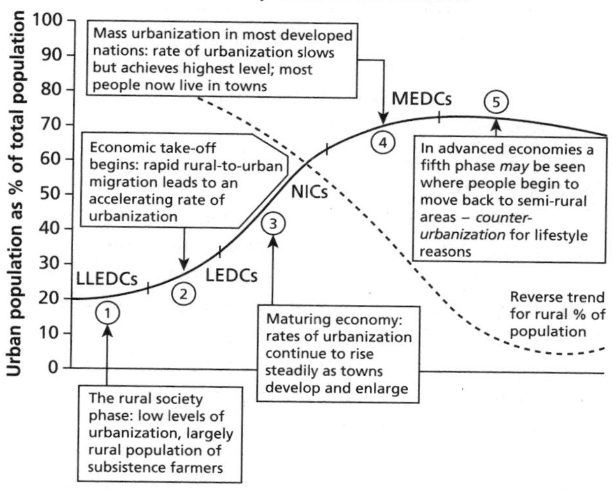

Source: *Warn, S.* Managing Changes in Human Environments. *Philip Allan Updates, 2001*

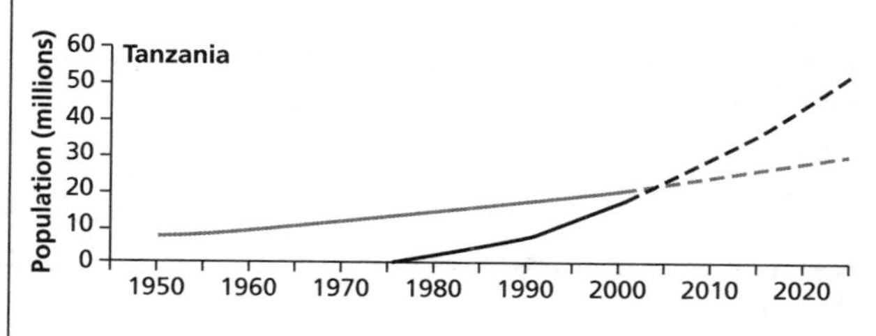

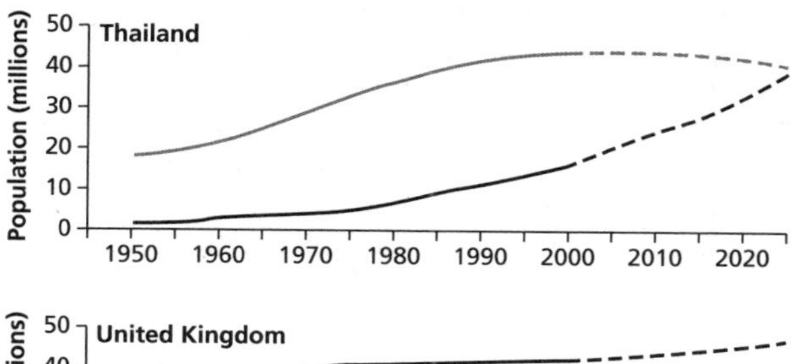

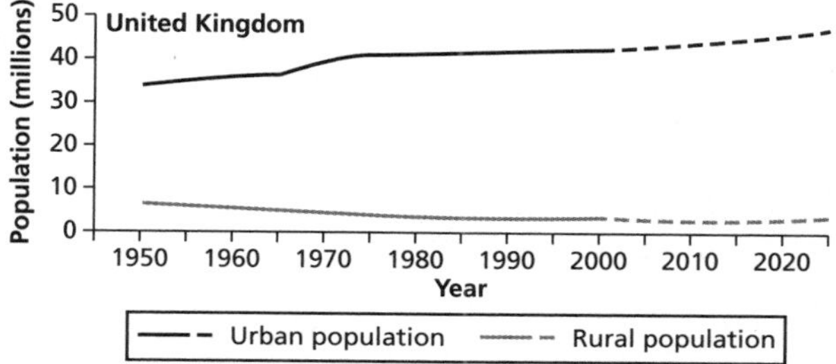

Trends in selected countries

Inward and outward movements

CENTRIPETAL MOVEMENTS

Rural–urban migration

Rural–urban migration refers to the movement of people away from the countryside to towns and cities. This is a very important process, especially in LEDCs and NICs. It occurs because people believe they will be better off in the urban areas than they are in the rural areas. As we saw in Part 1, reasons for this movement have been described using the concept of push and pull factors.

- Push factors are the negative features that cause a person to move away from a place (e.g. unemployment, low wages, natural hazards).
- Pull factors are the attractions (whether real or imagined) that exist at another place (e.g. better wages, more jobs, good schools).

Gentrification

Gentrification is the reinvestment of capital into inner-city areas. It refers mostly to an improvement of residential areas, although there is an economic dimension too. It is common in areas where there may be **brownfield sites** (abandoned, derelict or underused industrial buildings and land, which may be contaminated but has potential for redevelopment). Thus, as well as residential rehabilitation and upgrading, there is also commercial redevelopment. Gentrification may lead to the social displacement of poor people – as an area becomes gentrified, house prices rise and the poor are unable to afford the increased prices. As they move out, young upwardly mobile populations take their place.

Gentrification has occurred in many large old cities throughout the world, such as in New York (Greenwich Village and Brooklyn Heights), Toronto (Riverdale) and London (Fulham and Chelsea). It has also been observed in cities as diverse as Johannesburg, Tokyo and Sao Paulo.

Re-urbanization/urban renewal

Re-urbanization is a revitalization of urban areas and a movement of people back into these areas. A good example is the re-urbanization of Barcelona and the use of the 1992 Olympic Games to re-establish the city. "Urban renewal" refers to the rehabilitation of city areas that have fallen into decline (urban decay). A good example is the renewal of Manhattan in New York.

CENTRIFUGAL MOVEMENTS

Suburbanization

Suburbanization (see page 134) is the outward expansion of towns and cities, mainly in Europe, North America and Australia, largely thanks to improvements in transport systems. By the early 20th century, railways, electric tramways and buses were critical in the growth of middle-class, residential suburbs. Town extensions were really a form of suburban development along the lines of trams and trains. In addition, the price of farmland had declined dramatically and there was scope for urban expansion on a great scale.

The early 20th century was a period of optimism. Rising wages and living standards were matched by rising expectations. Housing was now available, affordable and of a quality unimaginable only a few decades earlier.

There were a number of reasons for this boom in private house-building:

- lower costs of living
- very low interest rates
- expansion of building societies
- willingness of local authorities to provide utilities, such as sewers, electricity, gas and water
- increased public transport.

Counter-urbanization

There are several reasons why people may wish to leave large urban areas and move to towns and villages in rural areas. These include:

- high land prices
- congestion
- pollution
- high crime rate
- a lack of community
- declining services.

In contrast, there is a perception that smaller settlements have a closer sense of community, better environments and a safer location.

Urban sprawl

Urban sprawl, the uncontrolled growth of urban areas at their edges, suggests that urban areas grow in an unchecked fashion. However, if there are Green Belts, urban sprawl is prevented as there are limits on how far the urban area can grow. Many of the world's largest cities, such as Tokyo, Seoul and Mexico City, have been characterized by urban sprawl.

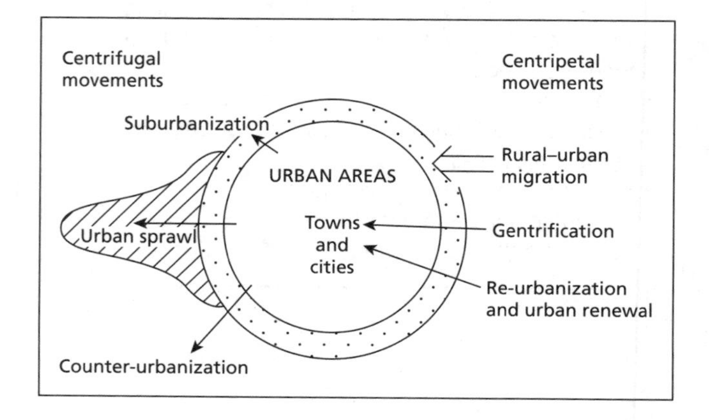

Megacities

WHAT ARE MEGACITIES?

Megacities are cities with a population of over 10 million people. The UN also calls them "metacities".

Megacities grow as a result of economic growth, rural–urban migration and high rates of natural increase. As the cities grow, they swallow up rural areas and nearby towns and cities. They become multi-nuclei centres. The world has never had so many very large settlements. Some of these cities have populations that are bigger than the population of entire countries – Mumbai, for example, has more people than Sweden and Norway combined!

Nevertheless, megacities contain between 4% and 7% of the world's total population, and grow at relatively slow rates, perhaps 1.5% per year. The first megacity was Tokyo, which now has a population of about 35 million (larger than Canada's population). By 2017, other megacities will include Mumbai, Delhi, Mexico City, Sao Paulo, New York, Dhaka, Jakarta and Lagos. Lagos has been growing at a very fast rate – 5% per annum – and is expected to increase at this rate until after 2020. Usually, very large cities grow more slowly than medium-sized cities.

By 2020, all but 4 of the world's megacities will be in developing regions, 12 of them in Asia alone. The impact of megacities on their region is huge. They are likely to require new forms of planning and management to cope with such large sizes.

The scale of environmental impacts is likely to be great. Rapid economic growth and urbanization in China has had a negative impact on the urban environment. China contains 16 of the 20 most polluted cities in the world and, after the USA, is the largest producer of greenhouse gases.

Megacities are important for the generation of wealth. In MEDCs, urban areas generate over 80% of national economic output; in LEDCs, the figure is over 40%. On the other hand, there are some aspects of megacities, such as crime and environmental issues, where they appear less than attractive.

	Population (millions)	Murders per 100,000	% of income spent on food	Persons per room	% of houses with water/ electricity	Telephones per 100 people	% of children in secondary school	Infant deaths per 1000 live births	Noise levels (1–10)	Traffic flow in rush hour (mph)	Quality of life score
Tokyo	35	1.4	18	0.9	100	44	97	5	4	28.0	81
Mexico City	19.4	27.8	41	1.9	94	6	62	36	6	8.0	38
New York	17.4	12.8	16	0.5	99	56	95	10	8	8.7	70
Sao Paulo	17.2	26.0	50	0.8	100	16	67	37	6	15.0	50
Osaka	16.8	1.7	18	0.6	98	42	97	5	4	22.4	81
Seoul	15.8	1.2	34	2.0	100	22	90	12	7	13.8	58
Moscow	13.2	7.0	33	1.3	100	39	100	20	6	31.5	64
Mumbai	12.9	3.2	57	4.2	85	5	49	59	5	10.4	35
Kolkata	12.8	1.1	60	3.0	57	2	49	46	4	13.3	34
Buenos Aires	12.4	7.6	40	1.3	86	14	51	21	3	29.8	55

Some measures of the quality of life in megacities

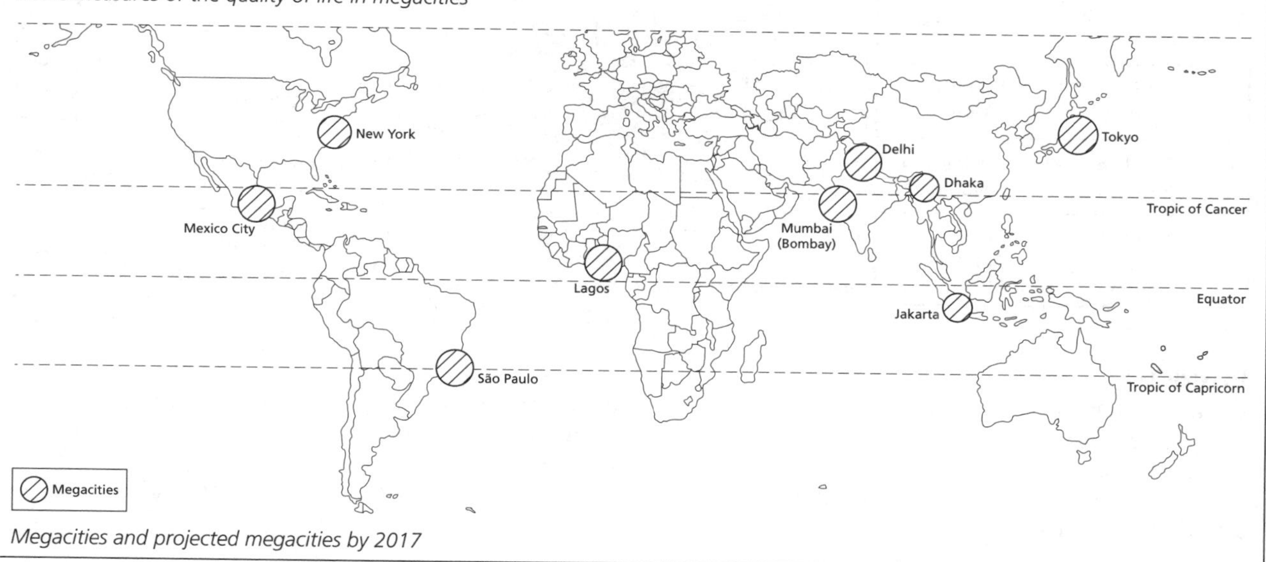

Megacities and projected megacities by 2017

Residential patterns in rich countries

THE LOCATION OF RESIDENTIAL AREAS

In most MEDC cities, there is a clear pattern of residential location. The highest residential densities are found in inner-city areas and are associated with terraced housing from the 19th century. Usually, residential density in the city centre is low due to high land values. However, with increasing distance away from the city centre, residential density decreases. This reflects the greater availability of land in the suburbs. Traditionally, poorer households were located in the inner city, close to jobs, whereas high-quality housing is located further out. However, densities in suburban areas have increased over the last 30 years due to decentralization and the development of edge-of-town estates.

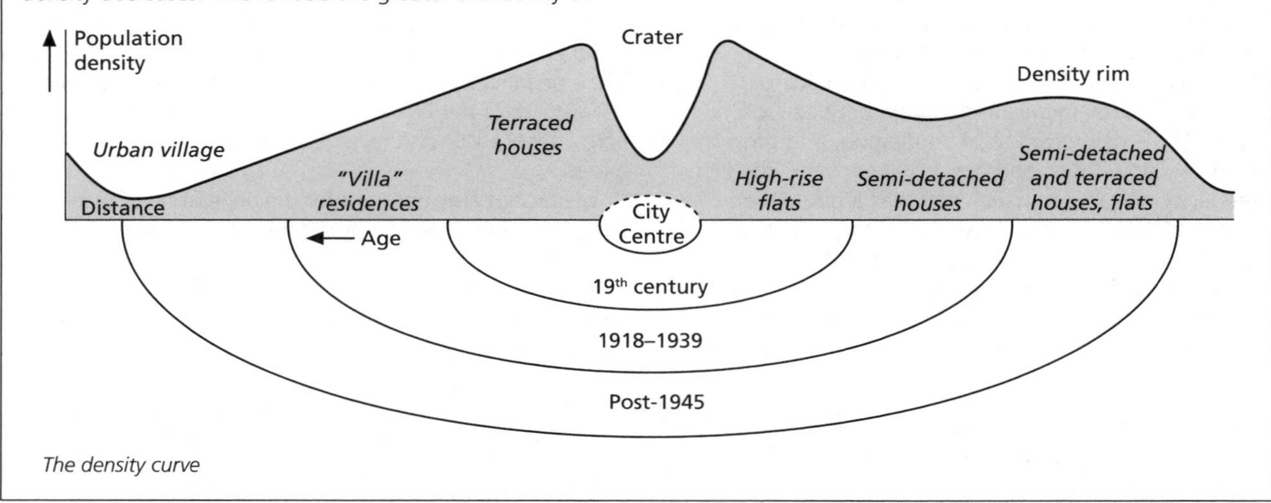

The density curve

THE FAMILY LIFE CYCLE

Housing choice is also partly related to life cycle and income. A person is likely to move around different zones of a city, depending on their age and their need for a house of a certain size. This is true for those in rented accommodation as well as for homeowners. Residential patterns are influenced by banks, building societies, local authorities, housing associations and free choice.

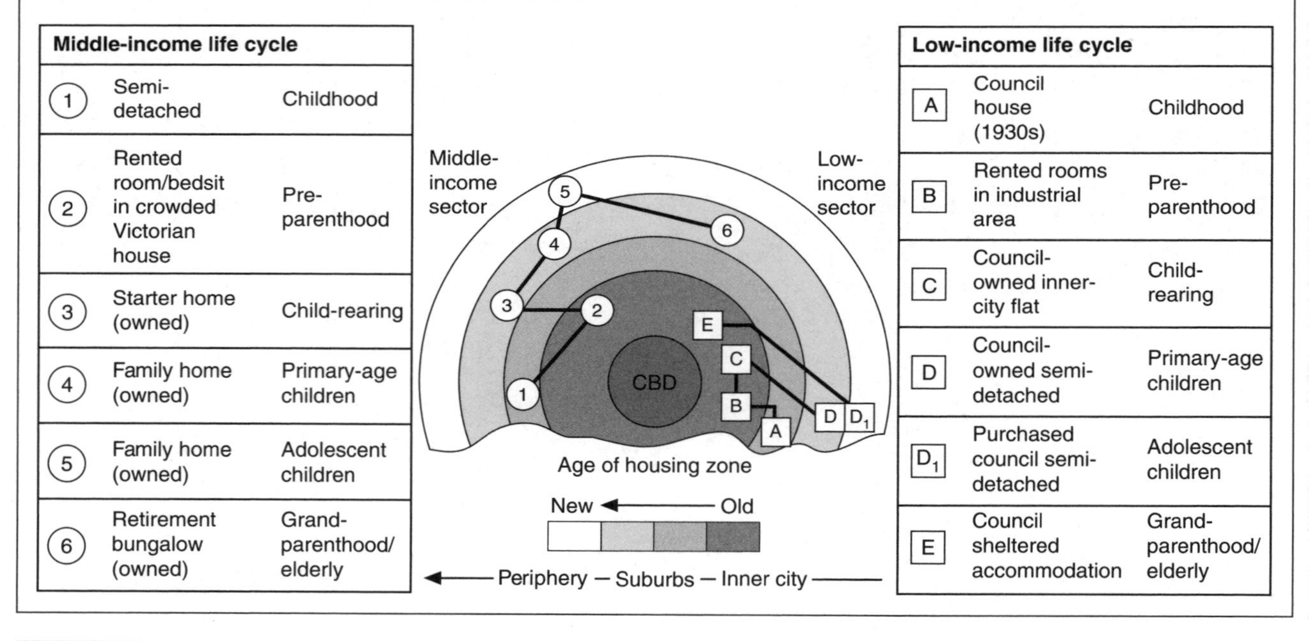

Middle-income life cycle		
1	Semi-detached	Childhood
2	Rented room/bedsit in crowded Victorian house	Pre-parenthood
3	Starter home (owned)	Child-rearing
4	Family home (owned)	Primary-age children
5	Family home (owned)	Adolescent children
6	Retirement bungalow (owned)	Grand-parenthood/ elderly

Low-income life cycle		
A	Council house (1930s)	Childhood
B	Rented rooms in industrial area	Pre-parenthood
C	Council-owned inner-city flat	Child-rearing
D	Council-owned semi-detached	Primary-age children
D_1	Purchased council semi-detached	Adolescent children
E	Council sheltered accommodation	Grand-parenthood/ elderly

ETHNICITY

In many cities there are clearly defined ethnic or racial areas. Famous examples include Harlem in New York, Watts in Los Angeles, and in Belfast (Falls, for example is a Catholic area; Shankill a Protestant area). Sometimes a population group chooses to live apart from the dominant population group in order to maintain their cultural integrity (positive segregation), while at other times the minority is excluded from society and is unable to afford the housing in more affluent areas (negative segregation).

Urban poverty and deprivation

QUALITY OF LIFE

Within most cities there is considerable variation in the quality of life. This raises questions about equality of opportunity and social justice. In MEDCs and LEDCs, there are areas that are labelled as "poor" and these are zones of deprivation, poverty and exclusion. In MEDCs, these are often inner-city areas or ghettos, whereas in LEDCs it is frequently shanty towns that exhibit the worst conditions. The factors associated with deprivation are varied, but they result in a cycle of urban deprivation and a poor quality of life.

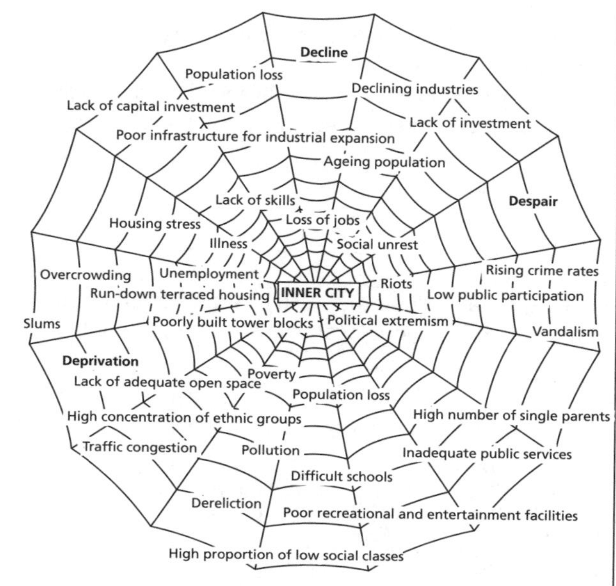

The inner city's web of decline, deprivation and despair

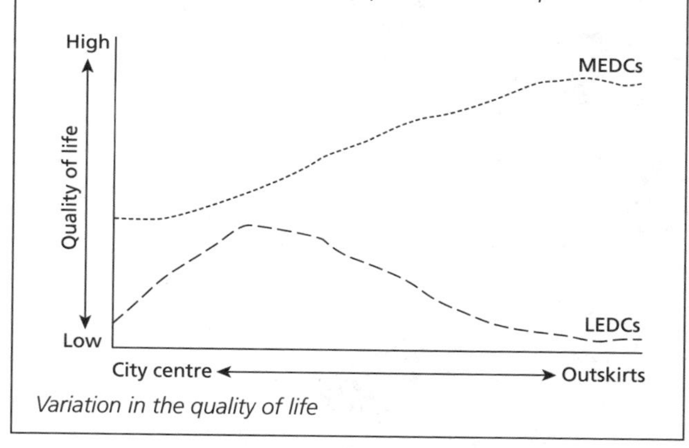

Variation in the quality of life

EXTENSION

Visit
http://www.scribd.com/doc/2336001/Measuring-Urban-Poverty-in-India for an article on urban poverty in India.
An excellent source of data and case studies on poverty in the UK is **http://www.jrf.org.uk/**.

MEASURING DEPRIVATION

A number of indices are used to measure deprivation. These include:

- physical measures – such as quality of housing, levels of pollution, incidence of crime, vandalism, graffiti
- social indicators – including crime (reported and fear of); levels of health and access to healthcare; standards of education; proportion of population on subsidized benefits (unemployment, disability, free school meals); proportion of lone-parent families
- economic indices – access to employment; unemployment and underemployment; levels of income
- political measures – opportunities to vote and to take part in community organization.

SLUMS AND SQUATTER SETTLEMENTS

The total number of slum dwellers in the world stood at about 924 million people in 2001. This represents about 32% of the world's total urban population, *but* 78.2% of the urban population in LEDCs. Slums are typically located in areas that planners do not want – steep slopes, floodplains, edge-of-town locations and/or close to major industrial complexes.

Slums have the most intolerable of urban housing conditions, which frequently include:

- insecurity of tenure
- lack of basic services, especially water and sanitation
- inadequate and sometimes unsafe building structures
- overcrowding
- location on hazardous land
- high concentrations of poverty and of social and economic deprivation, which may include broken families, unemployment, and economic, physical and social exclusion.

Furthermore, slum dwellers have limited access to credit and to formal job markets due to stigmatization, discrimination and geographic isolation.

On the positive side, slums are:

- the first stopping point for immigrants – they provide the low-cost and only affordable housing that will enable the immigrants to save for their eventual absorption into urban society
- the place of residence for low-income employees, thus serving to keep the wheels of the city turning in many different ways
- the base from which many informal entrepreneurs are able to operate, with clienteles extending to the rest of the city.

Most slum dwellers are people struggling to make an honest living, within the context of extensive urban poverty and formal unemployment.

Economic activities in cities (1)

LOCATION OF INDUSTRY IN URBAN AREAS

Models of urban land use have located manufacturing industry in inner-city areas, along major routeways and in industrial suburbs. This reflects the variety of manufacturing industries and their differing locational requirements. In these models, the location of industry is described but little explanation is given as to why it is there.

Industries found in cities include:

- those needing access to skilled labour, such as medical instruments; those needing access to the CBD, such as fashion accessories and clothes; and those which need the whole urban market for distribution, such as newspapers – these industries all having a central location
- port industries
- those located on radial routes, e.g. Samsung Electronics at Suwon, Korea
- those needing large amounts of land for the assembly, production or storage of goods, e.g. the Hyundai car works at Busan, Korea.

Large cities are attractive for industries for a number of reasons:

- Capital cities, such as Paris or Moscow, are the largest manufacturing centres of the nation.
- Cities are large markets.
- Port cities have excellent access to international markets.
- Cities are major centres of innovation, ideas and fashion.
- A variety of labour is readily available, including skilled and unskilled workers, decision-makers and innovators.

LAND USE IN NEW YORK

Industrial uses, warehouses and factories occupy 4% of the city's total lot area. They are found primarily in the South Bronx, along either side of Newtown Creek in Brooklyn and Queens, and along the western shores of Brooklyn and Staten Island. Riverfront locations are very important.

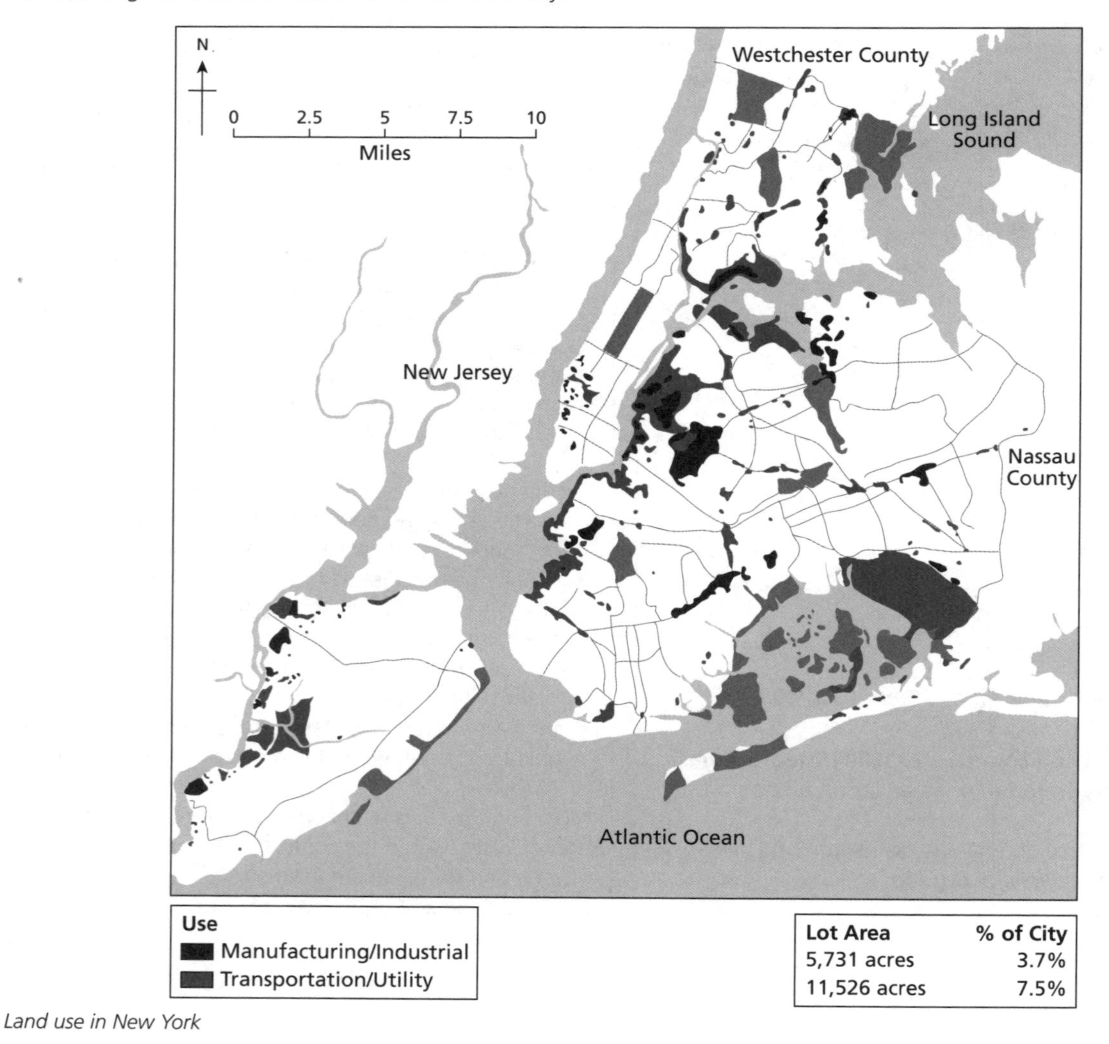

Land use in New York

Economic activities in cities (2)

THE CENTRAL BUSINESS DISTRICT (CBD)

The central business district (CBD) is the commercial and economic core of cities. It is the heart of the city, the area which is most accessible to public transport, and the location with the highest land values. It has a number of characteristic features and internal zoning. The following diagrams typify an MEDC city.

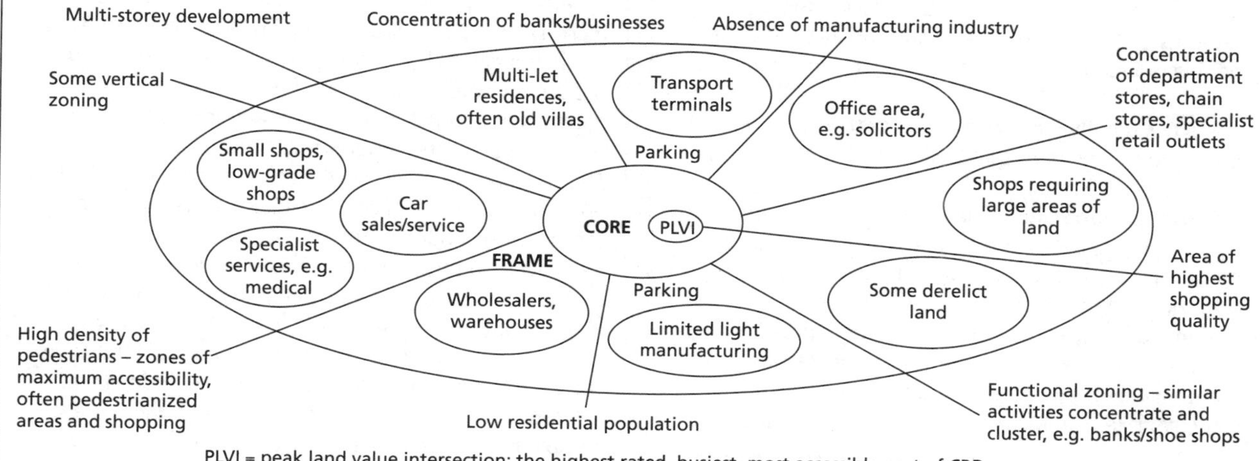

Core and frame elements of the CBD

Factors affecting CBD decline

Source: *Adapted from Warn, S.* Managing Changes in Human Environments. *Philip Allan Updates, 2001*

EXTENSION

Draw an annotated sketch diagram to show the main characteristics of this part of Seoul's CBD.

Shanty towns

SHANTY TOWNS

In most LEDC cities there is a considerable amount of informal or shanty housing. These are illegal settlements, generally built on unwanted land that may be unsafe. For example, some shanty towns in Rio de Janeiro are built on steep slopes and are subject to landslides, while the slums of Tegucigalpa in Honduras were built on a floodplain and badly affected by flooding following Hurricane Mitch in 1998.

There is great variety within shanty towns. "Slums of hope" and "slums of despair" are a further distinction of shanty towns.

- **Slums of hope** are the self-built houses where migrants are consolidating their position in the informal urban economy: housing is improving (e.g. in Mexico City, the *Colonias Paracondistas*).
- **Slums of despair** have little room for improvement because incomes are low, rents are high, leasing arrangements are insecure and there are environmental problems (e.g. in Mexico City, the *Ciudades Perdidas*).

SHANTY TOWNS IN RIO DE JANEIRO

The rise of the **favelas** as an urban feature of Rio has been rapid. The official definition of such settlements is "residential areas lacking formal organization or basic services, containing 60 or more families who are squatting illegally on the site". Between 1.7 and 2.5 million of Rio's 12 million inhabitants (nearly 20% of the total population) live in slums. The largest favela, Rocinha, has an estimated population of 80,000. Initially their destruction and the removal of the residents to *conjuntos habitacionais* in the suburbs was the significant policy. The clearance of favela sites for the building of high-class apartment blocks and condominiums served to maintain the status and value of the central area.

In 1990, a programme of electrification was started, as a means of improving conditions in the favelas. While long-established favelas, some dating back to 1940, have a mix of commercial services serving a more diverse socio-economic population, the worst conditions are still found in the most recent favelas. Here there is a complete absence of basic services, people have low incomes and there is high unemployment.

IMPROVING SHANTY TOWNS

Shanty towns can be improved by legalizing them and giving the residents security of tenure and by a variety of assisted self-help (ASH) measures. A good example is the upgrading of the shanty towns in Rio de Janeiro.

The Favela–Bairro Project (Favela–Neighbourhood Project) began in Rio in 1994. It aimed to recognize the favelas as neighbourhoods of the city in their own right and to provide the inhabitants with essential services. Approximately 120 medium-sized favelas (those with 500–2500 households) were chosen.

The primary phase of the project addressed the built environment, aiming to provide:

- paved and formally named roads
- water supply pipes and sewage/drainage systems
- crèches, leisure facilities and sports areas
- relocation for families who were currently living in high-risk areas, such as areas subject to frequent landslides
- channelled rivers to stop them changing course.

The second phase of the project aimed to bring the favela dwellers into the mainstream society and keep them out of crime. This is being done by:

- generating employment, for example by creating co-operatives of dressmakers, cleaners, construction workers, etc., and helping them to get established in the labour market
- improving education and providing relevant courses such as ICT
- giving residents access to credit, so that they can buy construction materials and improve their homes.
- The project has been used as a model of its type. The government is also helping people to become homeowners.

Nevertheless, there are some shanty towns in Rio de Janeiro, such as the *caxias* on the edge of the city, that remain very poor and isolated. The benefits of ASH and upgrading have done little to improve the standard of living of the residents.

EDGE TOWNS

An edge town is a new town development located on the edge of a city where there is land availability and good accessibility. An excellent example is Barra de Tijuca on the edge of Rio. It represents a movement away from more central areas by wealthy populations who can afford the high-quality housing built away from the CBD.

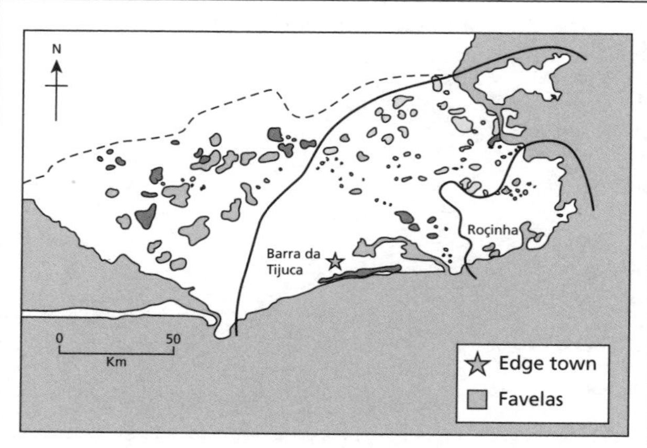

Location of favelas and edge towns in Rio de Janeiro

Urban microclimates

Structure of the air above the urban area

Greater amounts of dust mean increasing concentrations of hygroscopic particles; less water vapour because water is removed quickly via drains and sewers and because there is less vegetation to take in water and release it later; but more CO_2 and higher proportions of noxious fumes owing to combustion of imported fuels and discharge of waste gases by industry.

Structure of the urban surface

More heat-retaining materials with lower albedo (reflectivity) and better radiation-absorbing properties; rougher surfaces, with a great variety of perpendicular slopes facing different aspects; tall buildings can be very exposed, and the deep streets are sheltered and shaded.

The effect of city morphology on radiation received at the surface

(a) Isolated buildings

Isolated building

Sunny side heatedby insolation, reflected insolation, radiation and conduction

Heat stored and re-radiated

Shaded side

(b) Low buildings

Street collects reflected radiation

(c) High buildings

Very little radiation reaches street level. Radiation reflected off lower walls after reflection from near tops of buildings

Urban canopy layer below roof level

Prevailing wind

Urban boundary layer

Urban plume develops downwind

Urban canopy layer below roof level

Rural boundary layer

RURAL | SUBURBAN | URBAN | SUBURBAN | RURAL

The morphology of the urban heat island

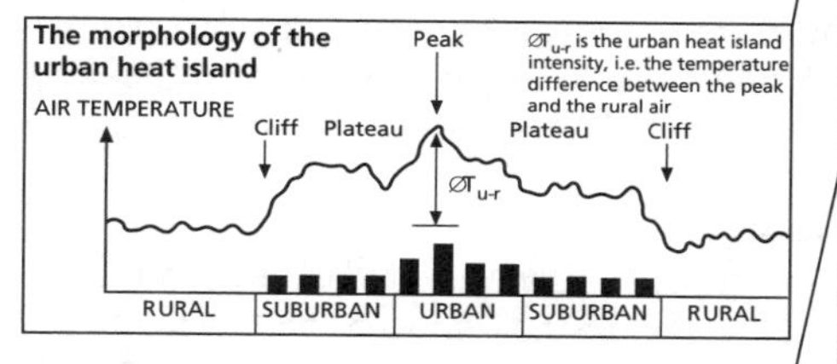

Airflow modified by a single building

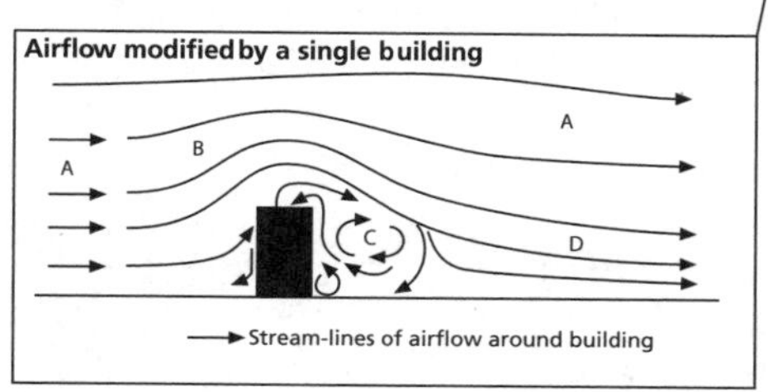

Resultant processes

1. Radiation and sunshine

Greater scattering of shorter-wave radiation by dust, but much higher absorption of longer waves owing to surfaces and CO_2. Hence more diffuse sky radiation with considerable local contrasts owing to variable screening by tall buildings in shaded, narrow streets. Reduced visibility arising from industrial haze.

2. Clouds and fogs

Higher incidence of thicker cloud covers in summer and radiation fogs or smogs in winter because of increased convection and air pollution respectively. Concentrations of hygroscopic particles accelerate the onset of condensation (see 5 below). Day temperatures are, on average, 0.6 °C warmer.

3. Temperatures

Stronger heat energy retention and release, including fuel combustion, gives significant temperature increases from suburbs into the centre of built-up areas, creating heat "islands". These can be up to 8 °C warmer during winter nights. Snow in rural areas increases albedo, thereby increasing the differences between urban and rural. Heating from below increases air mass instability overhead, notably during summer afternoons and evenings. Big local contrasts between sunny and shaded surfaces, especially in the spring.

4. Pressure and winds

Severe gusting and turbulence around tall buildings, causing strong local pressure gradients from windward to leeward walls. Deep, narrow streets much calmer unless aligned with prevailing winds to funnel flows along them – the "canyon effect"

5. Humidity

Decreases in relative humidity occur in inner cities owing to lack of available moisture and higher temperatures there. Partly countered in very cold, stable conditions by early onset of condensation in low-lying districts and industrial zones (see 2 above).

6. Precipitation

Perceptibly more intense storms, particularly during hot summer evenings and nights owing to greater instability and stronger convection above built-up areas. Probably higher incidence of thunder in appropriate locations. Less snow cover in urban areas even when left uncleared (e.g. road clearing).

THE URBAN HEAT ISLAND

Urban areas are generally warmer than those of the surrounding countryside. Temperatures are on average 2–4 °C higher in urban areas. This creates an urban heat island. It can be explained by heat and pollution release.

- Lower wind speeds due to the height of buildings and urban surface roughness.
- Urban pollution and photochemical smog can trap outgoing radiant energy.
- Burning of fossil fuels for domestic and commercial use can exceed energy inputs from the sun.
- Buildings have a higher capacity to retain and conduct heat and a lower albedo.
- Reduction in thermal energy required for evaporation and evapotranspiration due to the surface character, rapid drainage and generally lower wind speeds.
- Reduction of heat diffusion due to changes in airflow patterns as the result of urban surface roughness.

Environmental and social stress

HOUSING

Provision of enough quality housing is a major problem in LEDCs. There are at least four aspects to the management of housing stock:

- quality of housing – with proper water, sanitation, electricity and space
- quantity of housing – having enough units to meet demand
- availability and affordability of housing
- housing tenure (ownership or rental).

There are problems with much of the housing in many LEDC cities. Problems include lack access to water, adequate sanitation, a reliable and safe power supply, adequate roofs, solid foundations, secure tenure (i.e. the residents are at risk of eviction).

TRANSPORT

MEDC	LEDC
• Increased number of motor vehicles	• Lower private car ownership
• Increased dependence on cars as public transport declines	• Less dependence on the car, but growing
• Major concentration of economic activities in CBDs	• Many cars are poorly maintained and are high polluters
• Inadequate provision of roads and parking	• Growing centralization and development of CBDs, increasing traffic in urban areas
• Frequent roadworks	• Heavy reliance on affordable public transport
• Roads overwhelmed by sheer volume of traffic	• Shorter journeys, but getting longer
• Urban sprawl, resulting in low-density built-up areas and increasingly long journeys to work	• Rapid growth, resulting in enormous urban sprawl and longer journeys
• Development of out-of-town retail and employment, leading to cross-city commuting	• Emergence of out-of-town developments as economic development occurs (e.g. Bogota, Colombia)

Traffic problems in MEDC and LEDC cities

MANAGING ENVIRONMENTAL PROBLEMS

There are a range of environmental problems in urban areas. These vary over time as economic development progresses. The greatest concentration of environmental problems occurs in cities experiencing rapid growth. This concentration of problems is referred to as the Brown Agenda. It has two main components:

- Issues caused by limited availability of land, water and services
- Problems such as toxic hazardous waste, pollution of water, air and soil, and industrial "accidents" such as Bhopal in 1985

The environmental problems that most cities have to deal with include:

- problems of water quality
- dereliction
- problems of air quality
- noise
- environmental health of the population.

INEQUALITIES AND SOCIAL PROBLEMS

Examples of the many social problems found in cities include:

- access to services for the underclass
- problems related to crime
- ethnic and religious divisions, causing social and economic polarization.

Some examples have been striking, such as religion in Belfast and Jerusalem, ethnicity in Bradford and Oldham; whereas others, such as crime, are more widespread.

Issues of crime

The majority of criminal activity is concentrated in the most urbanized and industrialized areas and, within these, the poorest working-class neighbourhoods. Some, such as fraud and sexual offences, are relatively more common in lower-density neighbourhoods with lots of open spaces and a limited police presence.

Category	Indicator	Subgroup at risk
Demographic	Age Sex Marital status Ethnic status Family status	Young Male Single Minority group Broken home
Socio-economic	Family size Income Occupation Employment	Large Low Unskilled Unemployed
Living conditions	Housing Density Tenure Permanence	Substandard Overcrowded Rented Low

Common attributes of known offenders

The city as a system

SYSTEMS

A system is a simplified way of looking at how things work. Systems generally include factors (inputs), processes (throughputs) and results (outputs). Many aspects of geography, such as farming, industry, cities, coasts and rivers, have been described in terms of a systems approach.

SUSTAINABLE URBAN MANAGEMENT STRATEGY

An approach to urban management that seeks to maintain and improve the quality of life for current and future urban dwellers. Aspects of management may be social (housing quality, crime), economic (jobs, income) and environmental (air, water, land and resources).

THE CITY SYSTEM

Large cities are often considered unsustainable because they consume huge amounts of resources and produce vast amounts of waste. Sustainable urban development meets the needs of the present generation without compromising the needs of future generations. The Rogers model (*Cities for a Small Planet*) compares a sustainable city with that of an unsustainable one. In the sustainable city, inputs are smaller and there is more recycling.

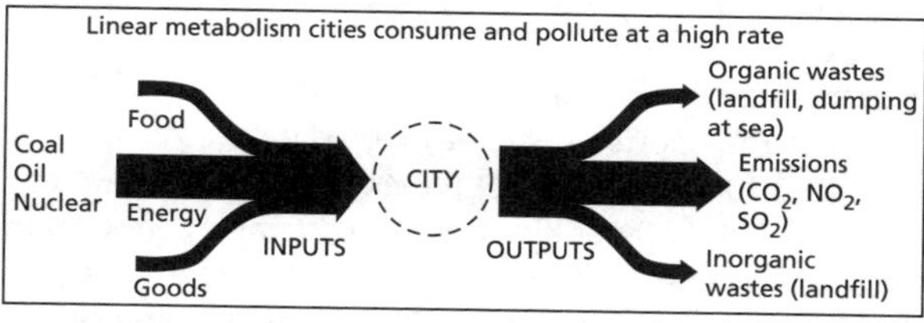

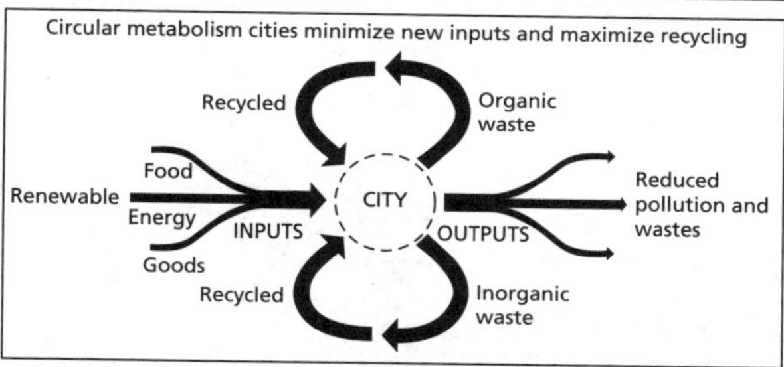

Source: *Rogers, R.* Cities for a Small Planet. *Faber & Faber, 1997*

Compact cities minimize the amount of distance travelled, use less space, require less infrastructure (pipes, cables, roads, etc.), are easier to provide a public transport network for, and reduce urban sprawl. But if the compact city covers too large an area it becomes congested, overcrowded, overpriced and polluted. It then becomes unsustainable.

To achieve sustainability, a number of options are available:

- reducing the use of fossil fuel, e.g. by promoting public transport
- keeping waste production to within levels that can be treated locally
- providing sufficient green spaces
- reusing and reclaiming land, e.g. brownfield sites
- encouraging active involvement of the local community
- conservation of non-renewable resources
- using renewable resources.

SUSTAINABLE URBAN DEVELOPMENT IN LEDCS SUSTAINABLE FUTURES REQUIRE:

- use of appropriate technology, materials and design
- acceptable minimum standards of living
- social acceptability of projects
- widespread public participation.

The main dimensions of sustainable development are:

- provision of adequate shelter for all
- improvement of human settlement management
- sustainable land use planning and management
- integrated provision of environmental infrastructure: water, sanitation, drainage and solid waste management
- sustainable energy and transport systems
- settlement planning in disaster-prone areas
- sustainable construction industry activities
- meeting the urban health challenge.

There are a range of successful local programmes for the urban environment:

Reducing pollution

- The *Hoy no circula* (car-free day), launched in Mexico City in 1989, saw air pollution fall by 21% in the first year.
- In Cubatao, Brazil, local and national government, and some businesses, have combined to reduce air pollution and enforce stricter regulations.

Integrated transport and land use

- Singapore's integrated transport and land use strategy has sought to decentralize development to regional and subregional centres that are served by mass rapid transit (MRT).
- See also the example of Curitiba, Brazil (page 131).

Recycling

- In Shanghai, a wide-ranging programme was established in 1957. It now employs 30,000 people retrieving and reselling reclaimed and recycled products, including 3600 advisors working with factories on sorting and retrieving waste.
- In Curitiba, 70% of households separate recyclable rubbish, and in squatter settlements food and bus fares are exchanged for garbage.

EXTENSION

Visit
http://www.environmentandurbanization.org
for up-to-date examples and case studies.

The sustainable city

SUSTAINABLE DEVELOPMENT IN CURITIBA

Curitiba, a city in south-west Brazil, is an excellent model for sustainable urban development. It has experienced rapid population growth: from 300,000 in 1950 to 1.8 million in 2007, but has managed to avoid all the problems normally associated with such expansion. This success is largely due to innovative planning:

- Public transport is preferred over private cars.
- The environment is used rather than changed.
- Cheap, low-technology solutions are used rather than high-technology ones.
- Development occurs through the participation of citizens (bottom-up development) rather than via centralized planning (top-down development).

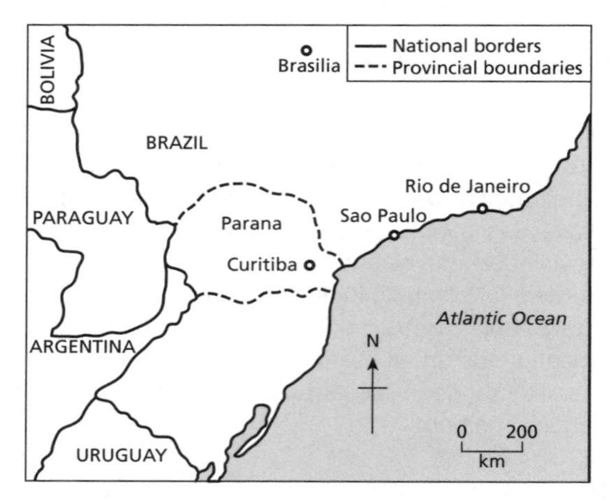

The location of Curitiba in Brazil

Sustainable solutions to flooding in Curitiba

Problems (1950s/60s)	Solutions (late 1960s onwards)
• Many streams had been covered to form underground canals, which restricted water flow.	• Natural drainage was preserved – these natural floodplains are used as parks.
• Houses and other buildings had been built too close to rivers.	• Certain low-lying areas are off-limits.
• New buildings were built on poorly drained land on the periphery of the city.	• Parks have been extensively planted with trees; existing buildings have been converted into new sports and leisure facilities.
• Increase in roads and concrete surfaces accelerated runoff.	• Bus routes and bicycle paths integrate the parks into the urban life of the city.

POLLUTION MANAGEMENT

The main way of reducing pollution has been to reduce the number of cars on the road. This has been done by having an integrated transport network.

In Curitiba, Brazil, the road network and public transport system have structural axes. These allow the city to expand, but keep shops, workplaces and homes closely linked.

There are five main axes of three parallel roadways: a central road with two express bus lanes – Curitiba's mass transport system is based on the bus.

- Inter-district and feeder bus routes complement the express bus lanes along the structural axes. Everything is geared towards the speed of journey and convenience of passengers.
- A single fare allows transfer from express routes to inter-district and local buses.
- Extra-wide doors allow passengers to board quickly.
- Double- and triple-length buses allow for rush-hour loads.
- The rationale for the bus system was economy and sustainability. A subway would have cost \$80–\$70 million per km; the express bus ways were only \$200,000 per km. The bus companies are paid by the kilometres of road they cover, not by the number of passengers. This ensures that all areas of the city are served.

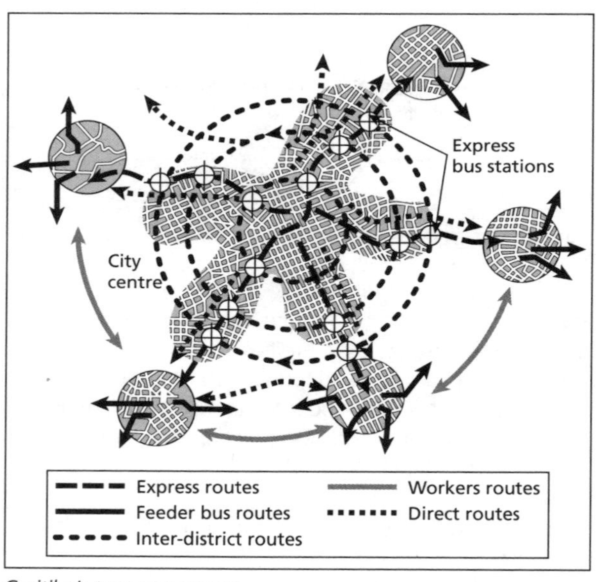

Curitiba's transport system

Sustainable strategies

THE URBAN ECOLOGICAL FOOTPRINT

According to the Global Development Research Centre, the urban ecological footprint is the land area required to sustain a population of any size. All the resources which people use for their daily needs, such as food, water and electricity, must be produced using raw natural resources. The urban ecological footprint measures the amount of arable land and aquatic resources that are needed to continuously sustain a population, based on its consumption levels at a given point in time. To the fullest extent possible, this measurement incorporates water and energy use, uses of land for infrastructure and different forms of agriculture, forests and all other forms of energy and material "inputs" that people require in their day-to-day lives. It also accounts for the land area required for waste assimilation.

Tokyo's ecological footprint

According to the Earth Council, a biologically productive area of 1.7 ha is available per capita for basic existence. This means that, for sustainable living, the people in Tokyo alone need an area of 45,220,000 ha – which is 1.2 times the land area of the whole of Japan. If mountains and other regions are discarded and only habitable land included, then this becomes 3.6 times the land area of Japan.

Tokyo is a city where the land is used several times at several levels. The difference between very high-density cities (compact cities with much vertical development) and extended cities (cities with suburban sprawl, like those in Australia and the USA, for example) would be three or four times greater.

Compact cities such as Tokyo have a large population living in a very small and dense area of land, freeing land area for other purposes. They also require reduced amounts of infrastructure and resources – it is easier to provide services, utilities and infrastructure to a population concentrated in a small area than is the case when people are spread over a large area.

SUSTAINABLE HOUSING

There are many problems with much of the housing in Mexico City. Many lack access to water, adequate sanitation, a reliable and safe power supply, adequate roofs, solid foundations, secure tenure, i.e. the residents are at risk of eviction.

There are a variety of possible solutions to the housing problems of many LEDC cities such as Mexico City (see page 130). Solutions include:

- government support for low-income, self-built housing
- subsidies for home-building
- flexible loans to help shanty-town dwellers
- slum upgrading in central areas
- improved private and public rental housing
- support for the informal sector/small businesses operating at home
- site and service schemes
- encouragement of community schemes
- construction of health and educational services.

CONTROL OF IN-MIGRATION – NEW CITIES IN KOREA AND MALAYSIA

There have been many attempts to reduce the importance of very large cities, such as London, Rio de Janeiro and Seoul. Developers have attempted to build new towns and new capital cities to deflect growth away from the main cities.

At the wealthier end of the scale are new towns and cities such as Brasilia, Canberra and, in Korea, Gongju-Yongi. Originally, Gongju-Yongi was planned to replace Seoul as Korea's capital by 2020. It is a US$54 billion scheme. Construction began in 2007. Seoul will in fact remain as the capital, but government offices will relocate to Gongju-Yongi. The new development is still necessary to ease chronic overcrowding, to aid redistribution of the state's wealth and to reduce the danger of a military attack from North Korea. Previous developments have concentrated huge amounts of money, power and up to half of Korea's population in Seoul.

Another impressive scheme is the Malaysian new town of Putrajaya. This is a totally new city situated 25 km to the south of Kuala Lumpur. Covering an area of 4931 ha, Putrajaya was established in 1995.

Putrajaya is a planned city, built according to a series of comprehensive policies and guidelines for land use, transportation system, utilities, infrastructure, housing, public amenities, information technology, parks and gardens.

The mission of the Putrajaya Corporation was to:

- provide an efficient and effective administration
- provide quality services to ensure customer satisfaction
- provide infrastructure and amenities conducive to creating an ideal environment for living and working.

Their functions include:

- the functions of a local government in the Putrajaya area
- to promote, stimulate, facilitate and undertake commercial, infrastructure and residential development in the area
- to promote, stimulate and undertake economic and social development in the area
- to control and coordinate the performance of the above activities in the area.

EXAM QUESTIONS ON PAPER 2 – OPTIONAL THEMES FOR HL AND SL

Key features

Timing: You have approximately 40 minutes for each question.

Choice: HL candidates must answer three questions and SL candidates must answer two questions. Each question must be chosen from a different theme.

Structure

One or both of the questions will have stimulus material in the form of a map, table, graph, photo or diagram. The question will normally have three parts:

a) This part is allocated about 4 marks and asks you to refer to the stimulus material using straightforward command terms such as *state*, *define*, *describe* and *identify*.

b) This part is allocated about 6 marks and asks you to draw on your own knowledge. It may or may not refer back to the stimulus material. The command terms are more difficult, such as *analyse*, *explain* and *suggest*.

c) This part is allocated 10 marks and asks for a more analytical approach. The command terms are more difficult than in previous parts and include terms such as *discuss*, *evaluate* and *justify*. This is where you should include examples and case studies.

Each question is worth 20 marks.

Option A: Freshwater – issues and conflicts

A1

Country	A Total renewable freshwater resources (m^3/person/year)	B Total water withdrawals (m^3/person/year)	C Water dependence %
Iceland	582,192	543	0
Congo	217,915	11	73
Canada	91,419	1,419	2
Norway	83,919	489	0
Uruguay	40,419	941	58
Bangladesh	8,089	576	91
Japan	3,365	696	0
Sudan	1,879	1,187	77
Egypt	794	1,013	97
Kuwait	8	198	99

Source: *FAO, UN, World Bank*

a) Referring to the table, define the terms:

i) total renewable freshwater resources

ii) water dependence. [2+2]

b) Referring to examples, explain the international variation in the amount of freshwater resources and withdrawals (columns A and B). [6]

c) Discuss the reasons why conflicts arise over freshwater supplies. [10]

A2 a) Describe the following terms:

i) watershed

ii) aquifer. [2+2]

b) Using only an annotated diagram, explain the hydrological cycle of a drainage basin. [6]

c) Referring to examples, examine the impacts of agriculture upon water quality. [10]

Option B: Oceans and their coastal margins

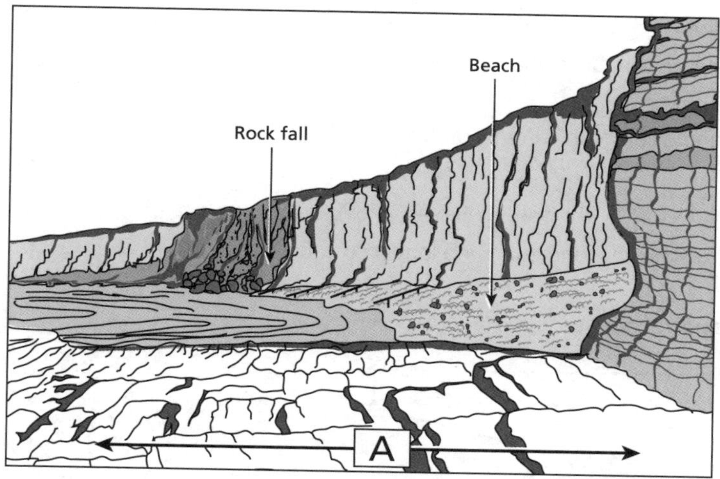

Source: Adapted from Allan Williams, *A Guide to Landforms*

B3 a) Identify feature **A** shown in the diagram and explain its formation. [1+3]

b) Explain the characteristics of this beach and its possible sources of sediment. [6]

c) Referring to examples, examine the reasons why some coastlines receive more protection than others. [10]

B4 a) Name and locate two abiotic oceanic resources and briefly decribe the zone where each is found. [4]

b) Explain the purpose of an exclusive economic zone (EEZ). [6]

c) Examine the causes of overfishing and evaluate the attempts to remedy this problem. [10]

Option C: Extreme environments

C5

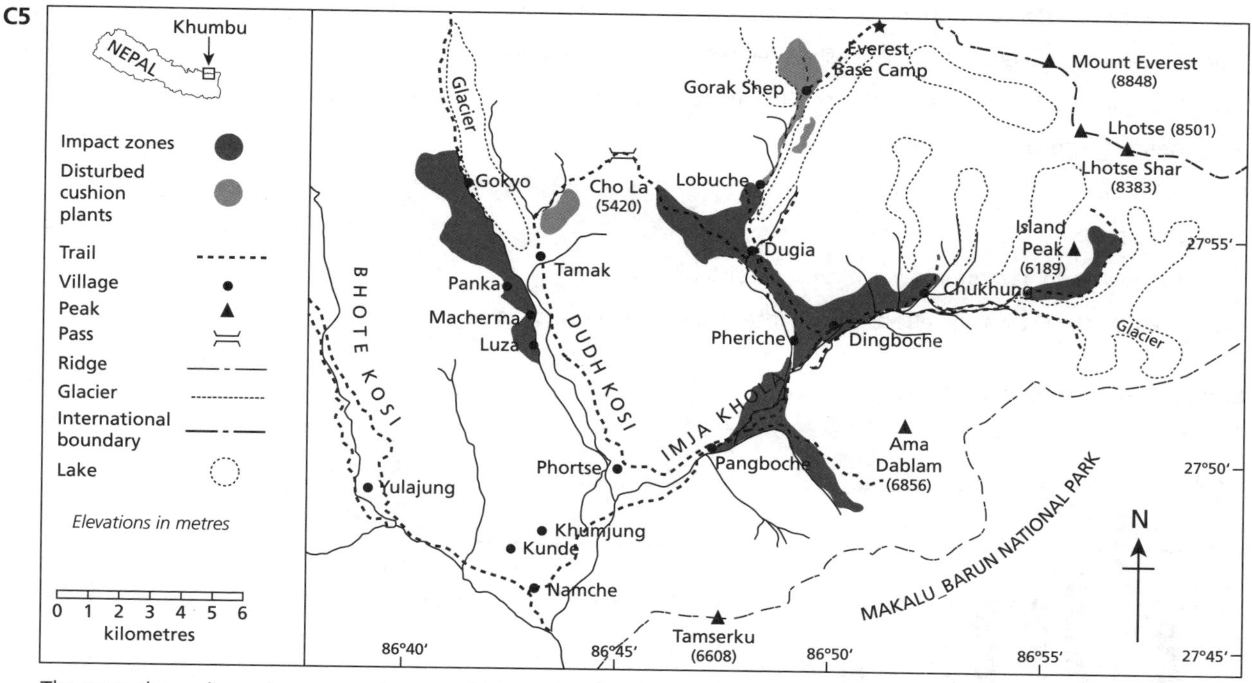

The map shows impact zones and areas of vegetation (cushion plants) disturbed by livestock and trekkers near Mount Everest in the Khumbu region of Nepal.

a) Describe and explain the pattern of all the impacts shown on the map. [4]

b) Explain the physical processes involved in mass movement in this type of region. [6]

c) Discuss the ways in which human activity may be sustainably managed in any one extreme environment. [10]

C6 a) Briefly describe the influence of continentality and cold ocean currents on the development of hot desert areas. [4]

b) Referring to arid areas, distinguish between the processes of weathering and erosion. [6]

c) Discuss the extent to which human activity in extreme environments is no longer sustainable. [10]

Option D: Hazards and disasters – risk assessment and response

D7

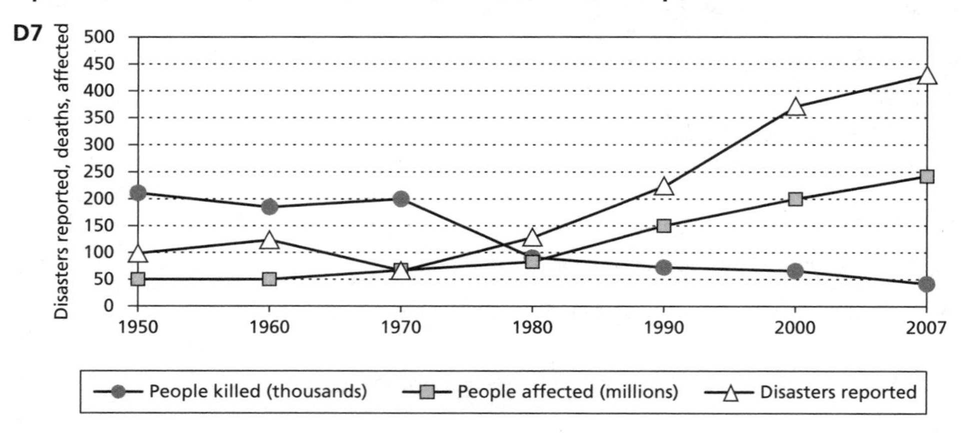

Global natural disasters

The graph shows the number of global disasters reported, the number of people killed (thousands) and the number of people affected (millions) by all disasters between 1950 and 2007.

a) Describe the trends shown on the graph. [4]

b) Explain the changes in the relationships between the three variables. [6]

c) Examine the reasons that people are attracted to regions threatened by one or more natural hazards. [10]

D8 a) Define the following terms:

i) hazard

ii) disaster. [2+2]

b) Compare two hazards in terms of their predictability. [6]

c) Examine the causes and effects of one recent human-induced disaster. [10]

Option E: Leisure, sport and tourism

E9

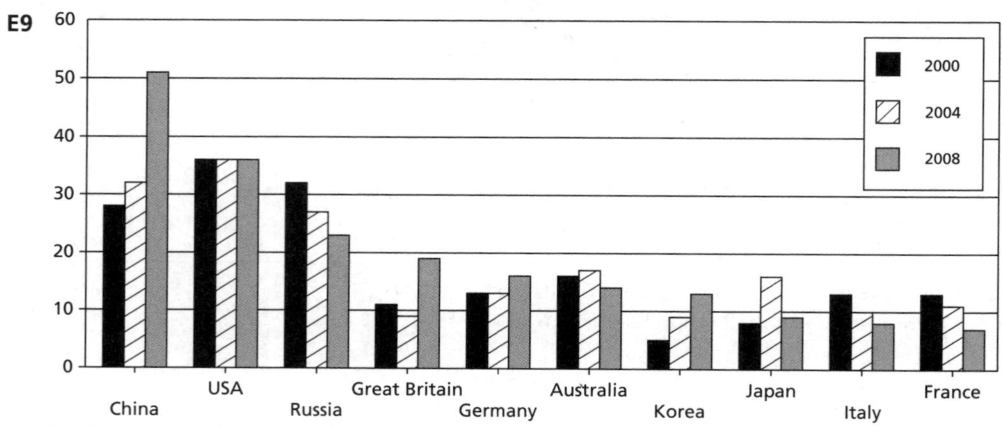

Gold medals awarded in summer Olympic Games in 2000, 2004 and 2008

Source: *www.databaseolympics.com*

The graph shows the number of gold medals awarded to the top 10 countries in the summer Olympic Games of 2000 (Sydney), 2004 (Athens) and 2008 (Beijing).

a) Describe the pattern and trends shown on the graph. [4]

b) Explain the factors which need to be considered when choosing a city to host a world sporting event. [6]

c) Discuss the international variation in the level of participation in world sporting events. [10]

E10 a) Define the following terms:

i) leisure

ii) sport [2+2]

b) Using only an annotated sketch map, describe and explain the location of leisure facilities in and around a named urban area. [6]

c) Referring to at least one example, discuss the strategies that may be used to maintain the carrying capacity of a popular tourist attraction in a rural area. [10]

Option F: The geography of food and health

F11

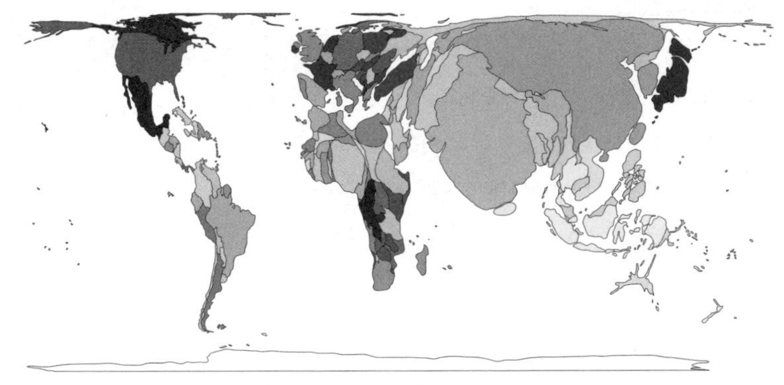

Map A

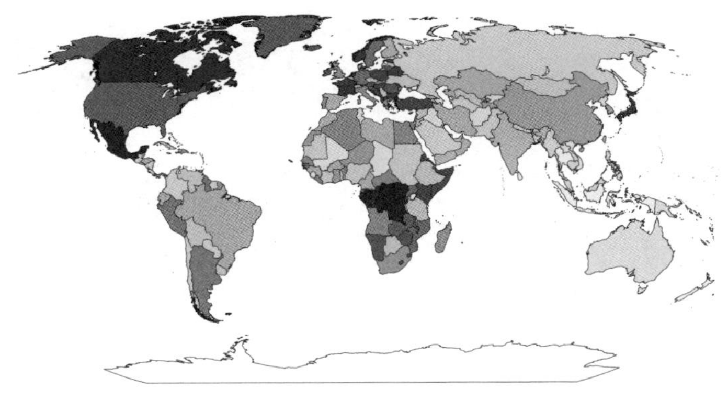

Map B

www.worldmapper.org

The size of each country on Map A represents the proportion of the world's unhealthy population that lives there. Map B shows the true national boundaries. The shading is used to distinguish each country.

a) Describe the pattern of unhealthy population shown on Map A. [4]

b) Suggest reasons for the pattern described in (a). [6]

c) Discuss the difficulties in containing the spread of one disease. [10]

F12 a) Describe the global pattern of malnutrition. [4]

b) Describe three technological innovations and explain how they have increased agricultural production in recent years. [6]

c) Referring to at least one recent example, discuss the causes of famine. [10]

Option G: Urban environments

G13

Photo by: Luiz Arthur Leirão Viera

The photograph shows part of Favela Paraisópolis (Paradise) in Sao Paulo.

a) Describe and briefly explain the contrasting land uses shown in this photograph. [2]

b) Explain the characteristics of the informal economic sector in the urban economy. [6].

c) Referring to one or more cities, discuss the causes of environmental problems and the attempts made to overcome them. [10]

G14 a) Explain what is meant by functional zoning. [4]

b) Explain three distinctive characteristics of the central business district (CBD). [6]

c) Discuss the processes of re-urbanization and gentrification that have occurred in the last 30 years in many richer countries. [10]

Globalization

DEFINITIONS

- **Core and periphery:** the concept of a developed core surrounded by an undeveloped periphery. The concept can be applied at various scales.
- **Cultural imperialism:** the practice of promoting the culture or language of one nation in another. It is usually the case that the former is a large, economically or militarily powerful nation and the latter is a smaller, less affluent one.
- **Gross domestic product (GDP):** the value of all final goods and services produced within a nation in a given year. The measure is easy to compute and gives a precise measure of the value of output.
- **Globalization:** "the growing interdependence of countries worldwide through the increasing volume and variety of cross-border transactions in goods and services and of international capital flows, and through the more rapid and widespread diffusion of technology" (IMF).
- **Globalization index:** the A.T. Kearney index is one of several measures of globalization. It tracks changes in the four key components of global integration: trade and investment flows; movement of people across borders; volumes of international telephone traffic and internet usage; and participation in international organizations (A.T. Kearney, *Foreign Policy*).
- **Glocalization:** a term that was invented in order to emphasize that the globalization of a product is more likely to succeed when the product or service is adapted specifically to each locality or culture it is marketed in. The increasing presence of McDonald's restaurants worldwide is an example of globalization, whereas the changes in the menus of the restaurant chain that are designed to appeal to local tastes are an example of glocalization.
- **Gross national income (GNI):** (the term is now used in preference to gross national product, GNP) the total value of goods and services produced within a country, together with the balance of income and payments from or to other countries.
- **Outsourcing:** the concept of taking internal company functions and paying an outside firm to handle them. Outsourcing is done to save money, improve quality, or free company resources for other activities.
- **Time–space convergence:** the reduction in the time taken to travel between two places due to improvements in transportation or communication technology.
- **Transnational corporation (TNC):** a firm that owns or controls productive operations in more than one country through foreign direct investment.

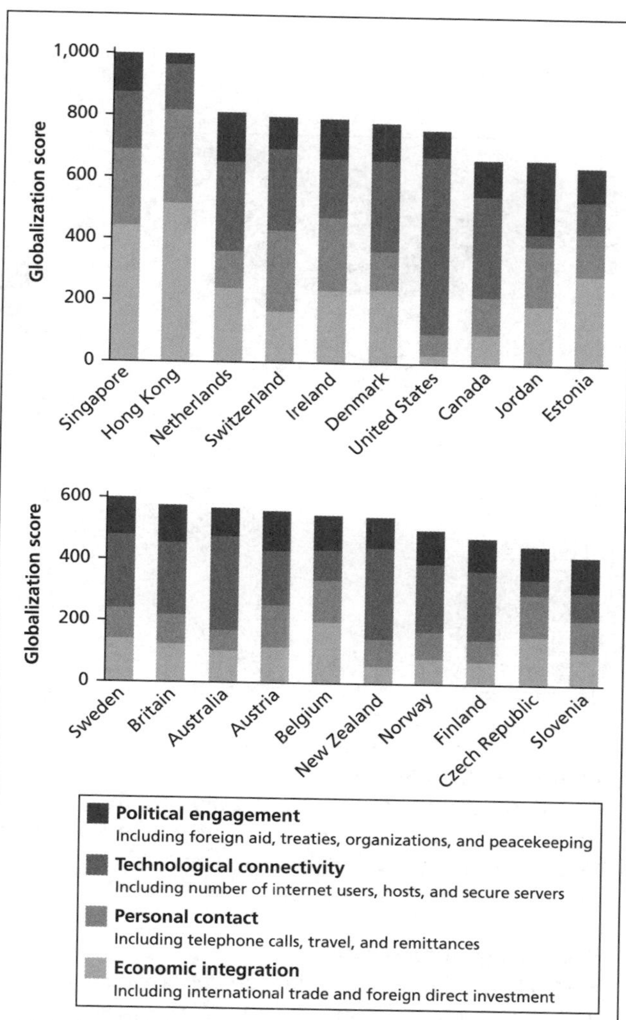

Globalization index

MEASURING GLOBAL INTERACTIONS

The globalization index tracks and assesses changes in four key components of global integration (see above). The 72 countries ranked in the 2007 globalization index account for 97% of the world's GDP and 88% of the world's population. Major regions of the world, including developed and developing countries, are covered to provide a comprehensive and comparative view of global integration.

Economic integration combines data on trade and foreign direct investment (FDI) inflows and outflows, international travel and tourism, international telephone calls, and cross-border remittances.

Technological connectivity counts the number of internet users and internet hosts. Political engagement includes each country's memberships in a variety of representative international organizations.

The resulting data for each given variable are then "normalized" through a process that assigns the value of 1 to the highest data, with all other data points valued as fractions of 1. The base year (1998 in this case) is assigned a value of 100. The given variable's scale factor for each subsequent year is the percentage growth or decline in the GDP – or population-weighted score of the highest data point, relative to 100. Globalization index scores for every country and year are derived by summing all the indicator scores.

Index of globalization

THE KOF INDEX OF GLOBALIZATION

The KOF index of globalization was introduced in 2002 and covers the economic, social and political dimensions of globalization. KOF defines globalization as: "the process of creating networks of connections among actors at multi-continental distances, mediated through a variety of flows including people, information and ideas, capital and goods. Globalization is conceptualized as a process that erodes national boundaries, integrates national economies, cultures, technologies and governance and produces complex relations of mutual interdependence."

More specifically, the three dimensions of the KOF index are defined as:

- **economic globalization**, characterized as long-distance flows of goods, capital and services, as well as information and perceptions that accompany market exchanges
- **political globalization**, characterized by a diffusion of government policies
- **social globalization**, expressed as the spread of ideas, information, images and people.

In addition to the indices measuring these dimensions, KOF calculates an overall index of globalization and sub-indices referring to actual economic flows, economic restrictions, data on information flows, data on personal contact and data on cultural proximity. The 2008 index introduced an updated version of the original index, employing more recent data than had been available previously.

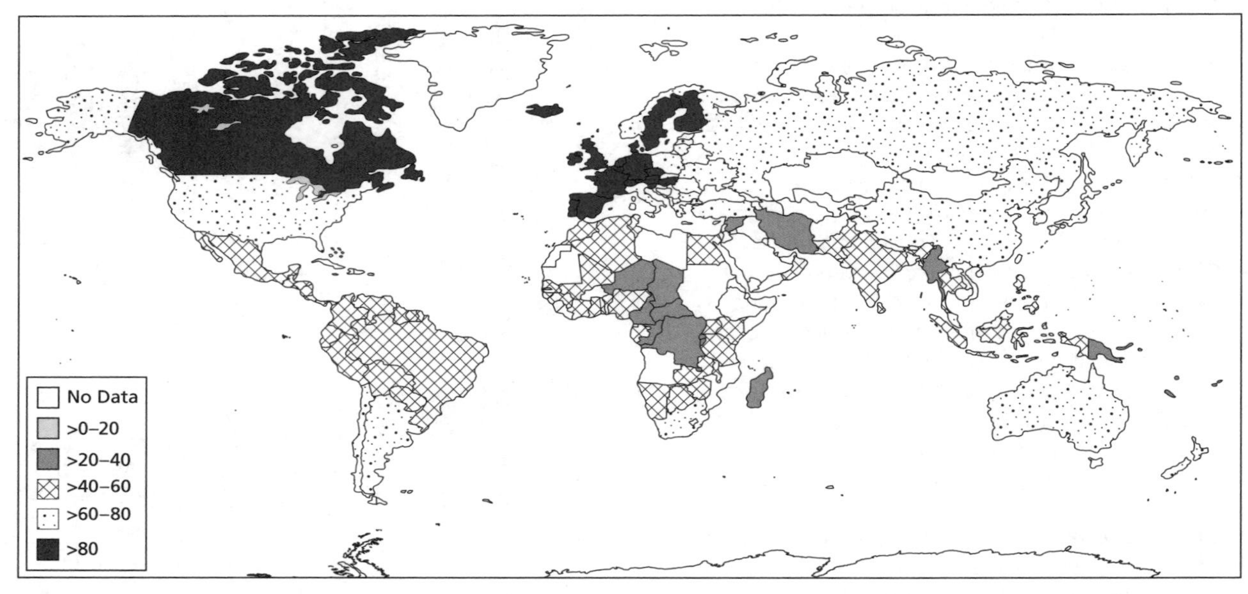

2005 KOF map

Economic globalization

Broadly speaking, economic globalization has two dimensions. First, actual economic flows, which are usually taken to be measures of globalization; and, second, restrictions to trade and capital.

Political globalization

Political globalization uses the number of embassies and high commissions in a country, the number of international organizations to which the country is a member and the number of UN peace missions a country has participated in.

Social globalization

The KOF index classifies social globalization in three categories. The first covers personal contacts, the second includes data on information flows and the third measures cultural proximity.

- **Personal contacts** includes international telecom traffic (outgoing traffic in minutes per subscriber) and the degree of tourism (incoming and outgoing) a country's population is exposed to. Government and workers' transfers received and paid (as a percentage of GDP) measure whether and to what extent countries interact.
- **Information flows** include the number of internet users, cable television subscribers, number of radios (all per 1000 people), and international newspapers traded (as a percentage of GDP).
- **Cultural proximity** is arguably the dimension of globalization most difficult to grasp. According to one geographer, cultural globalization mostly refers to the domination of US cultural products. KOF includes the number of McDonald's restaurants located in a country. In a similar vein, it also uses the number of Ikea stores per country.

Global core and periphery

WORLD SYSTEMS ANALYSIS

World systems analysis is identified with Immanuel Wallerstein (1974) and is a way of looking at economic, social and political development. It treats the whole world as a single unit. Any analysis of development must be seen as part of the overall capitalist world economy, not on a country-by-country approach. Wallerstein argued that an approach which looked at individual countries in isolation was too simplistic and suffered from **developmentalism**. The developmentalism school assumed that:

- each country was economically and politically free (autonomous)
- all countries follow the same route to development.

As such they were **ethnocentric**, believing that what happened in North America and Europe was best and would automatically happen elsewhere.

According to Wallerstein, the capitalist world system has three main characteristics:

- a global market
- many countries, which allow political and economic competition
- three tiers of countries.

The tiers are defined as the **core**, largely MEDCs; the **periphery**, which can be identified with LEDCs; and the **semi-periphery**. The semi-periphery is a political label. It refers to those countries where there are class struggles and social change, such as Latin America in the 1980s and eastern Europe in the late 1980s and early 1990s.

Wallerstein argued that capitalist development led to cycles of growth and stagnation. One of these cycles is a long-term economic cycle known as a Kondratieff cycle. This identifies cycles of depression at roughly 50- to 60-year intervals. The last two were in the 1920s–30s and the late 1980s. Stagnation is important for the restructuring of the world system and allows the semi-periphery to become involved in the development process.

Capitalism, according to the world systems approach, includes feudalism and socialism. They are extreme variations on the division of labour. As the world develops and changes, there will either be a swing towards a more socialist system or a transition towards a more unequal (feudal) system.

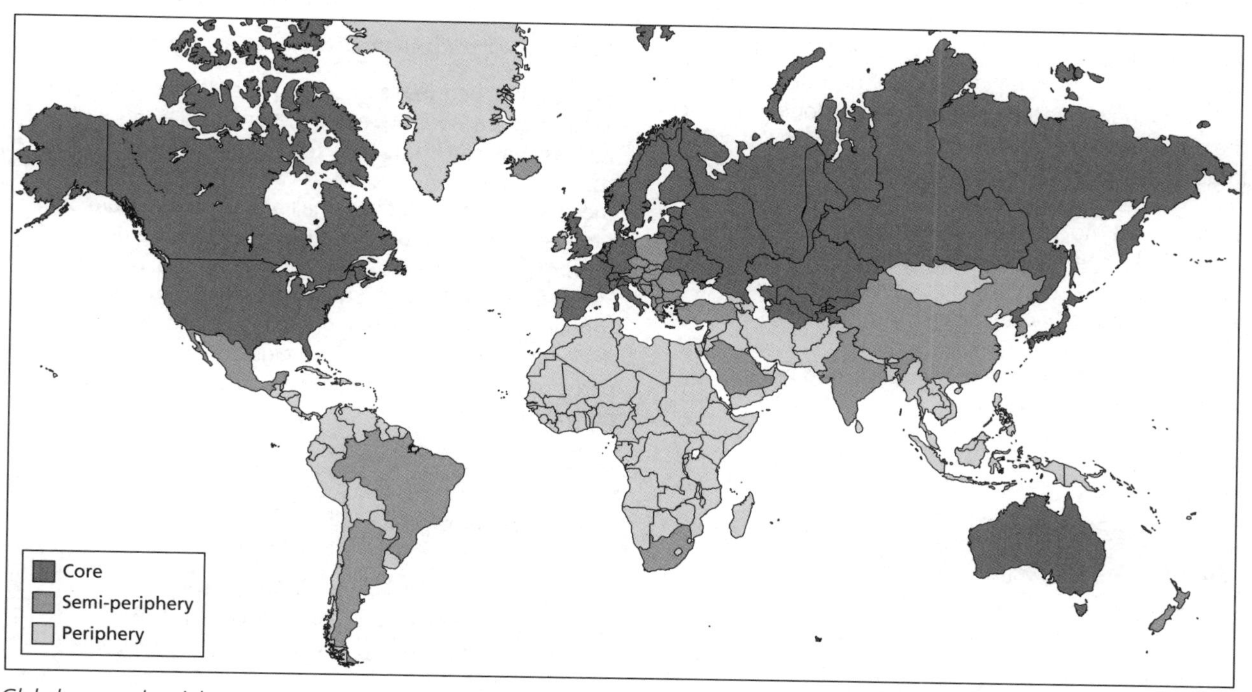

Global core and periphery

Time–space convergence

DISTANCE DECAY

The **frictional effect of distance** or **distance decay** suggests that areas that are closer together are more likely to interact, whereas areas further away are less likely to interact with each other. However, there has been a reduction in the frictional effect of distance as improvements in transport have allowed greater distances to be covered in the same amount of time. In addition, improvements in ICT have brought places on different sides of the world together almost instantaneously.

TYPES OF TRANSPORT

Transport costs are made up of **operating costs** and the **profit rate** of the carrier. Operating costs include:

- variable costs such as fuel and wages
- overhead costs, which include equipment, terminal facilities, tracks and repairs
- indirect costs such as insurance.

Some modes of transport are more competitive over a certain distance. For example, ocean transport is very competitive over long distances. This is due to very low operating costs. However, over short distances it is not competitive. This is because of the high overhead costs of ports. By contrast, the operating costs of road transport are very high but the overhead costs are low. This makes road transport very competitive over short distances but not over long distances. To compete over longer distances the vehicles need to carry much greater loads. Articulated lorries are able to spread the costs over a greater load. The same feature can be seen in other forms of transport. Some aircraft, notably wide-bodied jumbos, are getting larger. Tankers have increased in size. Very large crude carriers (VLCCs – ships built for a specific purpose) are more competitive because they can carry a greater load.

Factors affecting the type of transport used include:

- the item to be transported
- the cost of transporting it
- the speed at which it needs to be transported.

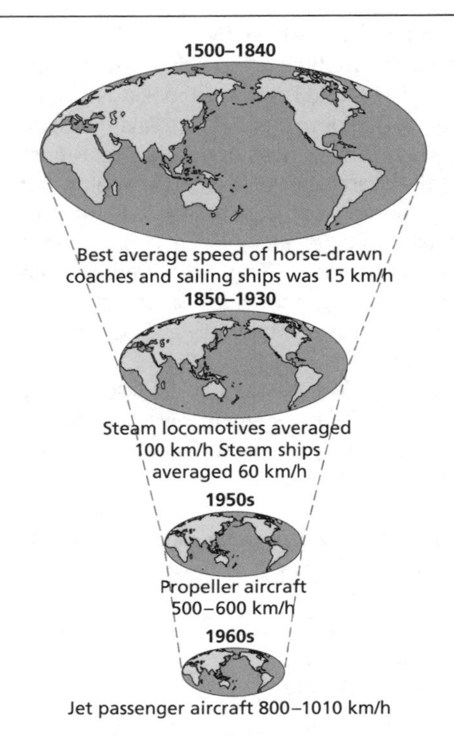

A shrinking world: the effects of changing transport on time–distance

For example, perishable goods, such as flowers and fruit, need to be transported rapidly, whereas bulky goods, such as coal, can be transported by the cheapest means possible. Economies of scale are also important. It is cheaper to carry bulk than small amounts; therefore bulk carriers are increasingly used. **Bulk carriers** are designed to carry cargoes in bulk, such as iron ore, coal or wheat. By contrast, **container carriers** are ships designed to carry containers. They are equipped with specialized handling devices for carrying expensive freight, such as machine parts, or high-value manufactures, such as electronic equipment.

	Advantages	Disadvantages
Water (ocean)	• Cheaper over long distances • No cost in building the route • Good for bulky, low-cost goods, e.g. coals, ores, grains • Costs spread over a large cargo	• Slow • Very limited routes to deep-water ports • Ships expensive to build/maintain • Environmental problems, especially pollution • Ports take up great space
Air	• Faster over long distances • Limited congestion • Good for high-value transport such as people, hi-tech industries and urgent cargo	• Lots of land needed for airports • Noise and visual pollution • Very expensive to build and maintain • No flexibility of routes • Very expensive • Can carry only small loads

Advantages and disadvantages of sea and air transport

In general, whilst aircraft have become faster, ocean tankers have become larger. There have also been increases in the size of planes. For example, the Airbus A380, the world's largest passenger plane, can carry about 555 people – more than the Boeing 747 jumbo. On the other hand, some very fast planes, such as Concorde, have been taken out of circulation.

Extension and density of networks

TELEPHONE CALLS

The map below shows the annual flow of intercontinental calls by fixed landline telephones (not cellphones) in 2007. The greatest volume of traffic is between North America and Europe, followed by North America and South-East Asia. There are also large flows between North America and the Caribbean and Latin America. There are relatively few flows between Africa and the other continents.

A number of reasons can help explain these patterns.

- Population size – countries with small populations, such as Greenland, are likely to generate a limited number of calls.
- Population density – within the USA, for example, there is a small flow to and from Alaska but a very large flow to and from the north-east USA.
- Wealth – countries that are wealthy, such as Japan and the USA, can afford more phones compared with poorer countries in Africa.
- Trading partners – countries within a trading bloc, such as the EU, are likely to generate large volumes of calls.
- TNC or MNC activities – companies which have offices and factories in different countries are likely to create large volumes of calls between those countries.
- Migration – there is likely to be a high volume of calls between the area a migrant moves to and their home country. However, the origin may be relatively poor and have relatively few phones.
- Colonial history – it is likely that there will be political and historical ties between a former colonial power and its former colonies. The UK and the former British Empire is a good example.
- Language – it is likely that the volume of calls will be greater among countries that share the same language.

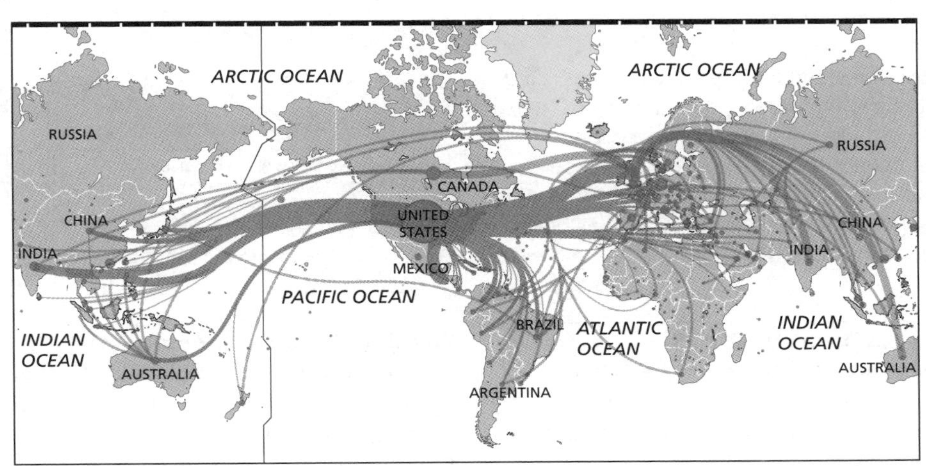

Global telephone traffic

THE INTERNET

How does the map of telephone traffic compare with the world internet map?

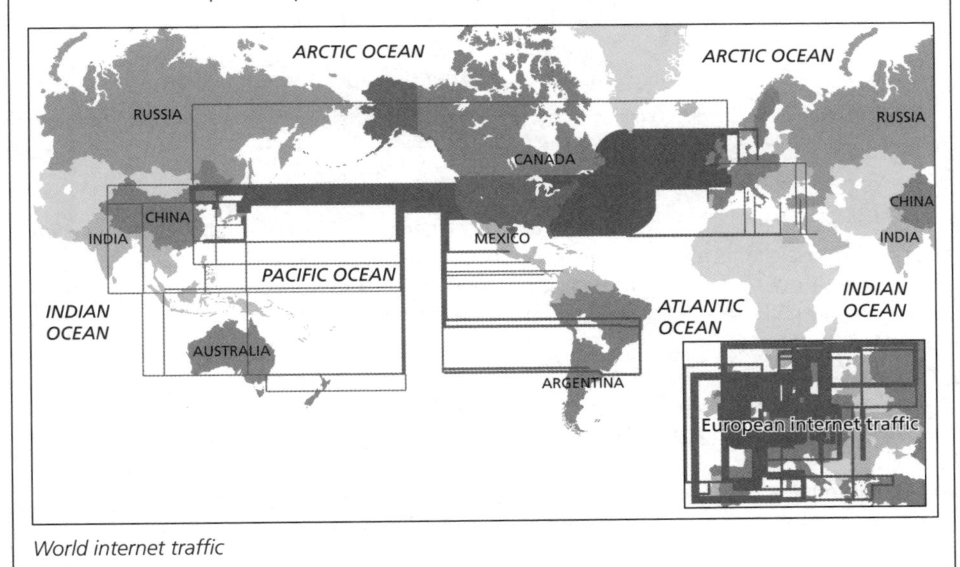

World internet traffic

EXTENSION

Visit www.telegeography.com/maps/index.php to see detailed maps on this topic.

The role of ICT

THE DIGITAL DIVIDE

The internet is the fastest growing tool of communications ever. Radio took 38 years to reach its first 50 million users; television took 13 years and the internet just four years.

The digital divide refers to the inequalities in opportunities between individuals, households, businesses and nations to access ICT. The digital divide also occurs between urban and rural areas, and between different regions of a country. Examples include the following:

- Over 75% of internet users come from rich countries, which account for just 14% of the world's population.
- In Thailand, 90% of internet users live in urban areas.
- In Chile, 74% of internet users are under 35 years of age.
- In Ethiopia, 86% of internet users are male.
- In the UK, 30% of internet users have salaries of over $120,000.
- In the UK, over 50% of internet users have degrees.

Instead of reducing inequalities between people, the digital divide may well have reinforced them. There is a widening gap between rich and poor countries.

Within rich countries, such as the USA, internet users are more likely to be white, middle class and male. There are many people who do not have access to ICT and therefore cannot benefit from the knowledge-based economy. To date there has been little action from rich countries to ensure that the benefits of ICT are extended to people in poorer countries, regions and areas.

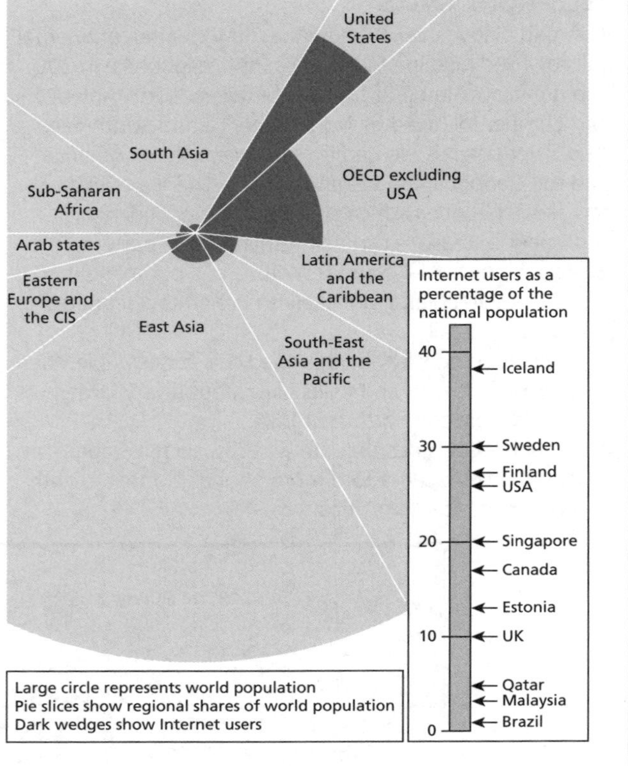

Global variations in internet use

HOUSEHOLDS AND INDIVIDUALS

The digital divide may be defined as inequality in the ICT network infrastructure and distribution of the IT knowledge, skills and resources necessary to access online services and information among different sections of a modern society.

The digital divide among households appears mainly to depend on two factors: income and education. The higher the income and the level of education, the more likely it is that more individuals will have access to information and communication technologies (ICTs).

The digital access index (DAI)

The DAI measures the overall ability of individuals in a country to access and use ICTs. It consists of eight variables grouped in five categories:

- infrastructure – combined fixed and mobile teledensity
- affordability – internet access price as a percentage of per capita GNI
- knowledge – represented by adult literacy and combined enrolment up to tertiary schools
- quality – represented by international internet bandwidth in bits per capita and percentage of broadband customers
- usage by internet users per 100 population.

CONTRASTING FORTUNES

India

The number of internet users in India has reached 42 million. Of these, the number of "active users" has risen to more than 21 million. India's population is over 1,130 million, so only 3.7% of the population has access to the internet. "Active user" defines users who have used the internet at least once in the previous 30 days.

Young people are the main drivers of internet usage in India. College students and those below the age of 35 are the biggest segment on the internet.

The reasons for the low uptake of ICT in India are simple. Poverty is the main one – people cannot afford the luxury of computers. In addition, not all areas have electricity; rural areas, and shanty towns in particular, have limited access. Third, the distances in India are so vast that trying to connect all areas to the web is almost impossible, as well as vastly expensive. Moreover, India has other issues to deal with – housing, health, food supply, water supply – access to the internet has much to compete with.

Iceland

In Iceland, some 258,000 people out of a population of 299,076 are internet users. That is a staggering 86.3% of the population. Unlike India, Iceland is a rich country and a sparsely populated one. Almost half of the country's population live in the Reykjavik region. Being able to communicate by ICT is extremely useful in a country where the road network is limited and travel in winter is difficult.

Financial flows (1)

THE GLOBAL ECONOMY

- Exports and imports of goods and services in 2005 exceeded $26 trillion or 58% of total global output, up from 44% in 1980. Developing economies still account for less than one-third of global trade, but their share has been increasing steadily.
- Gross private capital flows across national borders exceeded 32% of global output in 2005, up from 9% in 1980. Foreign direct investment (see page 000) and cross-border investment flows to developing economies have soared, despite occasional setbacks.

Many factors have accelerated the pace of globalization:

- Barriers to international trade and investment are coming down.
- Technological progress has dramatically cut transportation and communications costs.
- Some previously non-tradable services can now be traded easily.

Globalization has created opportunities and challenges for developing countries. The experiences of China, India and Korea, for example, show that integration into the global economy is necessary for long-term growth and poverty reduction. Nevertheless, there are concerns over equality of opportunity and the unequal distribution of benefits. Many poor countries and poor people in many countries have not been able to take full advantage of the opportunities brought by globalization or to participate in its benefits.

Expanding trade

Between 1990 and 2005, growth in trade outpaced growth in the overall global economy. The market share of low- and middle-income economies increased from about 16% in 1990 to almost 30% in 2005, although the sub-Saharan share lagged at around 1.5%. Trade between developing economies now accounts for about 8% of world merchandise exports. Between 1990 and 2005, merchandise exports between developing economies grew at an average annual rate of 13%, compared with less than 6% for exports between high-income economies.

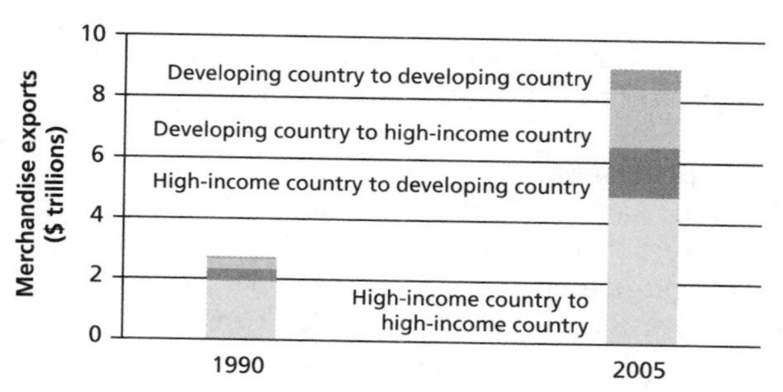

Changes in trade

Tariff barriers affect exports to developing economies disproportionately – tariffs are higher than those affecting exports to high-income economies. The simple mean tariff rate averages 9% in developing economies but less than 4% in high-income economies.

Expanding flows of private financial resources

International private financial flows have increased rapidly. Between 1990 and 2005, total gross capital flows tripled as a share of world GDP, and high-income economies still account for the majority of this finance. Financial flows to developing economies have also increased rapidly, although from a much lower base. Foreign direct investment (FDI) remains the largest component.

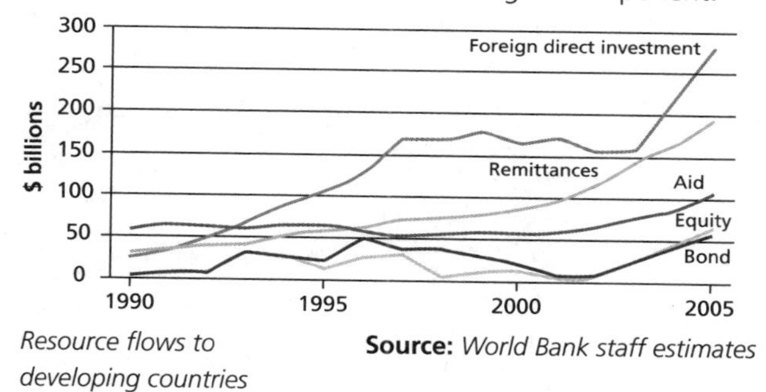

Resource flows to developing countries

Source: *World Bank staff estimates*

2008 FINANCIAL CRISIS

Until the 1970s many western governments aimed to keep unemployment rates low by expanding public spending or cutting taxes. However, government involvement in the financial markets declined from 1970–2000 during an era of deregulation. The financial crisis of 2008–09 saw governments becoming involved once more, and in some cases, taking hold of the financial markets. The era of easy credit was over – at least for the short term.

ECONOMIC GIANTS

- The USA and Europe account for over 50% of the global financial assets, with the USA alone representing over $47 trillion.
- Japan accounts for over $17 trillion, while the emerging Asia is close to $10 trillion in financial assets.

EXTENSION

Visit
http://www1.worldbank.org/economicpolicy/globalization/documents/table6-7.pdf for data on global financial flows.

Financial flows (2)

CROSS-BORDER INVESTMENTS

The map of cross-border investments is complex. Nevertheless, the dominance of flows between the USA, the EU and Japan are clear. Emerging Asia has overtaken Russia and eastern Europe. Singapore, Hong Kong and Taiwan have greater financial assets than all of Latin America.

The World Bank

The World Bank, or the International Bank for Reconstruction and Development (IBRD), was established in 1947 to provide aid to poor countries in terms of loans and technical assistance. Initially it provided loans for capital projects, such as infrastructure developments, but from 1980 onwards it provided help for debt repayment.

The World Trade Organization

The World Trade Organization (WTO) was formed in 1995. It covers trade in manufactured goods, raw materials, agricultural services and intellectual property rights. It has over 150 members, including China who joined in 2001. The WTO monitors whether countries are following free trade rules. Critics say that it is biased in favour of TNCs and against small producers.

The International Monetary Fund

The IMF is an international organization of 185 member countries. It was established to promote international monetary cooperation, exchange stability and orderly exchange arrangements; to foster economic growth and high levels of employment; and to provide temporary financial assistance to countries to help ease balance of payments adjustment.

The work of the IMF is of three main types:

- Surveillance involves the monitoring of economic and financial developments, and the provision of policy advice, aimed especially at crisis prevention.
- The IMF also lends to countries with balance of payments difficulties, to provide temporary financing and to support policies aimed at correcting the underlying problems; loans to low-income countries are also aimed especially at poverty reduction.
- The IMF provides countries with technical assistance and training in its areas of expertise.

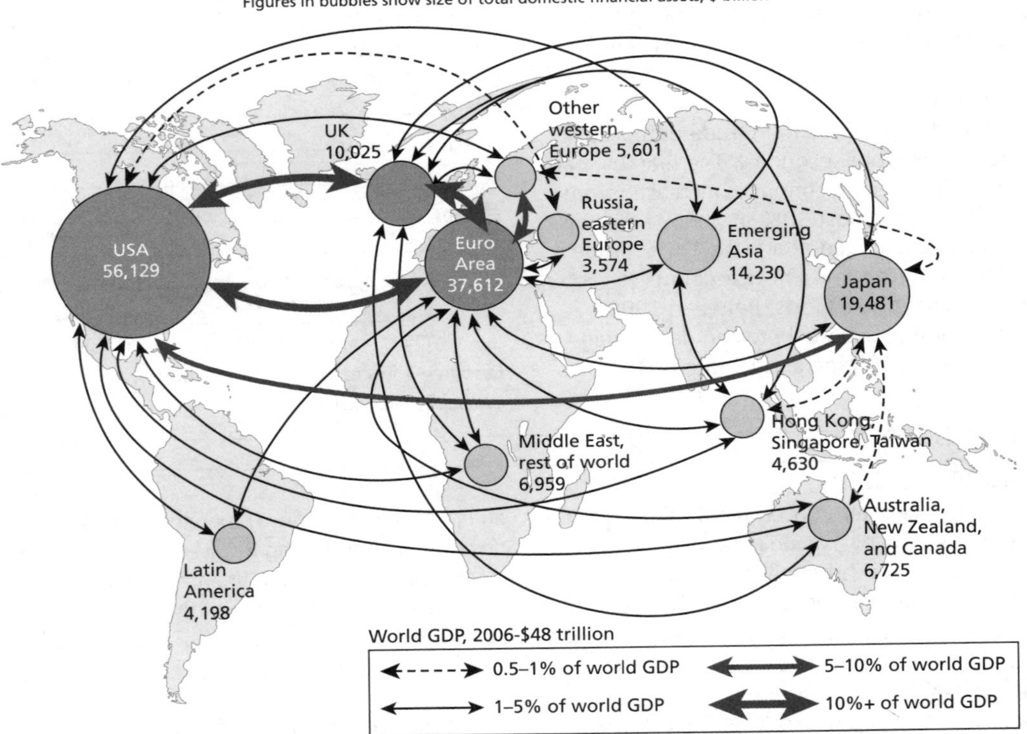

*Includes total value of cross-border investments in equity and debt securities, lending and deposits, and foreign direct investment.

Cross-border investments, 2006

Source: *McKinsey Global Institute Cross-Border Investments Database*

Financial flows (3)

A HISTORY OF FINANCIAL FLOWS

Neither the concept nor the phenomenon of financial globalization can be considered new. Cross-country capital movements have a long and well-documented history, dating back to the Renaissance period in Italy. Then, financial flows were limited among a small number of source and recipient countries. With expansion of trade, however, international financial systems expanded. For example, at a much later stage, as the Industrial Revolution spread out of Britain, the international financial markets increased in importance.

As economic activity expanded to the New World, international financial transactions supported it and international financial centres developed in the USA. Towards the end of the 19th century, France and Germany succeeded in developing international financial centres of their own. Paris and Berlin emerged as major financial centres.

However, following the first world war, trade barriers were erected and currencies were devalued in a competitive manner. The Great Depression of the 1930s and the second world war created crises and instability in the global economy. This was a period of economic and financial reverses. Many Latin American economies defaulted on their foreign loans. Financial flows shrank to just 1.5% of national income.

After the Second World War, most countries had restrictions over foreign investment. Cross-country capital movements reached and remained at their historical low levels in the 1950s and failed to pick up during the 1960s. Only the OECD economies and NICs participated in the slowly developing global financial markets.

The contemporary era of financial globalization began with the oil shock of 1973. The large current account surpluses earned by the members of OPEC (the Organization of Petroleum Exporting Countries) could not be invested in rich countries immediately, because of the restrictions. A good part of the surpluses was recycled to developing economies.

Since the mid-1980s, middle- and high-income countries liberalized their trade policy regimes and tried to integrate with the global economy. Several developing economies were highly successful in integrating with the global economy through trade.

In a financially integrated world, capital movements easily and rapidly take place from where capital is to where it is needed. The distribution of global capital among the recipient economies is highly uneven. Some economies, such as China and those in east Asia and Latin America, have easy access and receive large amounts of global capital resources, while others, such as those in south Asia (India being an exception in this group), have limited access. Many, for example the African economies, have not been able to attract any global capital.

Low-income developing economies receive very little net global capital, while some does go to the middle-income developing economies. The lion's share of global capital flows are attracted by a top 12 of recipient countries, namely Argentina, Brazil, Chile, China, India, Indonesia, Korea (Republic of), Malaysia, Mexico, Russian Federation, Thailand and Turkey.

STATE OF THE WORLD'S FINANCIAL FLOWS, MID-2000s

In 2005, the world's financial assets reached a record \$140 trillion worth of stocks, bonds and other financial assets, more than three times as large as the total worldwide GDP. This was an increase of \$7 trillion from a year earlier. Eurozone countries added \$3.3 trillion of assets in 2005, reflecting a 6% annual growth rate over 10 years.

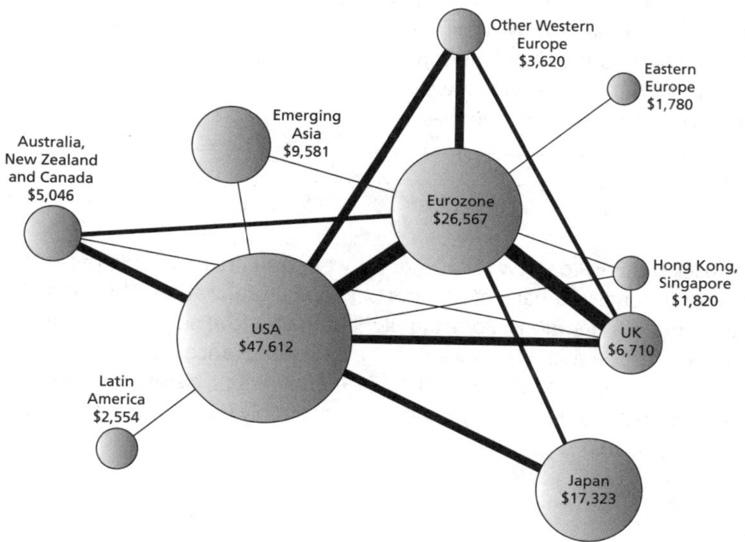

A worldwide web – how financial assets link different corners of the world

Source: *Wall Street Journal.* **Data source:** McKinsey & Co

Financial flows (4)

FOREIGN DIRECT INVESTMENT

The map of global FDI shows a very varied pattern. The countries with the highest FDI with respect to GDP are Ireland, Germany, Angola and Nicaragua. Much of Europe and South America have high levels of FDI, whereas Africa, the Middle East and south-west Asia have relatively low levels of FDI. However, when the total amount of FDI is considered (as a proportion of the world total) a different pattern emerges (see table). Highest investments are in rich countries and emerging economies, while lowest investment is in poor countries and countries that are politically isolated from the rich countries.

Highest (% of world total)	Lowest (% of world total)
USA 16.75	Kenya 0.01
UK 7.54	Sri Lanka 0.02
China 5.79	Iran 0.02
France 5.22	Kuwait 0.03
Belgium 4.78	Cuba 0.04

Foreign direct investment inflows, 2007–11

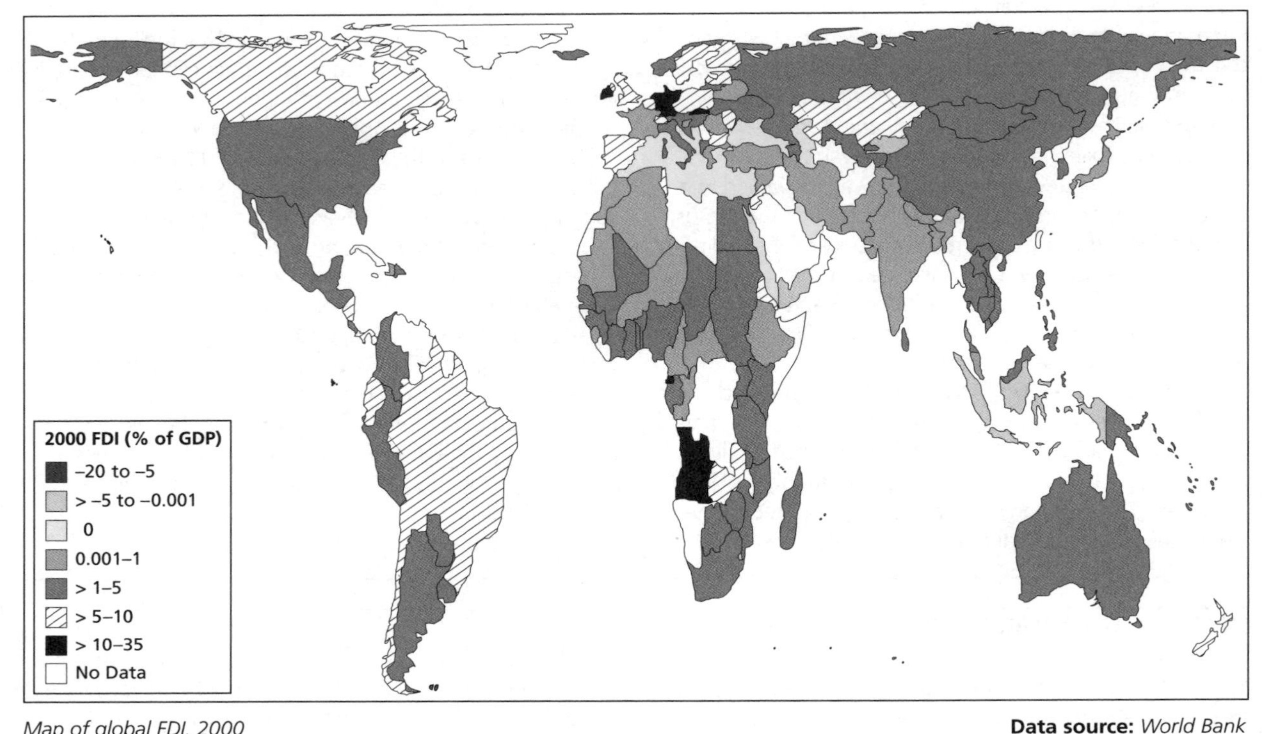

Map of global FDI, 2000

Data source: *World Bank*

Nevertheless, there is evidence of change. From a low initial level of less than $25 billion in 1990, net inflows of FDI to developing countries increased tenfold by 2005. The top 10 receivers of FDI net inflows accounted for about two-thirds of total FDI inflows among developing economies in 2005. FDI inflows are dominant in Latin America and the Caribbean, and in east Asia and the Pacific. Meanwhile, some developing economies are increasingly investing overseas to expand their global operations.

EXTENSION

Choropleth maps

A choropleth is a map that uses shading to show relative density per unit area – people per km^2 is a common choropleth map. Choropleths can be used to represent percentage and per capita information. They produce a striking visual impact. Nevertheless there are important considerations. For example, the map above suggests uniform conditions throughout the USA or Australia. It exaggerates the role of boundaries e.g. between France and Spain. Data can only occur in one category. Groupings can be in arithmetic intervals (e.g. 0–4, 5–9, 10–14 etc.), geometric intervals (e.g. 1–2, 3–4, 5–8, 9–16, 17–32 etc.) or at "natural breaks", by dividing the data into roughly equal groupings and using statistical variations, such as mean and standard deviation.

Financial flows (5) – loans and debt

INTERNATIONAL AID

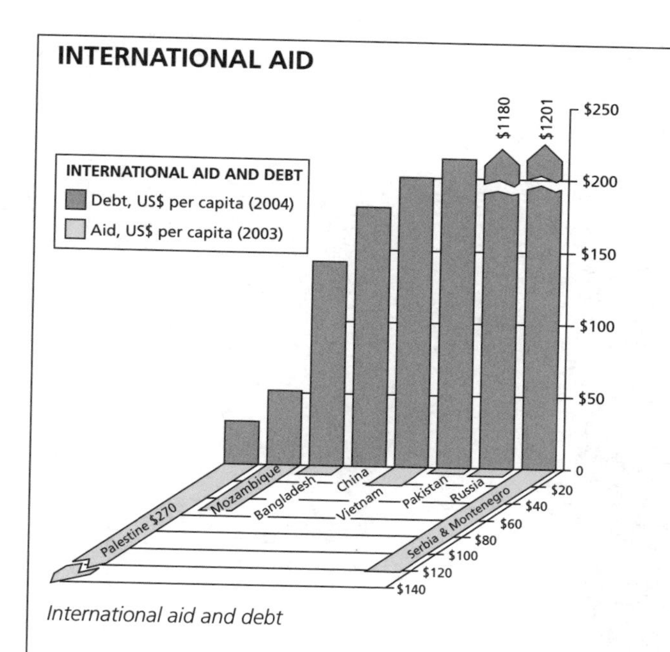

International aid and debt

The graph shows that there is a wide range in the level of debt per person. The highest debt is found Serbia and Montenegro, closely followed by Russia. There is also a high level of debt in poor countries such as Bangladesh and Mozambique. In addition, some rapidly developing countries such as China and Vietnam have high levels of debt.

The relationship between debt and aid is complex. Of the countries shown on the graph, Palestine, which has the lowest debt per person, receives the most aid, whereas Russia receives very little aid despite its massive debt. Mozambique appears to receive about the same amount of aid as its level of debt. In contrast, China appears to receive very little aid per person. It is important to remember that the statistics here will be affected by population size – the relatively small populations in Palestine and Serbia–Montenegro may inflate their figures compared with the countries that have very large populations, such as China, Pakistan, Bangladesh and Russia.

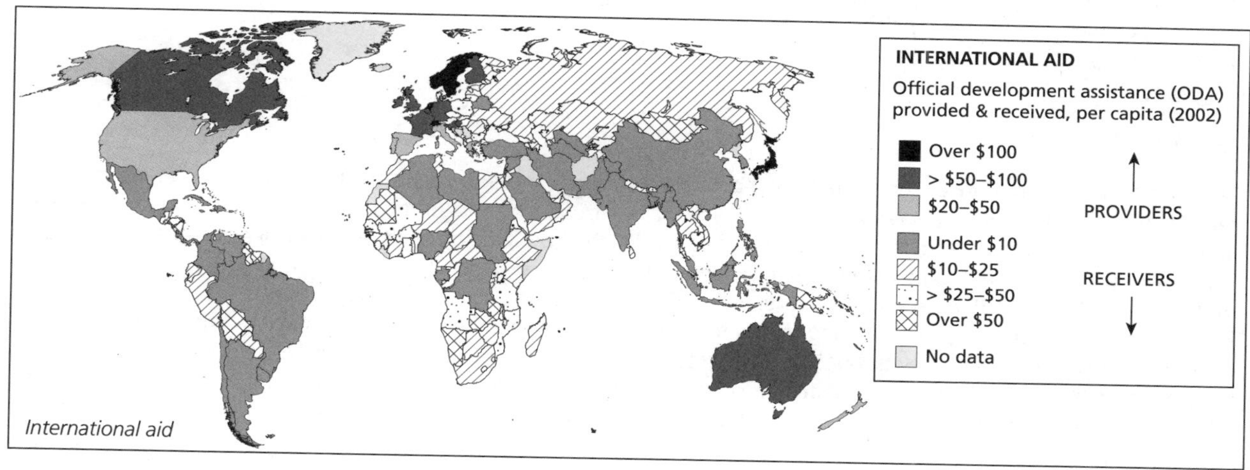

International aid

The world map shows that the main donors are the rich countries in North America, Europe, Australia, New Zealand and Japan. In contrast, the main recipients are in the poor countries. The highest levels of receipts would appear to be in much of sub-Saharan Africa, eastern Europe and Russia, and in South-East Asia.

The largest donors are the USA and Japan, although as a proportion of their GNI each donates less than 0.25%. France and the UK are the next largest donors, donating less than 0.5% of their GNI. The largest donors (in relation to GNI) are the Scandinavian countries, Norway, Denmark and Sweden.

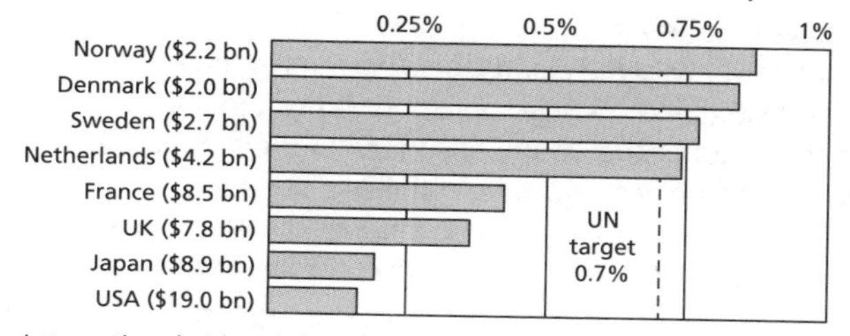

International aid and GNI

EXPANDING AID AND INCREASING EMPHASIS ON EFFECTIVE AID

Rich countries have committed to providing more and better aid, especially to the poorest economies that commit themselves to poverty reduction and good governance. After a period of decline and stagnation, aid flows began to rise, particularly after the Financing for Development conference in Monterrey, Mexico, in 2002. Total official development assistance (ODA) rose to a record high of $106.8 billion in 2005.

A large amount of aid is earmarked for special purposes such as debt relief, technical cooperation and administrative costs, and emergency relief and food aid.

Financial flows (6)

REMITTANCES

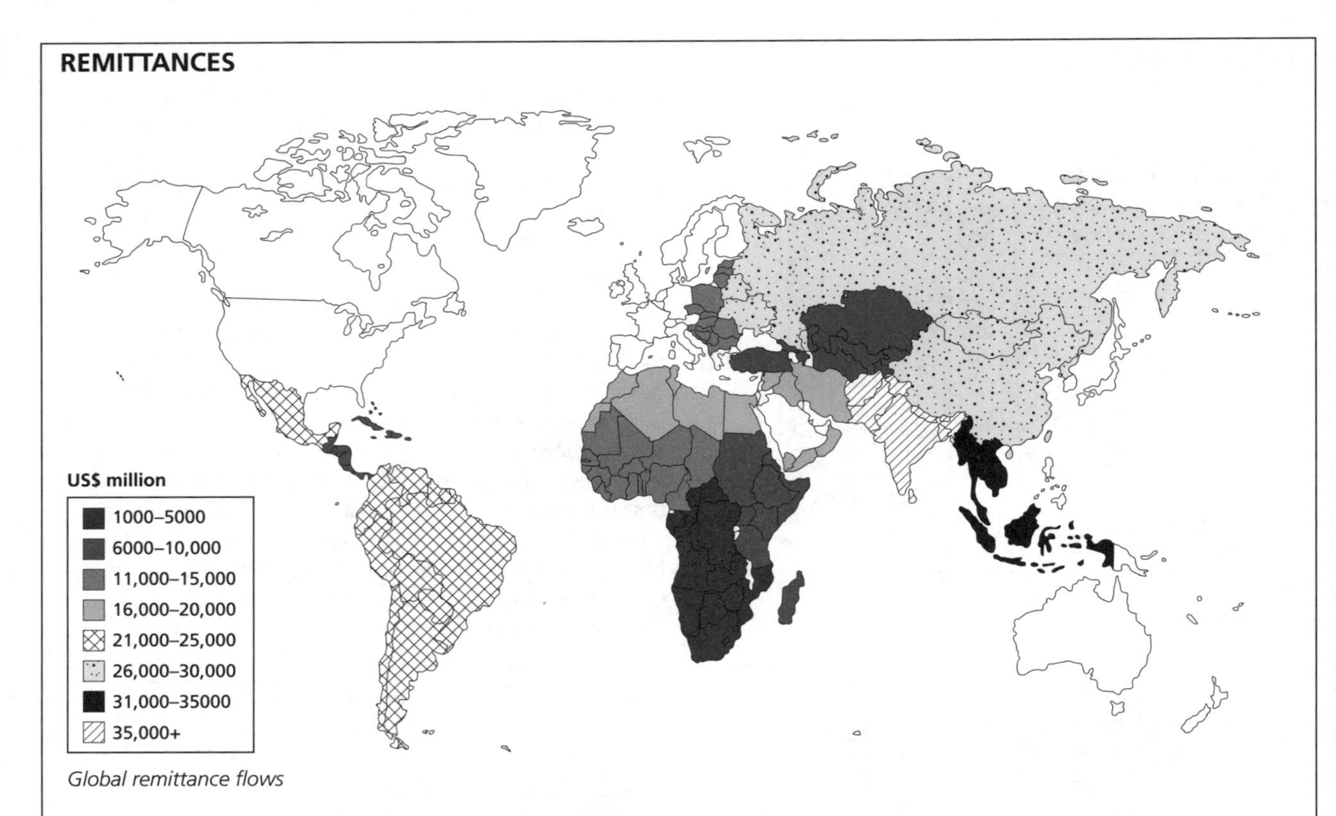

Global remittance flows

The map of global remittances shows that the region that receives the most in remittances is south Asia, in particular India, Pakistan and Bangladesh. In these countries the value of remittances is said to be greater than the amount of international aid that they receive. Countries in South-East Asia, such as Indonesia, Malaysia and Vietnam, receive a considerable amount of money through remittances. In contrast, most of Africa and the Caribbean receive a relatively small amount of remittances. Sub-Saharan Africa appears to be worst off. The pattern is different from the usual rich–poor divide in a number of ways; for example, the low value of remittances received in eastern Europe and in an arc of countries through Turkey to Kazakhstan makes this pattern unusual.

The main rich countries in North America, Europe, Japan, Australia and New Zealand do not show up on this map as they are the main source of income. Nevertheless, there could be some remittances between these countries.

THE VALUE OF REMITTANCES

The value of remittances to individual countries is impressive, with India and China each receiving over $20 billion in 2005. The source of Mexican money is without doubt the USA, while much of the remittances to the Philippines comes from the UK.

Some countries are very dependent on remittances. Nearly one-quarter of Haiti's GDP comes from remittances and in Jordan it is over one-fifth. In the Philippines, not only do remittances bring in a huge amount of income, about $13 billion, they also account for 13.5% of GDP. In India and China, the two largest recipients, remittances in 2005 accounted for 3.1% and 1.3% of GDP respectively.

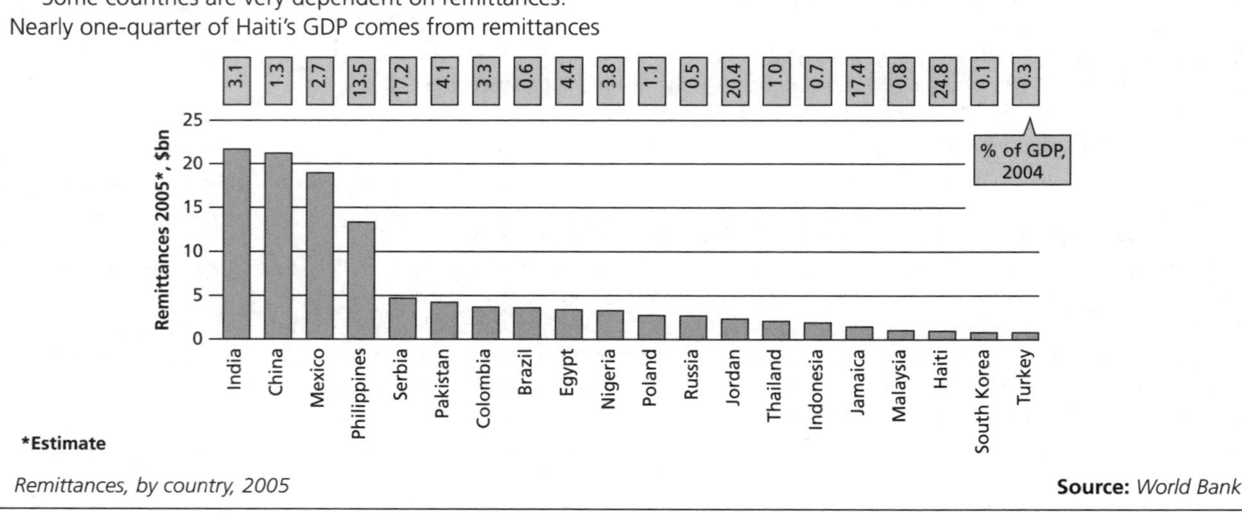

Remittances, by country, 2005

Source: *World Bank*

Labour flows

MEXICAN MIGRATION TO THE USA

The share of the USA's foreign-born population represented by Mexican immigrants doubled from 7.9% in 1970 to 15.6% in 1980, and then almost doubled again to 30.7% by 2006. In 2006, more than 11.5 million Mexican immigrants resided in the USA, accounting for one-tenth of the entire population of Mexico.

While Mexican immigrants are still settling in "traditional" destination states such as California and Texas, over the last 10 to 15 years, the foreign-born from Mexico, like other immigrant groups, have begun moving to "non-traditional" settlement areas. These include states in the south, such as Georgia and North Carolina, as well as Midwestern states, such as Nebraska and Ohio.

More than 83% of the Mexican-born reside in just 10 US states. In 2006, California had the largest number of foreign-born residents from Mexico, followed by Texas and Illinois. Nevertheless, more than 7 in 10 immigrants residing in the state of New Mexico were born in Mexico.

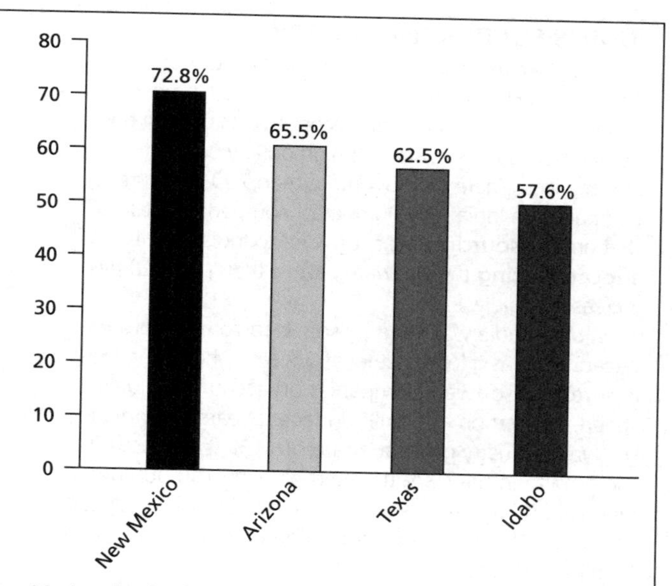

Mexican foreign-born as a percentage of all immigrants, by US state

Causes of movement

The migration of Mexicans to the USA is a classic example of push–pull factors. The negative push factors within Mexico include poor job opportunities, low wages, high unemployment and relatively low standards of living. In contrast, the "perceived" advantages of the USA include better job opportunities, better wages, better schools and healthcare, and an all-round improvement in standards of living.

Consequences

There are many consequences – both advantages and disadvantages – for both the source (Mexico) and the host (USA). For the USA, the migrants are a source of cheap labour and fill many of the jobs that US citizens do not want, especially unskilled low-paid jobs. On the other hand, there are tensions in areas with large numbers of migrants, especially in areas where unemployment among US citizens is above average.

For Mexico, the migrants are a major source of remittances. However, there is a drain of the younger, more skilled, more educated workforce, and this has a negative effect on the Mexican workforce.

DEMOGRAPHIC AND SOCIO-ECONOMIC OVERVIEW OF MEXICAN MIGRANTS IN THE USA

Of the total 11.5 million Mexican foreign-born in the USA in 2006, 27.9% entered the country in 2000 or later, 34.1% between 1990 and 1999, 20.4% between 1980 and 1989, 11.5% between 1970 and 1979, and the remaining 6.0% entered prior to 1970.

Three-quarters of Mexican immigrants in 2006 were adults of working age. Of the Mexican immigrants residing in the USA in 2006, 10.1% were minors (under age 18), 78.3% were of working age (between ages 18 and 54), and 11.6% were seniors (age 55 or older).

Of all Mexican immigrants residing in the country in 2006, 55.9% were men, while women accounted for 44.1%. In 2006, more than half of all unauthorized immigrants in the USA were from Mexico.

Nearly 75% of Mexican immigrants in 2006 were limited English proficient. Only 2.9% of the 11.4 million Mexican immigrants aged 5 and older reported speaking "English only", while 22.7% reported speaking English "very well". In contrast, 74.5% reported speaking English less than "very well", which is higher than the 52.4% reported among all foreign-born residents aged 5 and older.

Three in five Mexican immigrants had no high school degree. In 2006, 60.2% of the 8.9 million Mexican-born adults aged 25 and older had no high school diploma or the equivalent general education diploma (GED), compared to 32.0% among the 30.9 million foreign-born adults.

Employment characteristics

Mexican immigrant men were more likely to participate in the civilian labour force than foreign-born men overall and Mexican immigrant women. Forty per cent of Mexican-born men were employed in construction, extraction or transportation occupations. Both Mexican foreign-born men and women were significantly less likely to be employed as managers, scientists or engineers than foreign-born men and women overall, but they were more likely to be working in service or farming occupations.

The role of ICT in international outsourcing

INDIA'S SOFTWARE INDUSTRY

India's software export industry is worth more than $1 billion each year. It has become one of the most dynamic sectors of the Indian economy. Its growth has been based on low costs, but high quality products and services. There are now more than 700 software companies in India. The number of companies in MEDCs that are **outsourcing** their software to India (that is, **subcontracting** the software part of their product) has increased rapidly.

Initially, India was used by software companies because it was a low-cost location. Now, however, India is attracting software companies on account of quality, speed, innovation and skills. In recent years the Indian software industry grew at a rate of 46%, twice as fast as the growth in the USA. It employs nearly 150,000 people in India, and its exports are worth over 25 billion rupees each year. The Indian domestic market is worth a further 17 billion rupees.

The software industry is an attractive one for many countries.

- It demands high skills.
- It does not damage the environment.
- It is a growth industry.
- There is a great deal of investment money available.

A number of factors explain why India has done so well and how it has outshone competition from China, the Philippines and eastern Europe. These include the availability of a huge pool of relatively low-cost, technically qualified software professionals, high quality levels and a time zone advantage with both the USA and Europe. In addition, there have been attempts to improve India's telecommunications.

In the USA and Europe, there has been a growing shortage of software engineers. After the USA, India has the largest number of English-speaking scientific manpower. Combined with the trend towards outsourcing (subcontracting), there has been an ever-increasing market for Indian software. There is no shortage of software entrepreneurs and innovators. The sheer size of the workforce, its technical competence and relative low cost have been paramount in explaining the development of the software industry in India.

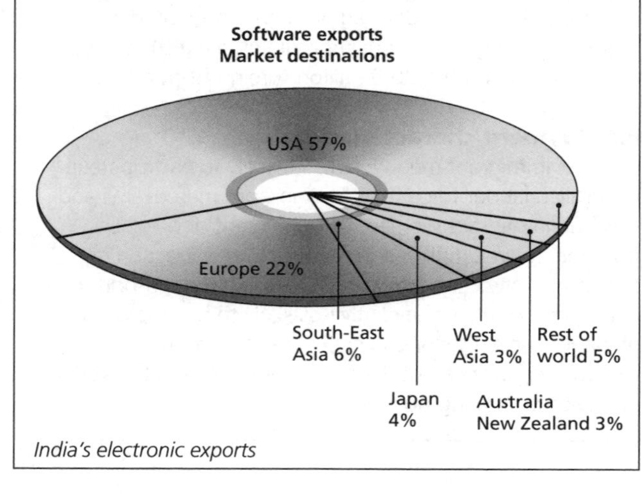

India's electronic exports

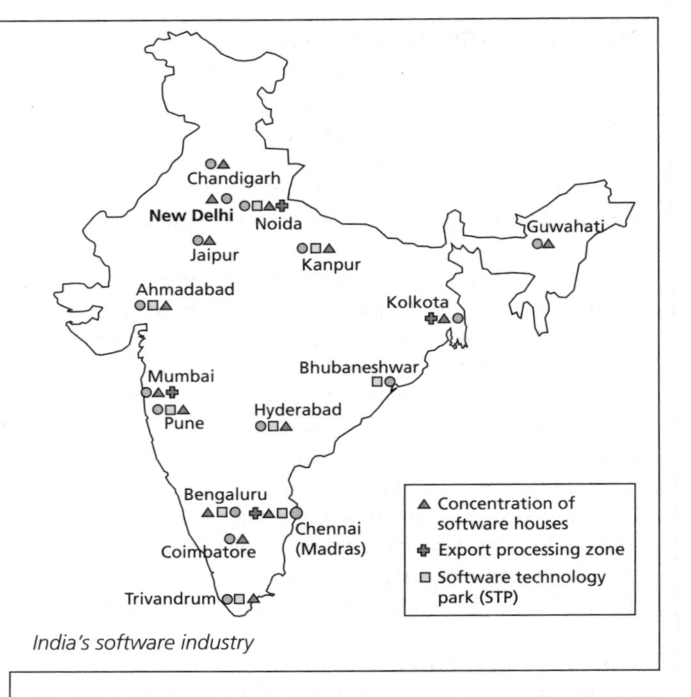

India's software industry

BENGALURU

Bangalore (now called Bengaluru) has been described as the "silicon plateau" of India. It is home to a cluster of high-technology firms – IBM, Hewlett Packard and Motorola. Bengaluru has attracted investors for a number of reasons.

- There is a skilled workforce.
- The city contains a number of research institutions and universities.
- Compared to the West, it offers low labour costs – a first-class graduate can be recruited for as little as 12,000 rupees a month (about £2800 a year).
- Bengaluru has low rainfall and pleasant temperatures, on account of its plateau location.
- India is an important base for western firms trying to enter the Asian markets.

One of the companies that has located there is Motorola, the US electronics and equipment company. In order to overcome the power cuts that plague India, Motorola has its own generator. It chose Bengaluru for a number of reasons.

- It is the hi-tech centre of India.
- Other US multinationals, such as Hewlett Packard and 3M, have located there.
- There is high quality but relatively cheap labour.
- It wanted a foothold in the expanding Indian market.

EXTENSION

Visit
http://news.bbc.co.uk/1/hi/world/south_asia/6107082.stm for an article on Bangalore/Bengaluru and to find out why not everyone is happy about the name change.

The effects of agro-industrialization on the environment

THE FOOD INDUSTRY

Food has gone global. Farming has become increasingly intensive, large scale and globalized in the drive for cheaper food. Advances in technology and communications have combined with falls in the costs of transport to transform the way in which food is sourced. The concentration of power in retailing and food processing has affected those at the other end of the scale, namely farmers in LEDCs and small farmers in MEDCs. Increasingly, modern farming methods are having a negative impact on the environment.

Since the 1950s, a revolution has taken place in the food industry. Every step in the process – how food is grown, harvested, processed, distributed, retailed and cooked – has changed. Until the Second World War, farmers were the major players in the food industry. After the war, they were given grants and subsidies, but these were merely to stop them going out of business. Many therefore intensified, increased efficiency and adopted labour-saving technologies such as agro-chemicals, machinery and high-yielding varieties (HYVs) of plants.

Improved yields and environmental impacts

In the last 50 years, wheat yields have increased from 2.6 to 8 tonnes per hectare, barley from 2.6 to 5.8 tonnes, and each cow produces twice as much milk.

Cleaning up the chemical pollution, repairing the habitats and coping with sickness caused by industrial farming costs up to £2.3 billion a year. It now costs water companies £135–200 million to remove pesticides and nitrates from drinking water.

Food processors usually want large quantities of uniform quality produce or animals at specific times. This is ideally suited to intensive farming methods, which favour synthetic chemicals and lead to land degradation and animal welfare problems.

Animals are reared on production lines. The spread of disease is a problem. In just two decades, new production methods have increased a dairy cow's average yield from 4000 litres to 5800 litres a year.

Cox's apples receive an average of 16 pesticide sprays. Lettuces imported to the UK from Spain, Turkey, Zimbabwe and Mexico are sprayed on average 11.7 times.

Air pollution and greenhouse gas emissions from farming cost more than £1.1 billion annually. About 10% of the UK's greenhouse gas emissions come from the methane from livestock digestion and manure, and nitrous oxide from fertilized land.

In the last 60 years farmers in the UK have ripped up about 190,000 miles of hedgerow, and destroyed 97% of meadows and 60% of ancient woodland; farmland birds have suffered a catastrophic decline. Birds that depend on agricultural fields have fallen in numbers by as much as 50% since 1970.

It is estimated that a kilogram of blueberries imported by plane from New Zealand produces the same emissions as boiling a kettle 268 times. Intensive farming in the UK has led to soil erosion and soil loss and increased the risk of flooding in some areas by 14%. This has added up to £115 million to insurance bills.

WATER PROBLEMS AND GLOBAL FARMING IN KENYA

The shores of Lake Naivasha in the "Happy Valley" area of Kenya are now blighted. Environmentalists blame the water problems on pollution from pesticides, excessive use of water on the farms, and deforestation caused by migrant workers in the growing shanty towns foraging for fuel.

British and European-owned flower companies grow vast quantities of flowers and vegetables for export, but the official Kenyan water authority, regional bodies, human rights and development groups, as well as small-scale farmers, have accused flower companies near Mount Kenya of "stealing" water which would normally fill the river. Kenya's second largest river, the Ngiro, is a life-sustaining resource for nomadic farmers, but it also sustains big business for flower farms supplying UK supermarkets.

According to the head of the water authority, the 12 largest flower firms may be taking as much as 25% of water normally available to more than 100,000 small farmers.

The flower companies are thereby exporting Kenyan water – this is known as "virtual water". A flower is 90% water. Kenya is one of the driest countries in the world and is exporting water to some of the wettest. The flower companies are in direct competition with the peasant farmers for water and the biggest companies pay the same as the smallest peasant for water.

The greatest impact is being felt on the nomadic pastoralists in the semi-arid areas to the north and east of Mount Kenya. The flower farms have taken over land that the pastoralists used and there is now less water.

EXTENSION

Visit
www.fao.org/DOCREP/005/Y4383E/y4383e0d.htm for a detailed account of the effects of agro-industrialization on the valleys of Chincha and Mantaro, Peru.

EXTENSION

Visit
http://www.financialexpress.com/news/agroindustry-to-boom-in-global-mkt/294399/ for a discussion of the role of agro-industrialization.

Environmental degradation

MINING OF RAW MATERIALS

Mined materials are normally classified in four groups:

- metals, such as iron ore and copper
- industrial minerals, such as lime and soda ash
- construction materials, such as sand and gravel
- energy minerals, such as coal, oil and natural gas.

Construction minerals are the largest product of the mining industry, being found and extracted in almost every country.

The environmental impacts of mining are diverse. Habitat destruction is widespread, especially if opencast or strip mining is used. Disposal of waste rock and "tailings" may destroy vast expanses of ecosystems. Copper mining is especially polluting – to produce 9 million tonnes of copper (world production levels in the 1990s), about 990 million tonnes of waste rock are created. Even the production of 1 tonne of china clay (kaolin) creates 1 tonne of mica, 2 tonnes of undecomposed rock, and 6 tonnes of quartz sand. Smelting causes widespread deforestation. The Grande Carajas Project in Brazil removes up to 50,000 ha of tropical forest each year.

There is widespread pollution from many forms of mining. The pollution results from the extraction, transport and processing of the raw material, and affects air, soil and water. Water is affected by heavy metal pollution, acid mine drainage, eutrophication and deoxygenation. Moreover, dust can be an important local problem. The use of mercury to separate fine gold particles from other minerals in river bed sediments leads to contamination in many rivers. In Brazil, up to 100 tonnes of mercury have been introduced into rivers by gold prospectors. Mercury is highly toxic and accumulates in the higher levels of the food chain, and can enter the human food chain.

Derelict land that results from extraction produces landforms of various size, shape and origin. A major subdivision is between excavations and heaps. The latter can be visually intrusive and have a large environmental impact. Heaps include those composed of blast furnace slag, fly-ash from power stations, as well as spoils of natural materials (overmatter), such as the white cones associated with china clay workings, oil shale wastes, and colliery spoil heaps.

Some environmental problems associated with mining

Problem	Type of mining operation			
	Open pit and quarrying	Opencast (as in coal)	Underground	Dredging (as in tin or gold)
Habitat destruction	X	X	–	X
Dump failure/erosion	X	X	X	–
Subsidence	–	–		X
Water pollution	X	X	X	X
Noise	X	X	–	–

X Problem present

– Problem unlikely

** Can be associated with smelting which may not be at the site of ore/mineral extraction*

Source: *Middleton, N.* The Global Casino. *Edward Arnold, 1995*

ENVIRONMENTAL IMPACTS OF INCREASED AIR TRAVEL

Transport as a whole produces about 25% of the world's CO_2 discharges. Within transport, aviation accounts for about 13%. Surface transport, by contrast, produces 22%. CO_2 emissions from shipping are double those of aviation and are increasing at an alarming rate.

Airlines' emissions are especially damaging because the nitrogen oxides from jet-engine exhausts help create ozone, a greenhouse gas, and because the trails that aircraft leave behind them help make the clouds that can intensify the greenhouse effect.

A ban on night flights would significantly reduce the impact on climate. Warming is much greater when aircraft fly in the dark, because of the effects of condensation trails (contrails). Aircraft contrails enhance the greenhouse effect because they trap heat in the same way as clouds. During the day, contrails reflect sunlight back into space, which helps to keep the planet cool. Contrails are responsible for about half of the aviation industry's impact on climate. Although one in four flights occurred between 6 p.m. and 6 a.m., they contributed 60–80% of the warming that could be attributed to contrails. Winter flights had more effect than those in the summer, contributing 50% of the warming despite providing only 22% of traffic.

Polluting industries and relocation to LEDCs

RELOCATION OF POLLUTING INDUSTRIES

Maquiladora development in Mexico

Mexico has attracted many US-owned companies to build low-cost assembly plants in places such as Ciudad Juarez, Nuevo Laredo and Tijuana. These factories are called *maquiladora* operations, as they are foreign owned but employ local labour. Since 1989, over 2000 US firms have set up in Mexico's border cities. The main attractions are:

- low labour costs
- relaxed environmental legislation
- good access to US markets.

Although firms are required by Mexican law to transport hazardous substances back to the USA, illegal dumping in Mexico is common. Air and water pollution are increasingly common. Despite the environmental problems, many Mexicans are in favour of the *maquiladora*, as it brings investment, money and jobs to northern Mexico. Over 500,000 people are employed in these factories.

Environmentalists point to Mexico's poor record of enforcing environmental laws. They fear that Mexico may become a dumping ground for hazardous material and show that Mexico's rivers, such as the Rio Grande, and air are already heavily polluted.

A study that investigated the relationship between *maquiladoras*, air pollution and human health in Paso del Norte found that particulate emissions from *maquiladoras* undoubtedly have significant impacts on human health, in particular respiratory disease. However, it found that particulate emissions generate health damages of similar magnitudes regardless of the source, and *maquiladoras* are clearly *not* the region's leading source of particulates. Unpaved roads, vehicles and brick kilns were the main sources of particulate emissions. Given that vehicles and brick kilns emit far more combustion-related fine particulates than *maquiladoras*, they inflict more health damages. The study found no evidence that health damages attributable to *maquiladoras* disproportionately affected the poor. However, brick kilns were far more likely in poor areas.

RELOCATION OF WASTE

Some countries export their waste to others, notably MEDCs to LEDCs.

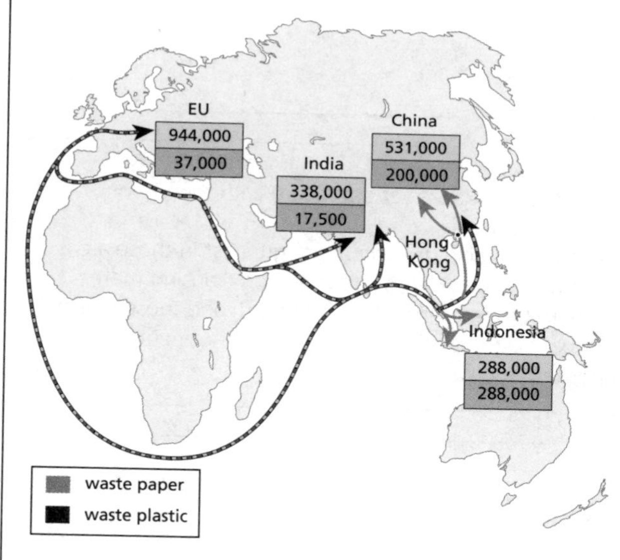

UK waste exports (tonnes)

China imports more than 3 million tonnes of waste plastic and 15 million tonnes of paper and cardboard each year. Containers arrive in the UK and other countries with goods exported from China, and load up with waste products for the journey back. A third of the UK's waste plastic and paper (200,000 tonnes of plastic rubbish and 500,000 tonnes of paper) is exported to China each year. Low wages and a large workforce mean that this waste can be sorted much more cheaply in China, despite the distance it has to be transported.

China is increasingly aware that this is not "responsible recycling" and that countries are exporting their pollution to them. They have begun to impose stricter laws on what types of waste can be imported.

A fairly new environmental problem is the dumping of old computer equipment. To make a new PC requires at least 10 times its weight in fossil fuels and chemicals. This can be as much as 240 kg of fossil fuels, 22 kg of chemicals and 1500 kg of clean water. Old PCs are often shipped to LEDCs for recycling of small quantities of copper, gold and silver. PCs are placed in baths of acid to strip metals from the circuit boards, a process highly damaging to the environment and to the workers who carry it out.

EXTENSION

The χ^2 test

The χ^2 is one of the most widely used tests of association. It is used to test whether an observed pattern (o) differs significantly from an expected pattern (ε). For example, a hypothesis might be set up stating that the UK exports its waste evenly to the four receiving areas. If so, each area would expect to receive about 523 000 tonnes of wate paper. Using the formula $\chi^2 = \Sigma(o - \varepsilon)^2/\varepsilon$ we get an answer of 508. Looking at the level of significance we can be 99.9% sure that there is a statistically significant difference in the volume of waste delivered to the four areas.

Transboundary pollution: acid rain

THE CAUSES

Acid rain – or, more precisely, acid deposition – is the increased acidity of rainfall and dry deposition, as a result of human activity. Rain is naturally acidic, owing to carbon dioxide in the atmosphere, with a pH of about 5.6. The pH of "acid rain" can be a low as 3.0.

The major causes of acid rain are the sulphur dioxide and nitrogen oxides produced when fossil fuels such as coal, oil and gas are burned. Sulphur dioxide and nitrogen oxides are released into the atmosphere, where they can be absorbed by the moisture and become weak sulphuric and nitric acids, sometimes with a pH of around 3. Most natural gas contains little or no sulphur and causes less pollution.

Coal-fired power stations are the major producers of sulphur dioxide, although all processes that burn coal and oil contribute. Vehicles, especially cars, are responsible for most of the nitrogen oxides in the atmosphere. Some come from the vehicle exhaust itself, but others form when the exhaust gases react with the air. Exhaust gases also react with strong sunlight to produce poisonous ozone gas which damages plant growth and, in some cases, human health.

DRY AND WET DEPOSITION

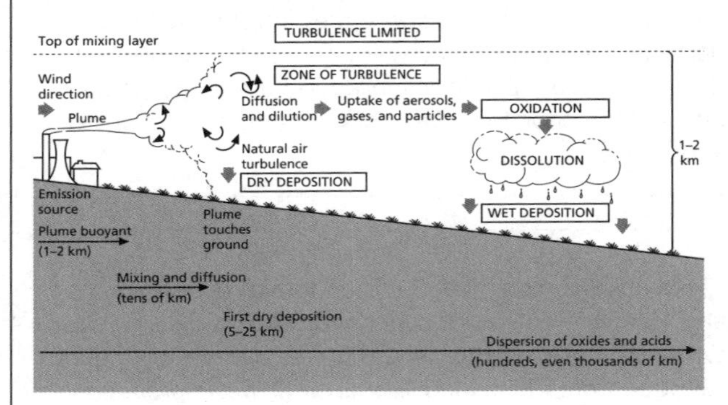

Dispersion and deposition

Dry deposition typically occurs close to the source of emission and causes damage to buildings and structures.

Wet deposition, by contrast, occurs when the acids are dissolved in precipitation, and may fall at great distances from the sources. Wet deposition has been called a "trans-frontier" pollution, as it crosses international boundaries with disregard.

THE EFFECTS

Acidification has a number of effects:

- Buildings are weathered.
- Metals, especially iron and aluminium, are mobilized by acidic water, and flushed into streams and lakes.
- Aluminium damages fish gills.
- Forest growth is severely affected.
- Soil acidity increases.
- There are links (as yet unproven) with the rise of senile dementia.

The effects of acid deposition are greatest in those areas which have high levels of precipitation (causing more acidity to be transferred to the ground) and those which have base-poor (acidic) rocks which cannot neutralize the deposited acidity.

THE SOLUTIONS

Various methods are used to try to reduce the damaging effects of acid deposition. One of these is to add powdered limestone to lakes to increase their pH values. However, the only really effective and practical long-term treatment is to curb the emissions of the offending gases. This can be achieved in a variety of ways:

- by reducing the amount of fossil fuel combustion
- by using less sulphur-rich fossil fuels
- by using alternative energy sources that do not produce nitrate or sulphate gases (e.g. hydropower or nuclear power)
- by removing the pollutants before they reach the atmosphere.

However, while victims and environmentalists stress the risks of acidification, industrialists stress the uncertainties. For example:

- rainfall is naturally acidic
- no single industry/country is the sole emitter of SO_2/NOx
- more cars have catalytic converters
- different types of coal have variable sulphur content.

Environmental awareness

There are many active players in the work of environmental awareness and conservation. These include:

- individuals (e.g. Gerald Durrell and the Durrell Wildlife Conservation Trust)
- groups (e.g. Greenpeace and the World Wide Fund for Nature)
- public servants, such as politicians and scientists (e.g. Al Gore, the former US vice president and author of *An Inconvenient Truth*; and Dian Fossey, made famous by the film *Gorillas in the Mist*).

In most cases there is a conflict between the need for economic development and the need for environmental conservation or management. Dian Fossey argued for the protection of the mountain habitats in Rwanda and Burundi that are home to the great silverback gorillas. On the other hand, population growth, civil conflict and the illegal trade in forest products led to a decline in forest cover and a reduction in the gorillas' habitat.

There is an urgent need for strategic thinking and planning, especially in some of the world's most valuable biomes, such as coral reefs. This needs to be done in a sustainable way, with the cooperation of the indigenous people.

THE ROLE OF GREENPEACE

Greenpeace is an international environmental organization founded in Vancouver, Canada in 1971. Its confrontational approach has secured it a high public profile, and helped develop strong support for the organization. It has tackled many issues, such as waste disposal, deforestation, nuclear power, harvesting of seal cubs and industrial pollution. Greenpeace's goal is "to ensure the continuing ability of the earth to nurture life in all its diversity". It has a presence in over 40 countries.

GREENPEACE

Greenpeace defines its mission as follows:

Greenpeace is a global campaigning organization that acts to change attitudes and behaviour, to protect and conserve the environment and to promote peace by:

- Catalyzing an energy revolution to address the number one threat facing our planet: climate change.
- Defending our oceans by challenging wasteful and destructive fishing, and creating a global network of marine reserves.
- Protecting the world's remaining ancient forests and the animal, plants and people that depend on them.
- Working for disarmament and peace by reducing dependence on finite resources and calling for the elimination of all nuclear weapons.
- Creating a toxic-free future with safer alternatives to hazardous chemicals in today's products and manufacturing.
- Supporting sustainable agriculture by encouraging socially and ecologically responsible farming practices.

Consequently, Greenpeace's main interests at present include:

- stopping climate change (global warming)
- preserving the oceans (including stopping whaling and seabed trawling
- saving ancient forests
- campaigning for peace and nuclear disarmament
- promoting sustainable farming (and opposing genetic engineering)
- eliminating toxic chemicals, including from electronic (E-) waste, many of which are cancerous (carcinogenic).

Greenpeace has been variously criticized, by governments, industrial and political lobbyists and other environmental groups, for being too radical, too mainstream (or not radical enough), for allegedly using methods bordering on ecoterrorism, for causing environmental damage, and for valuing non-human causes over human causes.

WORLD WIDE FUND FOR NATURE (WWF)

Formerly the World Wildlife Fund, the WWF was initially concerned with the protection of endangered species, but now includes all aspects of nature conservation, including landscapes (the environments in which species live). It has over 5 million supporters globally, and is increasingly concerned with the fight against environmental destruction.

The WWF is interested in climate change and global warming; forests; freshwater ecosystems; marine ecosystems; species and biodiversity; sustainability; agriculture; toxins; macroeconomic policies, and trade and investment.

The WWF works in recognized geographic areas, such as continents and countries, but also in ecoregions, large-scale geographic regions under threat from development. Examples include the Alps ecoregion and the Mekong ecoregion.

The WWF works with governments, NGOs, local peoples and businesses to find ways to protect the earth.

EXTENSION

Visit

www.greenpeace.org.uk/ to find out more about the work of Greenpeace in any selected country. Choose from the dropdown box to select the region of interest to you.

Homogenization of urban landscapes

UNIFORM URBAN LANDSCAPES?

Many urban landscapes look very similar. A stroll around Tokyo might include a visit to a McDonald's restaurant, just as a visit to Seoul could end up in Starbucks. Very tall towers are a feature of many cities, such as Toronto, Kuala Lumpur, Beijing and, of course, New York. Industrial estates and science parks are increasingly globalized, as TNCs outsource their activities to access cheap labour, vital raw materials, and potential markets.

Much appears to have changed about the city since the mid-1970s, with cities having undergone dramatic transformations in their physical appearance, economy, social composition, governance, shape and size.

So are urban areas around the world converging in form? Are we seeing a globalized urban pattern or do local and national characteristics still prevail?

Take Los Angeles, for example. In this city there is a dazzling array of sites in compartmentalized parts of the inner city: the Vietnamese shops and Hong Kong housing of Chinatown; the pseudo-Soho of artists' lofts and galleries; the wholesale markets; the urban homelessness in the Skid Row district; the enormous muraled barrio (shanty town) stretching eastwards towards east Los Angeles; the intentionally gentrifying South Park redevelopment zone. Many large cities have their Chinatowns and other ethnic/racial areas. Individual cities are anything but homogeneous. The point is that cities are increasingly globalized, increasingly heterogeneous, and that, as a result, cities are more similar now, because they are all diverse.

The post-industrial city is regarded as a more flexible, complex and divided city than its predecessor. The result is a patchwork city of different ethnic enclaves, economic areas and residential areas, where the boundaries between city and country (both physical and social) are difficult to define. Consequently, an array of new terms has emerged to describe the post-industrial city and its attendant spatial forms: the splintered city, the edgeless city, the urban galaxy, the spread city.

The Los Angeles school of geographers are very pessimistic about the development of their city, and see it as teetering on the verge of meltdown. They talk about the death of cities, ecological disaster, terrorism, inequality and dysfunction.

SEOUL – HOMOGENIZED CITY OR INDEPENDENT TRADER?

Seoul is a good example of the debate on the homogenization of urban landscapes. On the one hand, it fits the theory of a homogenized landscape – there are global firms (such as McDonald's) in Seoul, just as there are Korean firms such as Hyundai and Samsung located in other countries. The CBD is characterized by skyscrapers and international firms such as Barclays and Tesco. There are high-rise apartments and edge-of-town developments, and decentralization, such as at Gyeonggi-do and Pangyo on the south side of Seoul.

On the other hand, there has been a massive urban redevelopment centred on the restoration of the Cheong Gye Cheon River in downtown Seoul. This project has been not just the restoration of a river; it has a historic, cultural and touristic-economic value. Murals along the side of the river recount some of the most important events to occur in Seoul over the last 600 years and the river has become an important focus for Seoul residents and visitors – rather like Trafalgar Square in London – partly because it is stressing the individuality and uniqueness of Seoul, and of Korea.

The Cheong Gye Cheon

EXTENSION

Bias

When collecting data or information to use as "evidence" or to create a case study, it is important to consider how reliable the data are. It may be that printed material (including that on the web) is biased, so it reflects the viewpoint ofa particular group, and so is one-sided in its approach. (See the opposing views of Shell and some NGOs on page 185.)

There are a number of considerations here:

- Is the information valid? Is it authentic?
- Can it be substantiated by other sources?
- What are the advantages/disadvantages of this particular source?
- Who is the author of this source? Are they reliable? Are they an "expert"?
- What is the purpose of the source material?
- When and how was the source material collected?
- Was it done with a large organization with lots of capital available?

Cultural diffusion

CULTURE – A SUMMARY

- Culture denotes the systems of shared meanings which people who belong to the same community, group or nation use to help them interpret and make sense of the world.
- These systems of meanings include language, religion, custom and tradition, and ideas about "place".
- Cultures cannot be fixed, but shift and change historically.
- Culture is a process, rather than a thing, and is embodied in the material and social world.
- Cultures give us a sense of "who we are", "where we belong" – a sense of our own identity and identity with others.
- Cultures are, therefore, one of the principal means by which identities are formed.
- Cultures are not divorced from power relations.

A GLOBAL CULTURE?

It is commonly accepted that the world is changing fast, and the rate of this change is probably greater than ever before. New technologies, such as the internet and satellite communications, mean that the world is becoming more global and more interconnected. The increased speed of transport and communications, the increasing intersections between economies and cultures, the growth of international migration and the power of global financial markets, are among the factors that have changed everyday lives in recent decades.

Proponents of the idea of an emerging global culture suggest that different places and cultural practices around the world are converging and becoming ever more similar. A global culture might be the product of two very different processes:

- the export of supposedly "superior" cultural traits and products from advanced countries, and their worldwide adoption ("Westernization", "Americanization", "modernization")
- the mixing, or hybridization, of cultures through greater interconnections and time–space compression (the shrinking of the world through transport links and technological innovation), leading to a new universal cultural practice.

Music

Music lends itself to globalization because it is one of the few popular modes of cultural expression that is not dependent on written or spoken language for its primary impact. The production, distribution and consumption of music have a particular geography. The global music industry is dominated by TNCs, with the USA and the UK dominating domestically generated popular music. "World music" is now a significant component of the marketing strategies of these corporations, and exposes global audiences to local musical traditions from around the world. Migrations of people have also had cultural impacts on music, evidenced in increasingly "hybridized" forms.

Television

Until recently, television programmes tended to be produced primarily for domestic audiences within national boundaries, and could be subjected to rigorous governmental control. However, with the advent of cable, satellite and digital technologies, in addition to political and legal deregulation in many Western and developing states, several television channels are now globally disseminated, and to some extent circumvent national restrictions. The USA, France, Germany and the UK are major exporters of television programmes, while Brazil, Mexico, Egypt, Hong Kong and Spain are increasing their output.

Sport

Sports are forms of cultural expression that are becoming increasingly globalized, as well as increasingly commoditized. Football/soccer is the most obvious example, but similar trends can be observed in US Major League Baseball. The New York Yankees are a global icon; many Major League players hail from countries such as Cuba, the Dominican Republic, Puerto Rico and Costa Rica; the sport is becoming increasingly globalized through television coverage and its inclusion as an Olympic sport.

Tourism

Tourism is one of the most obvious forms of globalization. Once again, the geography of tourism is skewed, since it is dominated by people of all classes from rich countries. It can also be exploitative, particularly through the growth of international sex tourism and the dependency of some poor countries on the exploitation of women. However, it is a form of international cultural exchange that allows large numbers of people to experience other cultures and places. It also locks specific destinations into wider international cultural patterns.

EXTENSION

Visit

http://geography.about.com/od/culturalgeography/a/culturehearths.htm for a discussion on Culture Hearths and Cultural Diffusion and some useful links to other sites.

Consumer culture (1)

CONSUMERISM

Consumerism is the opposite (antithesis) and enemy of culture. Whereas culture is embodied in history, tradition and continuity, goods are manufactured for the profit they make. Consumerism represents the triumph of economic value over social worth. Everything can be bought and sold. Everything has its price.

"Every time you spend money, you're casting a vote for the kind of world you want."
Source: *Anna Lappe,* O Magazine, *June 2003*

MCDONALD'S RESTAURANTS

On an average day, over 30 million customers are served at one of more than 31,000 McDonald's restaurants in more than 100 countries. The world map shows that the first restaurants were located in the USA and Canada and then spread to Europe, Australia and Japan during the early 1970s. By the end of the 1970s, McDonald's were consolidating their position in Europe and New Zealand, and had opened restaurants in South America, namely in Brazil. The 1980s saw further expansion and consolidation in South America, Mexico, parts of Europe and South-East Asia. China, Russia and parts of the Arab world were reached only in the 1990s.

Over half these restaurants are in the USA, but the UK has over 600 outlets, Brazil over 250, China nearly 200, Thailand nearly 50. A promotional corporate statistic is that a new McDonald's restaurant opens somewhere in the world every three hours. Not only this, of course, but McDonald's are famed for their uniformity; the same decor, the same basic menu (with very small variations, including the McSpagetti in the Philippines!) and the same service style the world over. And yet McDonald's may not be just the force for cultural homogenization that this suggests. McDonald's has been localized, indigenized and incorporated into traditional cultural forms and practices. Exactly how this has happened varies across east Asia, for example. In Beijing, McDonald's has lost its American role as a place of fast and cheap food. Instead, it has become a middle-class consumption place, somewhere for a special family outing, somewhere where "customers linger for hours, relaxing, chatting, reading, enjoying the music". McDonald's here is seen as American, but Americana means something stylish, exotic and foreign, and as such actually results in the meanings and experiences of McDonald's in Beijing being very un-American! In contrast, in Japan, while there is a similar leisurely use of McDonald's, it is not a place of exotic social prestige, but a youth hangout, a place where someone in a business suit would be out of place. In Hong Kong, McDonald's was likewise marketed to the youth market. Today, McDonald's restaurants in Hong Kong are filled with people of all ages, few of whom are seeking an American cultural experience. The chain has become a local institution in the sense that it has blended into the urban landscape. McDonald's is not perceived as an exotic or alien institution. Hence the meanings and practices of McDonald's – an archetype of global homogenization – vary from place to place.

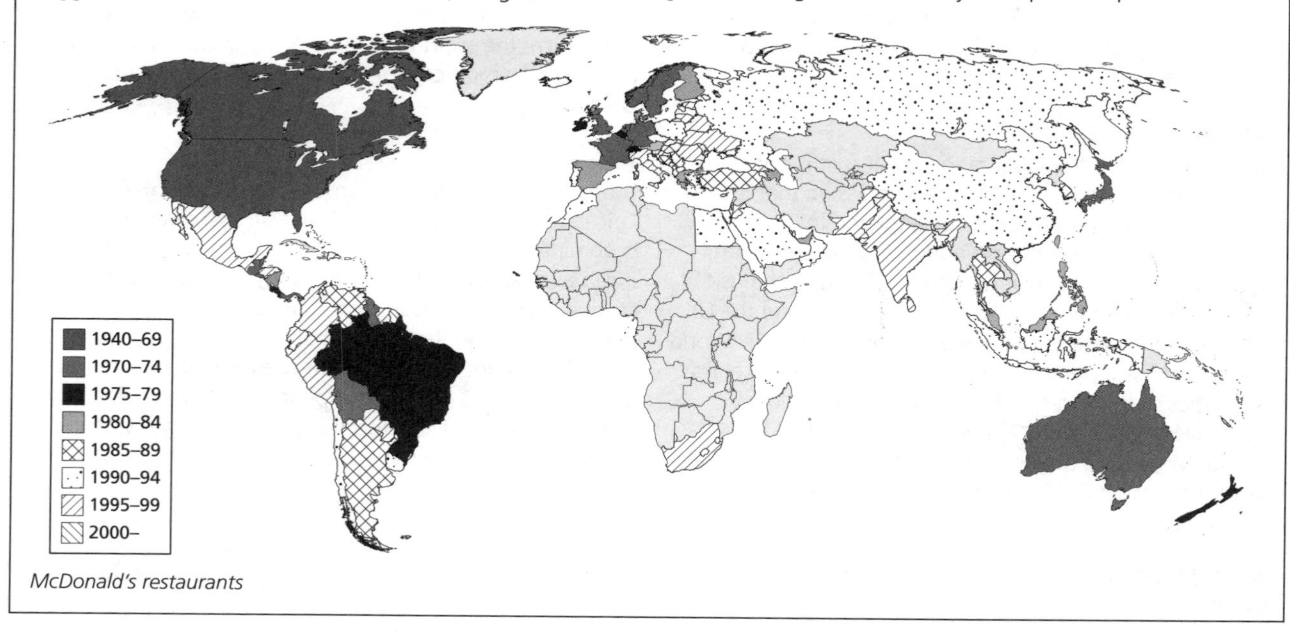

McDonald's restaurants

Consumer culture (2)

COCA-COLA

Founded in 1886 by pharmacist John Styth Pemberton in Atlanta, Georgia, The Coca-Cola Company is the world's leading manufacturer, marketer and distributor of non-alcoholic beverage concentrates and syrups, used to produce nearly 400 brands. The Coca-Cola Company continues to be based in Atlanta and employs 49,000 people worldwide, with operations in over 200 countries.

The biggest-selling soft drink in history, and one of the best-known products in the world, Coca-Cola was first offered as a soda fountain beverage in Atlanta.

Coca-Cola arrived in Britain in 1900. First sold regularly through soda fountain outlets including Selfridges and the London Coliseum in the early 1920s, Coca-Cola is now the most recognized trademark in the world. The word "Coca-Cola" itself is even thought to be the second most widely understood word in the world after "OK"!

Nowadays, the company is located in more than 200 countries, where its drinks are produced by *local* people with *local* resources. Coca-Cola produces brands that embrace distinct tastes and local preferences.

Coca-Cola workforce worldwide

EXTENSION

Visit

http://commons.wikimedia.org/wiki/Image:KFC_world_map1.png for a world map showing the worldwide distribution of KFC fast food restaurants.

EXTENSION

The nearest neighbour index

The nearest neighbour index provides a statistical value for the degree of clustering, regularity or randomness in a distribution pattern. The formula is $NNI = 2\overline{D}\sqrt{n/a}$ where $\overline{D}$ is the average distance between each point and its nearest neighbour ($\Sigma d/n$), n is the number of points being looked at, and a is the size of the area. The result varies between 0 (perfect clustering) and 2.15 (perfect regularity).

There are some coconsiderations to bear in mind:

- there may be sub-patterns within the overall pattern although the answer may sugggest a random pattern
- there may be controlling factors e.g. flood risk or soil type which influence the pattern.

Sociocultural integration

DIASPORAS

The term **diaspora** refers to the forced or voluntary dispersal of any population sharing common racial, ethnic or cultural identity, after leaving their settled territory and migrating to a new region. The Irish in New York and Boston are a good example, as are the Indians and Pakistanis in the UK. The Mexican labourers in the USA are another example of a diaspora.

THE IRISH DIASPORA

The Irish diaspora consists of Irish migrants and their descendants in countries such as the USA, the UK, Australia, Canada and those of continental Europe. The diaspora contains over 80 million people, more than 14 times the population of Ireland.

The USA was the most popular destination in the 19th century and Irish migration there reached a peak of 1.8 million in 1891. By 1951 the number of Irish in Britain had overtaken the US figure and by 1981 there were four times as many there as in the USA.

There have been major fluctuations in the figures since 1981 – there was a renewed increase in Irish migrations to the USA during the 1980s, a drop in the numbers going to Britain and a rise in numbers going to other EU countries. There were also high rates of return and an overall fall in absolute numbers of emigrants.

With improvements in Ireland's economic success and a fall in Irish birth rates (since the 1980s), the "bad old days" of high emigration are fast becoming part of Irish history. Nowadays fewer than 18,000 Irish people leave each year and many of these will return to Ireland again.

Britain

- Of all Irish-born people living abroad, 75% are in Britain.
- There are approx 1.7 million people in Britain who were born to Irish parents.
- The third-generation Irish community in Britain could be in the region of 6 million.

The USA

- Of the total US population, 10.8% claim Irish ancestry – the equivalent of seven times the population of Ireland itself.
- Irish-born people in the USA number 156,000.
- States with the largest Irish-American population are: California, New York, Pennsylvania, Florida, Illinois.
- Irish-Americans are the largest ancestral group in Washington DC, Delaware, Massachusetts and New Hampshire.

Canada

- First-generation Irish in Canada number approximately 28,500.
- 3.8 million say they are of Irish ancestry.

Argentina

- In the latter half of the 19th century, approximately 45,000 Irish arrived in Argentina – some 20,000 of whom settled, with most of the remainder moving back to the USA.
- Today in Latin America some 300,000 to 500,000 people are estimated to have some Irish ancestry, most of them living in Argentina, with lesser numbers in Central America, Uruguay and Brazil.

Australia

- Australia has the third largest Irish-born population outside Ireland.
- First-generation Irish in Australia number approximately 74,500.
- During the 18th and 19th centuries, 300,000 free emigrants and 45,000 prisoners sailed to Australia from Ireland.

New Zealand

- First-generation Irish in New Zealand number approximately 11,000.

IMPACT OF THE IRISH DIASPORA

Emigration has been a constant theme in the development of the Irish nation and has touched the lives of people in every part of Ireland. The economic and social prosperity of the country has been affected positively, through monies sent home from abroad, and negatively, through the loss of so many talented young Irish people. Irish emigrants have also had an enormous impact on the development of the countries in which they settled.

The term "Irish diaspora" is open to interpretation. One, preferred by the Irish government, is defined in legal terms: the Irish diaspora are those of Irish nationality who habitually reside outside of Ireland. This includes Irish citizens who have emigrated abroad and their children, who are Irish citizens by descent under Irish law. It also includes their grandchildren.

Irish Americans

In the USA, the Irish are largely perceived as hard workers. Most notably they are associated with the positions of police officers, firefighters, Roman Catholic Church leaders and politicians in the larger Eastern Seaboard metropolitan areas. Irish Americans number over 44 million, making them the second largest ethnic group in the country, after German Americans. The largest Irish American communities are in Chicago, Boston, New York City, Baltimore, Philadelphia, Kansas City and Savannah, Georgia. Each city has an annual St Patrick's Day parade, with Savannah having the largest. At state level, Texas has the largest number of Irish Americans. According to the 1990 US Census, Arkansas listed 9.5% of the population as Irish-descendent, primarily located in the south-east part of the state. In percentage terms, Boston is the most Irish city in the USA, and Massachusetts the most Irish state.

Cultural diffusion and indigenous groups: the Dani

This page looks at the cultural impact of globalization and tourism on indigenous people, specifically on the Dani tribes of Irian Jaya. For more information see Planet Geography by Stephen Codrington.

INTRODUCTION

Irian Jaya, on the west half of the island of New Guinea, was assumed to be an unoccupied Dutch territory until an American adventurer, Richard Archibold, discovered the Dani people in 1938. There are some 100,000 Dani people, consisting of 30 cultural groups, occupying the central highlands of Irian Jaya, mostly in the Baliem Valley, some 72km long and 30km wide.

PRE-CONTACT DANI LIFE

Housing

The Dani lived (and still do) in conical houses with poor ventilation. Many have died from pneumonoconiosis, a lung condition which results from domestic smoke inhalation. Life expectancy is just 38 years.

Clothing

Traditionally very little clothing is worn. The men wear a penis sheath, called a phallocrypt. Women wear only a grass skirt with beads.

Food

The Dani developed an elaborate drainage and irrigation system. Intensive sedentary agriculture provided sweet potato, the main source of carbohydrate. Pigs provided protein, but the diet of the Dani was poor. Pigs were also a status symbol and appeared to mark births, marriages and deaths.

Social organization

Monogamy was normal but polygamy occurred occasionally. Men usually married at age 20 and girls at age 12. Homosexuality was unknown before outside contact. After the birth of their second child, women ate tree sap to induce early menopause. When close relatives died, men commonly slashed their ear lobes off and women amputated a joint from their finger to show respect.

Ritual warfare

This was generally brief but ended in one or two deaths. Cannibalism, which involved eating the dead from battle, was practised by one Dani subgroup (the Yale).

Religion

The Dani saw their surroundings as living things – the moon was a man, the sun was a woman and rain was urine. The Dani saw the world as filled with spirits and the supernatural, which caused much anxiety. Bodies were preserved as mummies to appease the spirits.

OUTSIDE INFLUENCES ON DANI CULTURE

Missionaries

First contact with the outside world involved the Christian missionaries in 1954. They set up schools, churches and medical services, and by 1980 over 80% of the Dani people had been converted to Christianity. Missionary impacts were well accepted by the Dani and included:

- the burning of charms and fetishes
- adopting Christian names
- eliminating ritual warfare and cannibalism
- trading using money instead of cowrie shells
- raising the age of marriage of girls
- discouraging polygamy.

Government influences

Post-1989 Indonesian government influences were mostly resisted and included:

- the wearing of clothes and the use of Indonesian language
- rice cultivation instead of sweet potatoes as the staple crop
- Western-style houses, most of which were rejected
- expelling the missionaries, leaving the community without support.

The impact of tourism

Tourism is the third influence on Dani culture, after the missionaries and government. Tourists started coming in 1984 and were mainly trekkers from the USA, Germany and Australia. Most visited the Baliem Valley. The main attractions are the traditional culture. The main town, Wamena, contains a bustling market selling produce, crafts and artefacts. A number of characteristics can be observed:

- Tourists bring gifts and novelties for the Dani, who then expect them and develop a "cargo cult" attitude (i.e. they expect goods from contact with Westerners).
- The Dani have become exposed to new forms of dress – trekkers commonly give T-shirts to the people.
- The Dani work as guides and porters to gain cash – in this way they have become integrated into the cash economy.
- Severe leakage of tourist revenue occurs because most of the profits go to Indonesian businessmen – all the hotels and restaurants in Wamena are foreign-owned.
- Although the Baliem Valley has many natural attractions – salt wells, caves, lakes, preserved mummies, traditional hanging bridges and markets, for example – for many tourists the real attraction is the traditional "Stone Age" culture promoted in brochures – cannibalism, headhunting, mock wars and pig-slaughtering rituals.
- Some argue that Dani culture has become degraded to the level of a human zoo or a peepshow, and that tourism has begun to contaminate the culture which tourists wanted to preserve.
- The Dani people recognize that many tourists come in search of their traditional culture, which they have in some ways begun to abandon. In some villages the local people change into traditional clothing when outsiders arrive and pose for photographs to earn some income; they then return to their normal factory-made clothing.

Cultural imperialism

The world is becoming more uniform and standardized through a technological, commercial and cultural synchronization emanating from the West, and globalization is tied up with modernity.

Proponents of the cultural imperialism thesis date its inception to the industrial colonialism phase. It was during this phase that colonialism reached its zenith, peaking just prior to the First World War, when the British Empire reached its maximum territorial extent. However, the end of formal colonialism in the second half of the 20th century did not spell the end of cultural imperialism. Cultural imperialism has become an economic process as well as a political one. It is forged by TNCs that represent the interests of the elite, especially those of the USA.

ASPECTS OF GLOBAL CULTURAL IMPERIALISM

Language

There are around 6000 languages in the world, and this figure may drop to 3000 by 2100. Approximately 60% of these languages have fewer than 10,000 speakers; a quarter have fewer than 1000. English is becoming *the* world language. Although Mandarin is more widely spoken as a first language, the total number of English speakers if second-language speakers are taken into account is close to 1 billion. English is the medium of communication in many important fields, including air travel, finance and the internet. Two-thirds of all scientists write in English; 80% of the information stored in electronic retrieval systems is in English; 120 countries receive radio programmes in English; and at any given time over 200 million students are studying English as an additional language. It is an official language in much of Africa, the Pacific, and south and South-East Asia.

Tourism

Tourism is now the world's largest industry. The journey of many British people to the Costa del Sol, Spain, where they practise cultural traits such as drinking beer and eating fish and chips while lying on crowded beaches surrounded by tall buildings, is a stereotype which captures the essence of this type of standardization.

Global brands

Behind the growth in the influence of TNCs is the rise of global consumer culture built around world brands. McDonald's, for example, operates over 31,000 outlets in 119 countries. In 1997, it opened one outlet every four hours. Coca-Cola is sold in nearly every country. It is a transcultural item yet it is very much linked with US culture.

Media

National media systems are being superseded by global media complexes. Around 20 to 30 large TNCs dominate the global entertainment and media industry, all of which are from the West, and most of which are from the USA. These include giants such as Time-Warner, Disney, News Corporation, Universal Studios and the BBC.

Democracy

The spread of liberal democracy has been profound and is now practised in the vast majority of nation states across the planet. Underlying this diffusion is the western enlightenment belief that it is the most desirable form of governance.

CRITICISMS OF CULTURAL IMPERIALISM

It has been argued that the concept of cultural imperialism ascribes globalization with too much determining power. The power of locality, and of local culture, is thus overlooked. Moreover, a variation on the cultural imperialism argument sees the creation of a universalized hybrid culture. This type of culture is homogeneous but not entirely Western in nature. The impact of contact is not one-way. For example, the British drink tea because of the British imperial connection with India, and a number of words in the English language, such as *bungalow*, *shampoo*, *thug* and *pyjamas*, are borrowed from languages of the subcontinent. The influence of Black American and Hispanic dialects on rap, the most popular music globally at present, and the fact that football (soccer), which diffused through the British Empire, is thought to have been invented in China, are further examples of universalized hybrids.

EXTENSION

Visit
http://www.mcdonalds.com/corp/about/factsheets.html and find out about environmental responsibility.
Visit
http://www.mcdonalds.com/corp/about/mcd_faq.html for frequently asked questions.

Loss of sovereignty (1)

Some analysts believe that nations are far less important than they once were. They argue that the increasing flow of people, capital, goods and ideas across international boundaries illustrates the demise of the nation state. At the same time, the growth of trading blocs and TNCs heralds a new world order in which individual countries are less important than before.

GLOBALIZATION VERSUS REGIONALISM

While globalization of economic activity has certainly occurred, and there is evidence of a new international division of labour, political and cultural values have often created a new feeling of regional identity. Within major trading blocs such as the EU there are very strong nationalist tendencies, for example within Spain and the UK.

TNCS

Part of the reason for the decline of sovereignty in some countries is the sheer economic size and dominance of some TNCs, as shown in the figures below.

Rank	Company/country	Annual sales / GDP (US$ billions) (2007)
1	USA	13,811
2	Japan	4,376
3	Germany	3,297
4	China	3,280
5	UK	2,727

Rank	Company/country	Annual sales / GDP (US$ billions) (2007)
44	Singapore	161
59	Bangladesh	67
79	Kenya	29
109	Uganda	11
146	Zimbabwe	3

Rank	Company/country	Annual sales / GDP (US$ billions) (2007)
1	Wal-Mart, USA	379
2	Exxon-Mobil, USA	358
3	Royal Dutch Shell, Netherlands	355
4	Toyota, Japan	204
5	Chevron, USA	204

TRADING BLOCS

A **trading bloc** is an arrangement among a group of nations to allow free trade between member countries but to impose tariffs (charges) on other countries that may wish to trade with them. Examples of trading blocs include the European Union (EU), the Association of South-East Asian Nations (ASEAN), the North American Free Trade Agreement (NAFTA) and MERCOSUR, the common market of South America.

The European Union (EU)

In 1957 the six founder members of the European Economic Community desired closer union and greater economic and social progress. One of the main reasons for creating the EU was that trade had grown enormously since 1945. However, since then, plans to increase the number of countries in the EU and to extend into central and eastern Europe are no longer based on trade alone but increasingly on political grounds. The chances of conflict in an expanded EU are much lower than if the same countries are outside the EU.

REGULATORY BODIES

Much of the trade and money exchange that takes place is run by stock exchanges and the world's main banks. For example, Barclays Capital is the investment-banking sector of Barclays Bank. It deals with over £360 billion of investment through its 33 offices located worldwide. Its regional headquarters are located mostly in MEDCs such as London, Paris, Frankfurt, New York and Tokyo. Hong Kong is the exception, although it is an important financial centre, like most of the other cities on the list.

There is widespread criticism that many of the regulatory bodies have limited power, and that when faced with a powerful MEDC or TNC they capitulate.

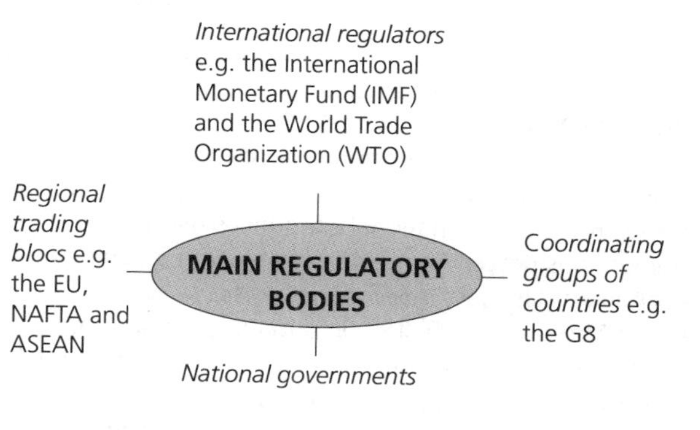

Loss of sovereignty (2)

TRANSNATIONAL CORPORATIONS

Transnational corporations (TNCs) or multinational enterprises (MNEs) are organizations that have operations in a large number of countries. Generally, research and development, and decision-making, are concentrated in the core areas of developed countries, while assembly and production are based in developing countries and depressed, peripheral regions.

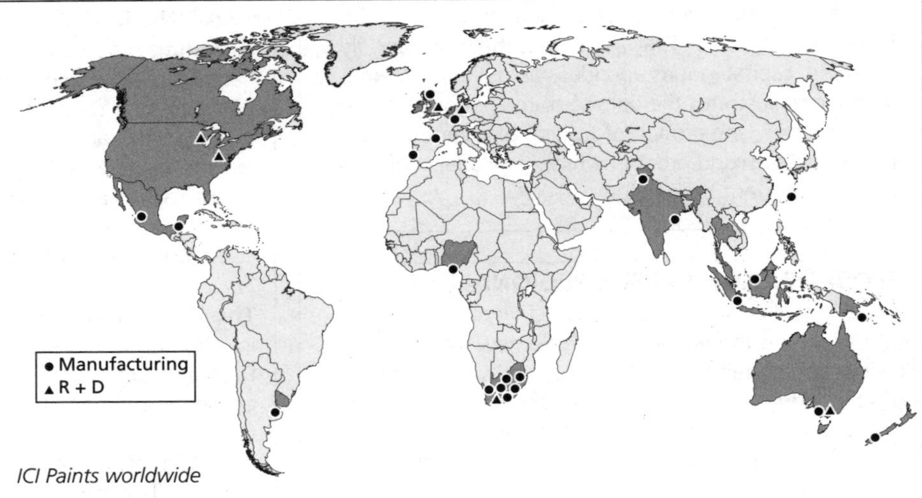

ICI Paints worldwide

Imperial Chemical Industries (ICI)

ICI was formed in 1926 and has its headquarters in the UK. It employs about 130,000 people worldwide and has sales of about £6500 million each year. ICI is often seen as one of the flagships of the British industry and its fortunes are seen as a barometer of the nation's fortunes.

The corporation is a vast conglomerate that makes almost the complete range of chemicals and chemical-related products, including fertilizers, paints, pharmaceuticals and plastics. Its sales and profits now depend on four main markets: the UK; western Europe; North America; and Australia and the Far East.

TNCS – THE BALANCE SHEET

TNCs provide a range of advantages and disadvantages for the host country. These include the following:

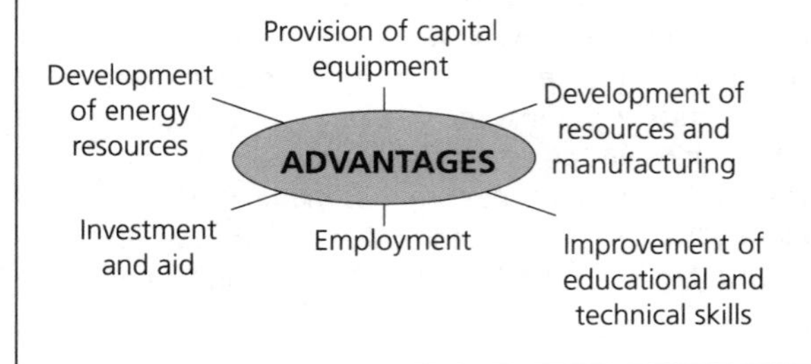

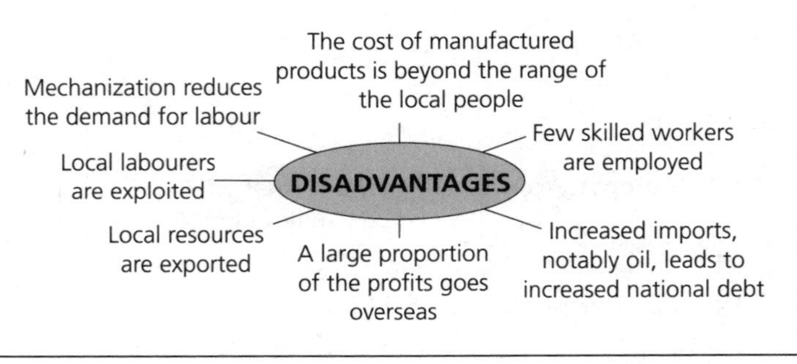

TNC POWER

The sheer scale of the economic transactions that TNCs make around the world and the effect they have on urban, regional and national economies gives them tremendous power. Thus TNCs have become planned economies with vast **internal markets**.

- Up to one-third of all trade is made up of internal transfers of TNCs. These transfers produce money for governments via taxes and levies.
- Economic power comes from the **ownership of assets**.
- Over 50 million people are employed by TNCs.
- Although many governments in developing countries own their own resources, TNCs still control the marketing and transport of goods.

TNCS AND THE WORLD'S ECONOMIC CRISES

Reduced demand and increased competition creates unfavourable economic conditions. In order to survive and prosper, TNCs have used three main strategies:

- **rationalization** – a slimming down of the workforce, which involves replacing people with machines
- **reorganization** – includes improvements in production, administration and marketing, such as an increase in the subcontracting of production
- **diversification** – refers to firms that have developed new products.

EXTENSION

Visit
http://www.globalpolicy.org/socecon/tncs for links and information on TNCs and global activites.

Responses

NATIONALISM

Nationalism refers to a political movement or a belief that holds that a nation, usually defined in terms of ethnicity or culture, has the right to an independent political development based on a shared history and common destiny. The concept of the nation state was a Western one but has spread throughout the world as a result of colonialism. Crucial to the development of the nation state was the creation of a national identity that cut across class. Nation-building involved a variety of factors, such as a common language, an education system, national communications networks, national symbols and promotion of national culture.

Ireland

Nationalism in Ireland shows many features. At one extreme were the political freedom-fighters or terrorists, and at the other were the members of the Gaelteacht, the Irish-speaking regions of Ireland.

In between these two extremes were a variety of programmes to develop the Irish sense of nationality. In schools, the curriculum delivered the Irish sagas in English, history and Irish lessons. The two main sports, Gaelic football and hurling, were unique to Ireland and the Irish diaspora. Economically, the campaign to buy "Guaranteed Irish" helped sales of Irish companies. The government's import substitution policies of the 1920s and 1930s helped reduce dependency on Britain. On the other hand, Ireland has been described as one of the most globalized countries in the world, given the amount of FDI it has attracted. Developing national identity and becoming an integral part of the global economy are hard to reconcile.

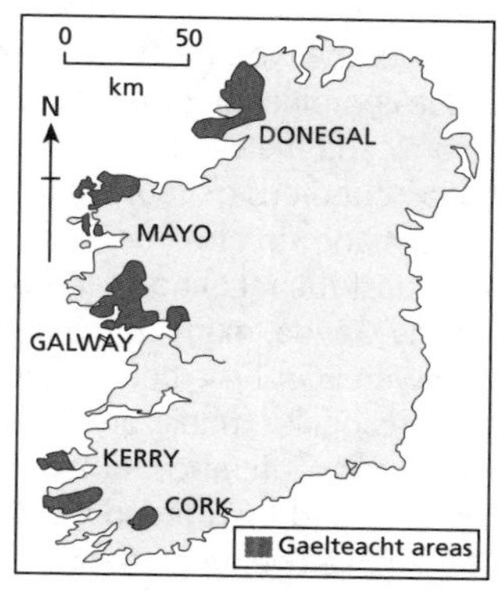

Gaelteacht areas of Ireland

ANTI-GLOBALIZATION

The anti-globalization movement (AGM) is a general term used to describe a wide variety of protestors, lobbyists and pressure groups. The AGM has attracted attention due to the protests it has mounted during international summits in places such as Seattle, Tokyo and Munich. The movement developed during the 1990s following the actions of the Zapatista National Liberation Army in Chiapas, Mexico. Some 3000 people took control of the main towns of Chiapas. Not only had they lost land in the process of development, protectionism for farmers had been removed and their livelihoods had suffered.

The AGM reached a global audience in 1999 during the WTO trade talks in Seattle, USA. Some 60,000 people arrived in the city, home to Starbucks and Microsoft, and protested. The trade talks were abandoned. In 2001 the World Social Forum was established in Brazil, involving large numbers of very diverse groups. The attacks on the World Trade Center (the Twin Towers attacks of 9/11) divided people. Some reacted by becoming more nationalistic and prepared to defend their national space; others saw it as the inevitable result of US global economic and political dominance.

Nevertheless, the AGM lacks focus. Some within the AGM are in favour of globalization – but at a slower pace. Others object to the economic and political power that some rich countries wield. Some see the work of organizations such as the World Bank as pedantic and stifling the needs of poor countries.

ATTEMPTS TO CONTROL MIGRATION

Australia's migration laws and regulations set the criteria and standards that foreign nationals must meet if they wish to travel to and remain in Australia for a period of time.

Australian Skilled Independent visa

The Skilled Independent visa is designed to provide work visas to individuals who have the qualifications or skills to fill Australia's skill shortages and contribute to the Australian economy.

Applicants for the Skilled Independent visa must be able to demonstrate that they possess a qualification or the skills of an occupation on the Skilled Occupation List (SOL).

Applicants must also undergo a character test. A person will fail this test where:

- they have a **substantial criminal record**
- they have, or have had, an **association** with an individual, group or organization suspected of having been, or being, involved in criminal conduct
- having regard to the person's **past and present criminal conduct**, the person is found not to be of good character
- there is a **significant risk** that the person will **engage in criminal conduct** in Australia, **harass, molest, intimidate or stalk** another person in Australia, vilify a segment of the Australian community, or **incite discord** in the Australian community or in a segment of that community, or **represent a danger** to the Australian community or a segment of that community.

Anti-globalization movements

PEOPLE'S GLOBAL ACTION

People's Global Action (PGA) is a network for spreading information and coordinating actions between grass-roots movements around the world. These diverse groups share an opposition to capitalism, and a commitment to direct action and civil disobedience as the most effective forms of struggle. PGA grew out of the international Zapatista gatherings in 1996 and 1997, and was formed as a portal for direct and unmediated contact between autonomous groups.

The first conference took place in 1998, when movements from all over the world met in Geneva and launched a worldwide coordination of resistance against the global market economy and the World Trade Organization (WTO). Later that year, hundreds of coordinated demonstrations, actions and street parties took place on all five continents, against the meeting of the G8 and the WTO. From Seattle and Genoa, many of the groups and movements involved with PGA have been a driving force behind the global anti-capitalist mobilizations.

A second international conference took place in Bangalore, India in 1999 and the third in Cochabamba, Bolivia in 2001. There have been regional conferences in Latin America, North America, Asia and Europe, and three caravans of movements: the Intercontinental Caravan, the Colombian Black Communities tour and the People's Caravan from Cochabamba to Colombia.

PGA is not an organization and has no members. However, PGA aims to be an organized network. There are contact points for each region, which are responsible for disseminating information and convening the international and regional conferences; an informal support group that helps with fundraising; a website; numerous email lists; and a secretariat.

PGA HALLMARKS

1 A very clear rejection of capitalism, imperialism and feudalism, and all trade agreements, institutions and governments that promote destructive globalization.
2 We reject all forms and systems of domination and discrimination including, but not limited to, patriarchy, racism and religious fundamentalism of all creeds. We embrace the full dignity of all human beings.
3 A confrontational attitude, since we do not think that lobbying can have a major impact in such biased and undemocratic organizations, in which transnational capital is the only real policy-maker.
4 A call to direct action and civil disobedience, support for social movements' struggles, advocating forms of resistance which maximize respect for life and oppressed people's rights, as well as the construction of local alternatives to global capitalism.
5 An organizational philosophy based on decentralization and autonomy.

The basis of unity and political analysis is expressed in the constantly evolving manifesto and hallmarks. Hallmark 4 was changed in Cochabamba to remove the word "non-violent". Non-violence has very different meanings in India (where it applies to respect for life) and in the West (where it applies also to respect for private property). The North American movement felt that the term could be understood to not allow for a diversity of tactics, or even contribute to the criminalization of part of the movement. The Latin American organizations said that "non-violence" seemed to imply a rejection of huge parts of the history of resistance.

Non-violence has to be understood as a guiding principle, relative to the particular political and cultural situation. Actions which are perfectly legitimate in one context can be unnecessarily violent (contributing to brutal social relations) in another.

PGA's detailed manifesto includes sections on each of the following:

- economic globalization power and the "race to the bottom"
- exploitation, labour and livelihoods
- gender oppression
- the indigenous peoples' fight for survival
- oppressed ethnic groups
- the onslaught on nature and agriculture
- culture, knowledge and technology
- education and youth
- militarization, migration and discrimination.

EXTENSION

Writing frames

There are a number of writing frames for answering an essay or a report. In general, the essay title and material to be included will suggest what type of structure should be used. However, for all questions you must examine the wording of the question and plan your answer. It is better to spend time thinking and planning, so that you do not waste any time writing about irrelevant material. Writing for 35 minutes on relevant material is better than 45 minutes on irrelevant material.

Read the question carefully and underline the command words (listed on page 193) and the topic to be discussed. There may be some technical words such as "and", "either…or". Questions with "and" in them generally ask for factual information and then require some interpretation. Often the interpretation is more important than the recall of fact. Questions stating "with the use of examples…" may allocate one third or half the marks for the examples used. If you do not answer the question you cannot get the marks.

There are three main types of essay. Descriptive essays are easy and require factual recall. Explanation requires you to give reasons and account for why a particular object is the way it is. Evaluation expects an opinion based on the evidence presented.

Migration and migration control

THE AGE OF MIGRATION

International migration has changed much in recent years. Four general trends can be identified:

- Migration is becoming more global in the sense that more countries are affected at the same time and the diversity of areas of origin is increasing.
- Migration is accelerating, with the number of movements growing in volume in all major regions.
- Migration is becoming more differentiated, with no one type of movement dominating a country's flows, but instead with combinations of permanent settlers, refugees, skilled labour, economic migrants, students, retirees, arranged brides and so on.
- Migration is being feminized, with women not only moving to join earlier male migrants but now playing a much fuller part in their own right, notably among labour migrants themselves, as well as often being dominant in refugee flows.

These trends have implications for policy-makers. There are new challenges for governments for providing for migrants, but there is also increased hostility in receiving countries. Increasing globalization and a growing diversity of migrants make it harder for governments to restrict migration.

MIGRATION CONTROL IN THE USA

Illegal immigration to the USA refers to the act of foreign nationals voluntarily residing in the USA in violation of US immigration and nationality law. Illegal immigration carries a civil penalty. Punishment can include fines, imprisonment and deportation.

It is estimated that there were between 11.5 and 12 million illegal immigrants in the USA in 2006. Their mode of illegal entry into the country is believed to break down as follows:

Visa overstay
A traveller is considered a "visa overstay" once he or she remains in the USA after the time of admission has expired. Visa overstayers tend to be somewhat more educated and better off financially than those who crossed the border illegally.

Fraudulent marriage
People have long used sham marriages as a way to enter the USA.

Border crossing
Each year, an estimated 200,000–400,000 illegal immigrants try to make the 24–48 km hike through the wilderness to reach cities in the USA. Often, the people who choose to sneak across the border employ expert criminal assistance – smugglers who promise a safe passage into the USA.

Entry by sea ports
In 1993, 283 Chinese immigrants attempted entry into the USA via a sea vessel. Ten of them arrived dead.

Slavery
Indian, Russian, Thai and Chinese women have been reported as having been brought to the USA under false pretences to be then used as sex slaves. As many as 50,000 people are illicitly trafficked into the USA annually, according to a 1999 CIA study.

Prostitution
Trafficking in women plagues the USA as much as it does developing nations. Organized prostitution networks have migrated from metropolitan areas to small cities and suburbs.

Indispensable
Foreign-born population in selected OECD countries
% of total population, 2005

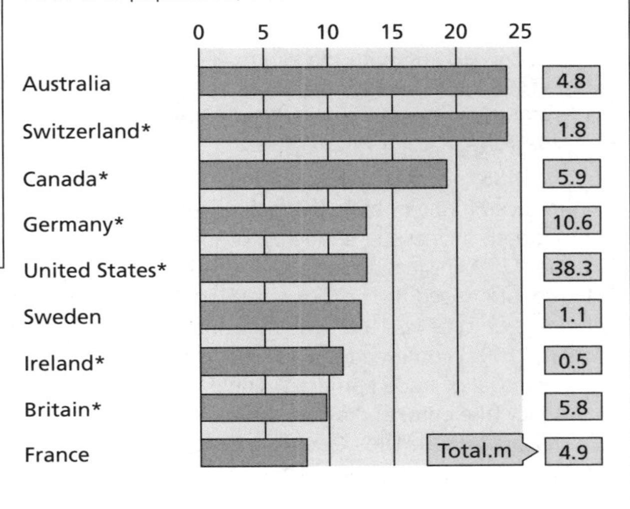

IMMIGRATION AND ENFORCEMENT

Illegal migration on the USA–Mexico border is concentrated around big border cities such as El Paso and San Diego, which have extensive border fencing and enhanced border patrols. Stricter enforcement of the border in cities has failed to curb illegal immigration significantly, instead pushing the flow into more remote regions and increasing the cost to taxpayers of each arrest from $300 in 1992 to $1700 in 2002. The cost to illegal immigrants has also increased: they now routinely hire coyotes, or smugglers, to help them get across.

In 2005, the US House of Representatives voted to build a separation barrier along parts of the border not already thus protected. A later vote in 2006 included a plan to blockade 860 miles (1380 km) of the border with vehicle barriers and triple-layer fencing, along with granting an "earned path to citizenship" to the 12 million illegal aliens in the USA and roughly doubling legal immigration (from their 1970s levels). In 2007, Congress approved a plan calling for more fencing along the Mexican border, with funds for approximately 700 miles (1100 km) of new fencing.

Globalization versus nationalism in the EU

GLOBALIZATION OR ANTI-GLOBALIZATION – THE CASE OF THE EU

The growth of the European Union to a union of 27 countries, with at least two more wishing to join, would appear to be a strong symbol of globalization. Member nations have given up some of their sovereignty and political power to a multinational government. The EU has moved beyond mere economic integration and has achieved some political, social and cultural integration.

Attempts at European integration developed after 1945, partly as an attempt to prevent a world war from ever occurring again. (Part of the desire to include Turkey in the EU is to integrate a large Islamic state into the union and thus reduce the possibility of war between Islam and the West.) The six countries that formed the European Coal and Steel Community (ECSC) went on to form the European Economic Community (EEC) in 1958. The UK attempted to join the EEC but was turned down in 1961 and 1967. In 1973 the EEC was expanded, and in the 1980s it expanded further still.

In 1986 the Single European Act introduced a rule of majority decisions – this greatly increased the powers of the Council of Ministers and the Parliament. It also introduced the goal of removing all barriers to trade by 1992. The Maastricht Treaty (1991) confirmed the agenda for the removal of trade barriers, a single currency (the euro) and a range of social regulations. The Maastricht Treaty established the European Union. In 2002 the euro was introduced into 12 of the then 15 members. The EU was expanded to 25 members in 2004.

Membership of the EU over time

However, there has been reaction against the growth of the EU and its imposition of economic, political and social regulations. The UK and Denmark, for example, opted out of the single currency, deciding to retain their own. During the 1980s, while France and Germany were pro-integration, the UK argued aggressively against loss of sovereignty. Ironically, within the UK there have been movements within Wales, Scotland and Northern Ireland for greater political and economic autonomy. In all three there is now a devolved parliament with responsibility for some decision-making. Within Northern Ireland there is power-sharing between the Democratic Ulster Party and Sinn Fein. The Scottish Nationalist Party and Plaid Cymru offer alternatives to the main UK political parties in Scotland and Wales. In Spain, Catalonia and Galicia have achieved significant autonomy, while the Basque Country has not, largely in response to the violence of the independence-seeking party, ETA.

As the EU has expanded, it has become more diverse. Economically, socially and culturally it is more varied and divided than ever before. This diversity means that integration is likely to be less complete than when there were just 6 (or 9 or even 15) countries. Being large may help economic prospects (a larger market, for example) and political ones (less chance of war), but national identity and regional cooperations are likely to become more important over time.

Globalization and glocalization

DEFINITIONS

Globalization is defined as the growing interdependence of countries worldwide through the increasing volume and variety of cross-border transactions in goods and services and of international capital flows, and through the more rapid and widespread diffusion of technology.

In contrast, **glocalization** is a term that was invented in order to emphasize that the globalization of a product is more likely to succeed when the product or service is adapted specifically to each locality or culture it is marketed in. The increasing presence of McDonald's restaurants worldwide is an example of globalization, while the changes in the menus of the restaurant chain that are designed to appeal to local tastes are an example of glocalization.

GLOBALIZATION AND GLOCALIZATION IN THE MANUFACTURING SECTOR

	Globalization	Glocalization
Organization	Worldwide	Concentrated in the Triad (EU, North America and Japan)
Locational requirements	Comparative advantage and economies of scale	Depressed regions of major international trade blocs
Labour and management	Foreign managers in senior ranks; spatial division of labour	Very difficult for foreign managers to reach senior ranks
Market	Production for world markets Geographically dispersed	Production for local or regional markets Geographically concentrated
	Export-orientated strategy	Export-orientated strategy

A comparison between globalization and glocalization in the manufacturing sector

Globalization aims at a worldwide intra-firm division of labour. In this strategy, activities are established in many sites spread over the world, based on a country's comparative advantages. A manufacturer striving for globalization aims to secure the supply of its inputs by locating production of these inputs at the most favourable locations. Thus, labour-intensive production of components will be situated in low-wage areas, while the production of high-tech and high value-added parts will require a skilled or well-educated workforce. In a European context, this would mean locating research facilities in core areas and assembly plants in peripheral areas.

Glocalization aims to establish a geographically concentrated inter-firm division of labour in the three main trading blocs: Japan and South-East Asia, the USA, and the EU – collectively these are known as the Triad. Manufacturers striving for glocalization are building their comparative advantage on close interaction with suppliers and dealers, as well as with other relevant actors, such as banks and governments. Two essential elements stand out in a firm's glocalization strategy:

- the decentralization of production to hierarchical networks of local subcontracting
- a high degree of control over supply and distribution.

The strategy for glocalization involves the attempt of a manufacturer to become accepted as a "local citizen" in a different trade bloc, while transferring as little control as possible over its strategic activities. Glocalization is first of all a political, and only in the second place a business location strategy. A manufacturer aiming for glocalization will localize activities in a different trade bloc area only if:

- it otherwise risks being treated as an "outsider" and so subject to trade or investment barriers and thus stands to lose market share, or
- the inevitable compromise in costs and control will allow it to produce competitively, i.e. there are suitable areas of low labour costs or regional assistance.

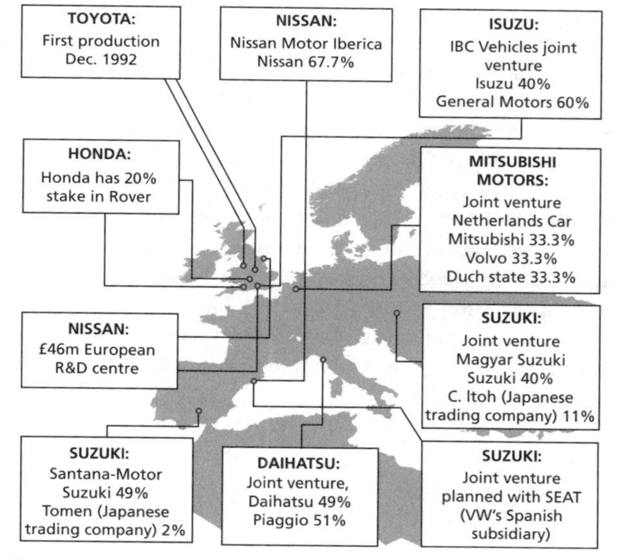

The Japanese presence in Europe

Adoption of globalization

AN UNEVEN PROCESS

Globalization is a very uneven process. For example, the diffusion of telecommunications and ICT has left vast numbers of people without access to either. These are mostly poor, rural people who have missed out on the advantages of globalization. Globalization is not a homogeneous process or feature – its outcomes vary markedly across the world. Indeed, the growth of globalization has led to increased inequalities between nations, regions, urban and rural areas, and within cities. Globalization has marginalized and excluded many people from its benefits.

Looking back at economic development between 1945 and 1980 (the modernization era), there are certain groups of people who benefited little economically. Economic growth was accompanied by exclusion. One group of people who have not been able to benefit from globalization are women. Of the 1.3 billion people in poverty, 70% are women. Along with women, landless labourers also failed to make much progress and were also unable to benefit from globalization.

Since 1970 many poor countries have become poorer. According to the United Nations Development Programme, between 1960 and 1991 the share of the world's richest 20% increased to 80%. The ratio of richest to poorest increased from 30:1 to 60:1. Such polarization means that there are increasingly more people unable to benefit from globalization. They are either excluded from it, or exploited by it. There are clear differences between urban and rural areas, with large urban areas much better off. Nevertheless, within urban areas, those in low-income areas, such as shanty towns and bustees, are much less able to benefit from globalization than the wealthy in rich areas.

Globalization has made some places worse off. This is because of what is termed "capital flight". This means that wealthy companies can decide to invest in some places, remove investment from others, close factories, and open in new locations on the basis of where they can take advantages in changes in tax regimes, pay negotiations, government incentives, and the availability of grants and loans. To an extent, some countries, especially poor countries, are at the mercy of wealthy TNCs. Who is part of the globalization process and who is out of the globalization process may therefore be determined by the TNC and the nation's government, rather than its people.

Factors influencing the adoption of globalization strategies

Thus the adoption of strategies for globalization may be influenced by many factors, including levels of wealth, landlessness, gender, TNC–government negotiations, and sociocultural factors, such as whether a product is acceptable to a local market.

MCDONALD'S AND GLOBALIZATION

On page 170, the globalization of McDonald's was illustrated. A number of questions could be asked. For example:

- Does the spread of fast food undermine local food producers?
- Does the spread of fast food undermine local dietary patterns?
- Does the spread of fast food create local environmental problems, such as water shortages?
- Does the spread of fast food create sociocultural problems, such as change in farming patterns and conflict between people of different generations?

On the other hand, some researchers believe that the extremes of globalization may be modified – or even adapted – to fit local conditions. For example, there is evidence to suggest that McDonald's has caused small but significant changes in Asian dietary patterns. The Japanese rarely ate food with their hands – this is now acceptable and commonplace. On the other hand, whereas a McDonald's outlet is a fast-food restaurant in the USA, in places such as Seoul, Taipei and Beijing, it is somewhere to go for a very leisurely meal.

McDonald's has opened restaurants in New Delhi and Mumbai (Bombay). Environmentalists believe that foreign investment in farming and the fast-food industry is destroying not only the Indian environment but also its tradition and way of life. Environmentalists believe that foreign fast-food chains encourage people to eat a diet based on meat, which the country cannot afford. They say that breeding large numbers of animals will make it difficult for India to feed itself. The animals which give the meat, milk and eggs are fed on grain. The same amount of grain would feed five times as many people if they ate it themselves. Similarly, 1700 litres of water are needed to produce half a kilo of chicken. That is 20 times the water that an average Indian family would need each day.

Globalized and glocalized production

FOOD PRODUCTION

Glocalized commercial production	Globalized production
Benefits	**Benefits**
Producer • Increased market access and sales • Possibly more farm-gate sales	*Producer* • Ability to produce foods cheaply and at a uniform standard
Consumer • Fresh food • Local products "in season" • Reduced air miles • Smaller carbon footprint	*Consumer* • Cheap food available year round • All types of products available year round • Competition between producers keeps main costs down
Local economy • Improved local farming economy • Multiplier effects, e.g. demand for fertilizers, vets, farm equipment	*Local economy* • May be able to provide large amounts of a single product to a major TNC • Specialization allows intensification and increased production
Costs	**Costs**
Producer • Increasing cost of oil makes cost of inputs higher • Greater emphasis on quality may make production less profitable	*Producer* • Increased air miles • Higher input costs, especially fertilizers and oil • Profit margins increasingly squeezed
Consumer • Higher cost of local farm products • Less choice "out of season"	*Consumer* • Increased costs are likely to be passed on to the consumer • Indirect costs such as pollution control, eutrophication of streams, soil erosion, declining water quality
Local economy • Cost of subsidies to maintain farming, e.g. payments to encourage farming in environmentally friendly ways	*Local economy* • Undercuts local farmers who may quit farming • Producers are vulnerable to changes in demand and are at the mercy of TNCs

Costs and benefits of globalized and glocalized food production

MULTINATIONAL COMPANIES AND LOCAL PRODUCTION

The impact of fast-food producers in India

The impact of multinational companies on local agriculture in India is considerable. Contract farming, where a company buys the entire crop of a large group of farmers, is common. For example, Pepsi Foods now controls about 1000 farmers in the state of Punjab, telling them what to grow and how to grow it. In 1989, the company opened a factory to make tomato paste for the Pizza Hut chain of restaurants around the world. Local farmers were given interest-free loans to buy foreign seeds and tools, and Pepsi buys the produce at fixed prices. Tomato yields trebled in four years and production quadrupled, making the paste competitive with that made in other countries. Pepsi Foods also makes potato crisps locally. However, the company now uses imported potatoes which, it claims, make better crisps – Indian potatoes have a high sugar content, which causes browning. Other foreign-food giants are now trying to get permission to grow imported varieties to make frozen French fries. Environmentalist critics claim that imported seeds reduce the use of local types. They also claim that contract farming will force farmers to rely on a limited number of crops, which will disrupt traditional farming practices.

Multinational companies claim that they are encouraging local farmers to provide fresh produce at the local market. However, much of this produce is sold to overseas fast-food outlets. Indian farmers now earn about $10 million worth of exports for the McDonald's chain. A similar experience arose in Brazil, where initially McDonald's dealt with just eight farmers. Within eight years, it was dealing with 7000 Brazilian suppliers.

Increased production would clearly benefit India's agricultural exports. Indeed, increasing agricultural exports, especially by promoting local food processing, has become a priority for India's government. One of the most promising areas for exports is fruit and vegetables. India is the world's largest producer of fruit and the second largest of vegetables, but its share of global exports is only about 1%. An estimated 25–30% of the fruit and vegetables grown in India goes to waste, and just 1% is commercially processed locally. This compares with 70% commercial processing in Brazil and Malaysia. Fast-food chains say that they have no desire to undermine India's traditional food. They claim that they are not trying to change eating habits and that no one is being forced to visit a fast-food chain. Why should people in a developing country be deprived of something that the rest of the world has?

Alternatives (1)

THE ROLE OF CIVIL SOCIETIES

A **civil society** refers to any organization or movement that works in the area between the household, the private sector and the state to negotiate matters of public concern. Civil societies include non-government organizations (NGOs), community groups, trade unions, academic institutions and faith-based organizations.

Global civil society is extraordinarily heterogeneous. Groups that comprise it can be liberal, democratic and peaceful, while others are illiberal, anti-democratic and violent. If organizations such as Oxfam International and Greenpeace are part of global civil society, arguably so too is al-Qaeda. Furthermore, even those global civil society groups that advocate progressive values – development NGOs, for example – may sometimes act in ways that run counter to those values.

The perception that global institutions, such as the World Bank and the IMF, are undemocratic and do not help all people equally has led to a global civil society movement that is attempting to regulate the global system from below. This has witnessed a massive rise in NGOs representing the needs of many "victims" of globalization. The statistics are impressive:

- A survey of NGOs in 22 nations showed that they employed 19 million workers, recruited 10 million volunteers and generated $1.1 trillion in revenue.
- In 1960, each country had citizens participating in 122 NGOs – by 1990, the number had increased to over 500.
- In western Europe, 66% of NGOs have been formed since 1970.
- There are over 2 million NGOs in the USA, 75% of which have been formed since 1968.
- In eastern Europe, 100,000 non-profit organizations appeared between 1989 and 1995.
- In Kenya, over 250 NGOs appear every year.
- In 1909, there were just 176 international NGOs; by 2000, there were over 29,000, 60% of which had been formed since the 1960s.

However, a note of caution is required. Evidence from South Africa suggests that many small-scale NGOs and local bottom-up development schemes fold after a short period of time. The figures must be treated with care.

A number of broad alliances have emerged within the NGOs, such as the global environmental movement, the anti-globalization movement and the global women's movement. Well-known individual NGOs include Greenpeace, The Fair Trade Network, Stop The War Coalition, Globalize Resistance, Oxfam, CAFOD, Amnesty International and Médecins Sans Frontières. Each of these have different aims and methods but all agree that major globalizing bodies such as the World Bank, the IMF and the G8 countries are pushing an agenda that favours rich western countries at the expense of others.

At an individual level, some people have decided to boycott GM crops. Others, during the recent increases in oil prices, have boycotted garages owned by Shell and BP. Others choose to do something positive – buying Fair Trade products is one way of helping producers in poor countries at the expense of large TNCs.

While the role of global civil society should not be overstated (it is generally much less powerful than governments, international organizations and the private sector), there are plenty of examples of where global civil society groups have been a force for progressive social change. The International Campaign against Landmines and the Jubilee 2000 campaign for debt relief are two of the best known and most successful. More generally, parts of global civil society have succeeded in putting new issues and ideas onto the international agenda, and in effecting changes in national and international policies. They have helped to improve the transparency and, to some extent, the accountability of global institutions, and to mobilize public awareness and political engagement.

Important areas where global civil society is trying to have an impact include:

- creating a more level playing field for the global South
- supporting free media and access to information
- making global civil society more accountable and transparent
- establishing a new relationship with global institutions.

EXTENSION

Visit

http://web.worldbank.org/WBSITE/EXTERNAL/TOPICS/CSO/0,,contentMDK:20127718~menuPK:288622~pagePK:220503~piPK:220476~theSitePK:228717,00.html

for the World Bank website on civil society organizations.

EXTENSION

Visit

http://www.msf.org.uk

for Medicins Sans Frontiers (UK). Select an issue or a country to investigate.

Alternatives (2)

CIVIL SOCIETIES – AN EXAMPLE FROM NIGERIA

Shell and Ogoniland: development for whom?

In 1979 Nigeria was at the peak of an oil boom. Oil brought in $25 billion and external debt was less than $10 billion. Within a few years, however, Nigeria had gone from boom to bust and has yet to recover. Shell is by far the largest oil company in Nigeria and has long been the focus of many protests. Shell is responsible for nearly half the country's output of 2 million barrels a day, and Nigeria is as dependent on oil as it ever was. Oil accounts for 80% of export earnings and 90% of government revenue. Additionally, Shell is the leading partner in a proposed liquefied natural gas (LNG) project. This promises to be the most important source of foreign exchange in Nigeria since the development of the oil industry.

The abbreviated text below is reproduced from the newspaper advertisements taken out by Greenpeace, The Body Shop International, Friends of the Earth and Chaos Communications. It raises a broad spread of issues:

- economic (should the public buy Shell products?)
- environmental (degradation and pollution)
- social (poor people unable to defend themselves)
- cultural (the chances for the Ogoni people to continue as farmers and fishermen).

DEAR SHELL, THIS IS THE TRUTH. AND IT STINKS.

For over thirty years, the activities of the Nigerian Government, Shell and other multinational oil companies have led to the widespread degradation and pollution of the region's lakes, rivers, land and air. The Ogoni are mostly farmers and fishermen, who need their land and water to live. The oil spills and pollution must be cleared up and the lands restored.

Shell must take responsibility for their part in this pollution. We believe that Shell has an obligation to operate to the highest environmental and social standards. We do not believe that Shell has done so in Nigeria.

Please heed the words of Ken Saro-Wiwa himself, writing from his prison cell before his execution on 10 November 1995: "I believe that only a boycott of Shell products and picketing of garages can call Shell to their responsibility to the Niger Delta. I remain hopeful that men and women of goodwill can come to the assistance of the poor deprived in Ogoni and other parts in the Niger Delta who are in no position to defend themselves against a multinational such as Shell."

THE BODY SHOP, WATERSMEAD, LITTLEHAMPTON, WEST SUSSEX
FRIENDS OF THE EARTH, 26–28 UNDERWOOD STREET, LONDON N1 7JQ
GREENPEACE, GREENPEACE HOUSE, CANONBURY VILLAS, LONDON N1 2PN

THIS MESSAGE WAS FUNDED BY THE BODY SHOP INTERNATIONAL, FRIENDS OF THE EARTH, GREENPEACE AND CHAOS COMMUNICATIONS LIMITED

Shell also took out an advertisement. These are some of its points.

CLEAR THINKING IN TROUBLED TIMES

There are certainly environmental problems in the area, but as the World Bank's Survey has confirmed, in addition to the oil industry, population growth, deforestation, soil erosion and overfarming are also major environmental problems there.

In fact, Shell and its partners are spending US$100 million this year alone on environment-related projects, and US$20 million on roads, health clinics, schools, scholarships, water schemes and agricultural support projects to help the people of the region. And, recognizing that solutions need to be based on facts, they are sponsoring a $4.5 million independent survey of the Niger Delta.

Some campaigning groups say that we should intervene in the political process in Nigeria. But even if we could, we must never do so. Politics is the business of governments and politicians. The world where companies use their economic influence to prop up or bring down governments would be a frightening and bleak one indeed.

We'll keep you in touch with the facts.

EXAM QUESTIONS ON PAPER 3 – HIGHER LEVEL EXTENSION – GLOBAL INTERACTIONS

Key features

Timing: You have 1 hour to do one question worth 25 marks. It is recommended that you write a short plan for your answer.

Choice: There is a choice of one out of three questions.

Structure

Part (a) uses straightforward terms such as *describe* and *explain*, whereas part (b) requires a more analytical approach. Examples and case studies should be used wherever appropriate.

Note: The term "global interaction" means a process of exchange between nations involving people, goods, services and ideas. Globalization includes global interactive processes and also their outcomes.

---

1 a) Explain the process of cultural diffusion. [10]

b) To what extent has global interaction reduced cultural diversity? [15]

2 a) Describe the characteristics and pattern of global interaction. [10]

b) Examine the economic benefits that derive from participation in the global economy. [15]

3 a) Explain the role of technological changes in transport and communications upon the process of globalization. [10]

b) Examine the environmental problems resulting from global interaction. [15]

4 a) Explain how and why globalization may be measured. [10]

b) Examine the reasons why some counties are more globalized than others. [15]

5 a) "Increasing globalization helps to expand opportunities for nations and, on average, helps workers in rich and poor countries alike." (World Bank Development Report)

Explain the argument being put forward in this statement. [10]

b) Analyse the changing role of the nation state in an increasingly globalized world. [15]

6 a) Explain the causes, impacts and responses to one major global pollution incident. [10]

b) Examine the growth of global concern over environmental issues arising from global interactions. [15]

THE RELATIVE IMPORTANCE OF ESSAYS AT HL AND SL

IB exams consist of a number of different approaches to assessment, including extended responses. The advice given here is directed towards conventional full-length essays, which are compulsory in Papers 1 and 3. In both cases, one essay carries a relatively heavy mark weighting, as shown below:

- Paper 1, Section B HL 6.25%, SL 10% of total marks
- Paper 3 (HL only) 20% of total marks. In this exam you will have one hour to answer the question, which appears as parts (a) and (b). These may be linked to the same topic, for example Economic Interactions, or may be independent. Either way, you should approach the two parts separately and assume that the examiner will not cross-credit them, i.e. transfer marks from one to the other if information is misplaced.

INTERPRETING THE ESSAY TITLE

1. Underline the keywords in the title.
2. Go through the checklist below to check each aspect against your essay title to see if it is relevant or not. This will ensure that you give the essay title its broadest interpretation. The title may be brief and leave you to think creatively and to comment on specific aspects of the subject which are not actually mentioned in the title but which are relevant to it. For example, if the question asks you to comment on the global variation in fertility rate, you would need to write about variations in time as well as space.

Checklist

Note that not all the items in this checklist will be relevant to your essay.

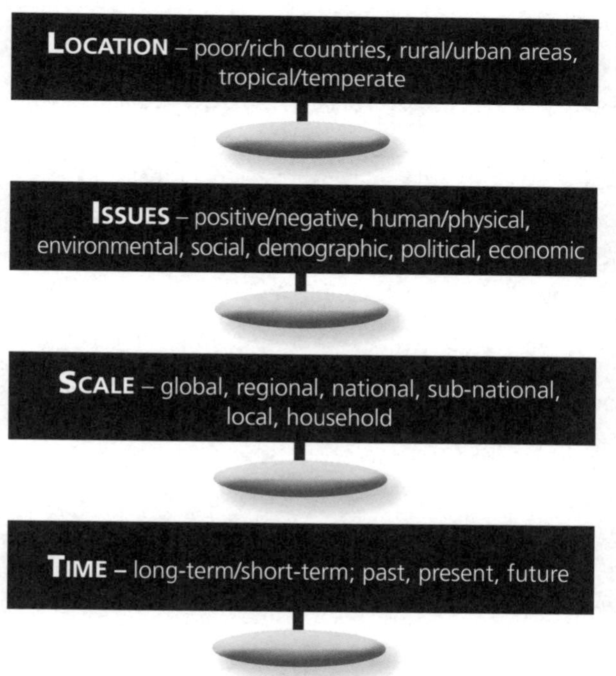

PLANNING

Planning is important. Reasons why you should plan your essay include:

- it allows you to order your thoughts before writing.
- you can return to the essay plan and insert new points as you get inspiration while writing.
- it presents a logical sequence of ideas that the reader can easily follow.
- examiners have little time and will credit a well-structured answer that is easy to follow.
- it allows you to focus on the question and make sure that the content is relevant.

STRUCTURE OF THE ESSAY

Introduction

The introductory paragraph gives an interpretation of the title, defines terms, indicates the slant or the direction of the argument and generally sets the scene.

The main body of the essay

Make sure that each paragraph in this part of your essay presents a distinct point or idea. The opening line of each paragraph should clearly indicate its content. The remainder of the paragraph elaborates on that point.

Examples, case studies and illustrations, such as sketch maps and diagrams, should appear in this section.

Conclusion

Here you should return to the essay title and provide an overview of your response. The conclusion should not contain new ideas; it should round off an argument and summarize the key features of the content.

THE LANGUAGE OF IB EXAMS

It is recommended that you become familiar with the command words and other terms listed and defined below. They are all found in IB geography exam questions – misinterpretation costs marks.

Analyse	break down in order to bring out the essential elements or structure
Annotate	add brief notes to a diagram or graph
Classify	arrange or order by class or categories
Compare	give an account of the similarities between two (or more) items or situations, referring to both (or all) of them throughout
Compare and contrast	give an account of similarities and differences between two (or more) items or situations, referring to both (or all) of them throughout
Construct	display information in a diagrammatic or logical form
Contrast	give an account of the differences between two (or more) items or situations, referring to both (or all) of them throughout
Define	give the precise meaning of, for example, a word, phrase, concept or physical quantity

Describe	give a detailed account
Determine	obtain the only possible answer
Discuss	offer a considered and balanced review that includes a range of arguments, factors or hypotheses. Opinions or conclusions should be presented clearly and supported by appropriate evidence
Distinguish	make clear the differences between two or more concepts/items
Draw	represent by means of a labelled, accurate diagram or graph, using a pencil. A ruler (straight edge) should be used for straight lines. Diagrams should be drawn to scale. Graphs should have points correctly plotted (if appropriate) and joined in a straight line or smooth curve
Estimate	obtain an approximate value
Evaluate	make an appraisal by weighing up the strengths and limitations
Examine	consider an argument or concept in a way that uncovers the assumptions and interrelationships of the issue
Explain	give a detailed account, including reasons or causes
Identify	find an answer from a number of possibilities
Justify	give valid reasons or evidence for an answer or conclusion
Label	add labels to a diagram
Outline	give a brief account or summary
State	give a specific name, value or other brief answer without explanation or calculation
Suggest	propose a solution, hypothesis or other possible answer
To what extent	consider the merits or otherwise of an argument or concept. Opinions and conclusions should be presented clearly and supported with empirical evidence and sound argument

Source: *Adapted from the Geography Subject Guide, IBO*

Exam-speak – common terms that confuse

Verbs

Referring to	mentioning or using
Influence	the effect of one thing upon another
Modify	change
Respond to	take action

Nouns

Outcome	consequence/result
Benefits/advantages	positive outcomes
Costs/disadvantages	negative outcomes
Impacts/effects	usually dramatic outcomes
Issues	important and controversial results
Problems	difficulties
Pressures/conflicts	undesirable competition
Challenges	difficulties which may be overcome
Opportunities	potential benefits
Trend	change over time (on a graph)
Pattern	distribution in space
Feature	a distinct part, e.g. a cliff is a coastal feature
Process	the actions or changes that occur between parts
Relationship	a two-way interaction

Adjectives

Global	the whole world
Regional	global regions, e.g. Asia-Pacific
National	belonging to one country
Local	the immediate area or district
Possible	likely to happen
Probable	very likely to happen
Economic	relates to business, finance, employment
Social	relates to human welfare e.g. housing and health
Cultural	relates to language, customs, religion and moral codes
Political	relates to the actions of governments
Demographic	relates to populations e.g. fertility rate
Environmental	relates to the physical environment

PAPERS 1 AND 2 MARKBANDS

Markband	Mark range Paper 1 Section B	Mark range Paper 2	Descriptor
A	0	0	No relevant knowledge; neither examples nor case studies; no evidence of application; the question has been completely misinterpreted or omitted; no evaluation; no appropriate skills
B	1–3	1–2	Little knowledge and/or understanding; largely superficial or of marginal relevance; or no irrelevant examples and case studies; very little application; important aspects of the question are ignored; no evaluation; very low level skills; little attempt at organization of material; no relevant terminology
C	4–6	3–4	Some relevant knowledge and understanding, but with some omissions; examples and case studies are included, but limited in detail; little attempt at application; answer partially addresses question; no evaluation; few or no maps or diagrams; little evidence of skills or organization of material; poor terminology
D	7–9	5–6	Relevant knowledge and understanding, but with some omissions; examples and case studies are included, occasionally generalized; some attempt at application; competent answer although not fully developed, and tends to be descriptive; no evaluation or unsubstantiated evaluation; basic maps or diagrams, but evidence of some skills; some indication of structure and organization of material; acceptable terminology
E	10–12	7–8	Generally accurate knowledge and understanding, but with some minor omissions; examples and case studies are well chosen, occasionally generalized; appropriate application; developed answer that covers most aspects of the question; beginning to show some attempt at evaluation of the issue, which may be unbalanced; acceptable maps and diagrams; appropriate structure and terminology
F	13–15	9–10	Accurate, specific, well-detailed knowledge and understanding; examples and case studies are well chosen and developed; detailed application; well-developed answer that covers most or all aspects of the question; good and well-balanced attempt at evaluation; appropriate and sound maps and diagrams; well-structured and organized responses; terminology sound

These markbands are to be used for Papers 1 and 2 at both standard level and higher level.

PAPER 3 MARKBANDS

Markband	Mark range Part (a) Maximum 10 marks	Descriptor	Mark range Part (b) Maximum 15 marks	Descriptor
A	0	No relevant knowledge or inappropriate; the question has been completely misinterpreted or omitted; no appropriate skills	0	No relevant knowledge or inappropriate; the question has been completely misinterpreted or omitted; no synthesis/evaluation; no appropriate skills
B	1–3	Little relevant knowledge and/or understanding; important aspects of the question are ignored; little attempt at organization of material	1–4	Little relevant knowledge and/or understanding; important aspects of the question are ignored; little attempt at synthesis/evaluation; little attempt at organization of material
C	4–6	Some relevant knowledge and understanding; answer partially addresses question; some indication of structure or organization	5–8	Some relevant knowledge and understanding; answer partially addresses question; basic synthesis; basic or unsubstantiated evaluation; some indication of structure or organization
D	7–8	Generally accurate knowledge and understanding; answer is developed, and covers most aspects of the question; appropriate structure with generally appropriate terminology	9–12	Generally accurate knowledge and understanding; answer is developed, and covers most aspects of the question; synthesis that may be partially undeveloped / evaluation that may be partially unsubstantiated; appropriate structure with generally appropriate terminology
E	9–10	Accurate, relevant knowledge and understanding; well-developed answer that covers most or all aspects of the question; well-structured response with sound terminology	13–15	Accurate, relevant knowledge and understanding; well-developed answer that covers most or all aspects of the question; clear, developed synthesis and substantiated evaluation; well-structured response with sound terminology

Source: *Adapted from the Geography Subject Guide, IBO*

These markbands are devised to assist examiners in grading answers. A best fit approach is adopted, which means that the student's answer should fulfil most but not all the requirements of any markbands.

PAPER 1 – THE CORE

Notes:

AOVP – any other valid point
OWTTE – or words to that effect
Explain includes describe.

Section A

1 *Populations in transition*

a Allow [1 × 3 marks] for a description of three changes and [1 mark] for quantification. The following would be suitable or AOVP: an increase in total population; improved female survival; increased life expectancy. Credit other valid changes. **[4]**

b This is a tendency for population to continue to grow [1 mark] even when birth rates are falling [1 mark]. Growth is sustained by the large number of young people with reproductive potential [1 mark]. **[3]**

c Advantages – Ageing societies are typical of rich countries. They can provide a cheap and amenable, semi-retired workforce; they can offer childcare for working families; they support the "grey" economy via health and social care; many have disposable income and have boosted the retirement and leisure industries.

Disadvantages – They do not have the potential as a future workforce and are costly to support in terms of health, housing, social care and pensions. An ageing society puts financial pressure on the workforce. Fertility may be boosted by promoting immigration, sometimes resulting in social conflict. Award a maximum of up to [4 marks] for advantages or disadvantages; the answer need not be evenly balanced, but both need mentioning. **[6]**

d The most likely diagram is one similar to the demographic transition model, with one line for birth rate [1 mark] above the line for death rate [1 mark] and the intervening space shown as natural increase [1 mark]. Add [1 mark] for the Y axis labelled correctly as BR/DR (‰) and [1 mark] for the X axis indicating time in years. **[5]**

2 *Disparities in wealth and development*

a 75% of global trading activity is between the high income countries (HICs) such as those in Europe and North America. The middle income countries, such as those in South-East Asia, have benefited from increasing access to world markets; but many of the low income countries have limited access to world trade, mainly due to the exclusive and protectionist policies of trading blocs. In addition, the commodities they export are generally of relatively low value and the terms of trade unfavourable. **[5]**

b High CDR is often linked to poor standard of living [1 mark], but not always. CDR depends on the age structure of the population [1 mark]; ageing populations have high CDRs [1 mark] and vice versa. E.g. India's CDR of 8‰ is lower than that of the UK at 10‰ [1 mark]. CDR is general rather than age-specific such as IMR. **[4]**

Award full marks for four valid points.

c Advantages of aid – It can relieve a crisis such as famine. It can be used for development projects. Disadvantages of aid – It leads to foreign dependence. Development projects sometimes use inappropriate technology or have foreign ties. Food aid depresses local market prices. **[4]**

Credit any two of each plus AOVP.

d Gender inequality means the low status of women (not men). Such a situation prevents the full utilization of human capital and restrains economic growth. Low status can limit a woman's education and restrict her role to childbearing. This not only results in high rates of maternal and child mortality, but adds further economic burdens to the state. Finally, the poor status of women compromises the well-being of the nation and its international recognition. **[6]**

A good answer will cover most of these points and refer to at least one country.

3 *Patterns in environmental quality and sustainability*

a There are numerous possibilities, but the most common are organic wastes from sewage, fertilizers and pesticides from agriculture, and heavy metals and acids from industrial processes and transport. **[2 + 2]**

b Answers should explain any two of the following strategies, used to prevent or arrest soil degradation: **[3 + 3]**

- mechanical techniques to prevent excessive surface runoff, including terracing, contour ploughing, bunding and filling gullies with brushwood
- mechanical techniques against wind, including windbreaks and mulching to improve soil cohesion
- afforestation to impede rapid surface runoff and soil loss on steep slopes
- maintenance of crop cover and use of organic fertilizers and mulches to improve and stabilize soil structure
- treatment and prevention of salinization by flushing out the salt, avoiding high amounts of evaporation that results from inefficient irrigation techniques.

Credit AOVP.

c The focus is on environmental sustainability and this should be defined or a clear understanding shown. Accept any two valid points concerning environmental protection and resource conservation [2 marks]. The scale should also be local, which means within one country [1 mark]. The chosen management strategy should be described and the explanation should touch upon need and/or suitability [2 marks]. **[5]**

d The diagram should show incoming shortwave (UV) radiation [1 mark], some outgoing longwave (infrared) radiation returning to space [1 mark], some of the outgoing radiation reflected and absorbed [1 mark], and a layer of greenhouse gases [1 mark]. **[6]**

Award [2 marks] for further detail and accuracy of the diagram.

4 ***Patterns in resource consumption***

a First, food insecurity of any kind is evidence of population demand exceeding food supply and is widespread throughout sub-Saharan Africa, where famines are a regular occurrence. Second, the exhaustion of resources can be exemplified through the global depletion of fossil fuels, loss of biodiversity and soil erosion. **[5]**

Award [2 marks] for a reference to food, [2 marks] for a reference to other resources and [1 mark] for a named example or AOVP.

b Any two regions may be chosen, but those at each end of the spectrum would be ideal – e.g. North America and Africa. A brief description of their respective footprints [2 marks] should be followed by an explanation regarding consumption. The total footprint of North America is 9.5 global hectares per person (GHP); but for Africa it is only 1 GHP. Explanation should focus on specific aspects of resource consumption and waste production, both of which are related to the level of economic development. **[5]**

c Overpopulation is when the number of people exceeds the carrying capacity/resources available [2 marks] under the current level of technology [1 mark] there. Award [2 marks] for examples. All points should be fully developed. **[5]**

d Advantages of nuclear power – It is cheap. It has very large reserves. It generates high amounts of electricity.

Disadvantages of nuclear power – It is a radioactive material. It is a hazard to dispose of. There is a risk of accidents. Its effects are persistent and diffuse. **[4]**

Credit any two of each plus AOVP.

Section B

1 Answers should show an understanding of the following terms – migration, resources, uneven distribution.

- Migration is the movement of people, involving a permanent (more than one year) change of residence. It can be internal or external (international) and voluntary or forced. It does not include temporary circulations such as commuting or transmigration.
- Resources can mean biological (wood, soil, animals, plants, food); physical (geology, minerals, water, fuel), or manmade (money, housing, infrastructure, technology).
- Uneven resource distribution may lead to people leaving areas of resource deficit and moving to areas where there are plenty of resources for the population (areas of optimum population or underpopulation).

The counter-argument may be put forward that both internal and international migration may result from factors other than resource imbalance. These include wars, civil unrest and natural disasters, none of which is directly concerned with the uneven distribution of resources (war being a possible exception).

Answers reaching markband E or above are expected to cover the arguments concerning uneven distribution, but also to address the other non-resource related factors. Examples and case studies are expected throughout the answer. **[15]**

2 The fundamental socio-economic motive for migration would include the search for a better quality of life. Push factors (unemployment, underemployment, lack of public utilities, poor medical and education services, poor housing and overcrowding) and pull factors, which are generally the opposite, should be mentioned. The discussion should include international labour movements, such as those from Mexico to the USA and also between the NICs. Internal migration driven by socio-economic motives would include rural-to-urban movement in poor countries and centrifugal movement of workers and retirees in the rich world. Temporary movements such as commuting and tourism are not relevant.

To achieve markbands E/F, factors other than socio-economic ones (e.g. political and environmental) need to be considered briefly in terms of relative importance, and migration should be discussed on more than one scale. **[15]**

3 Sustainable development is defined as "development that meets the needs of the present population without compromising the ability of future generations to meet their own needs". This involves environmental protection, reducing pollution and ensuring that population does not exceed resource supply or environmental carrying capacity. The definition also extends to quality of life and minimizing human stress.

The unprecedented growth rate of global populations has outstripped the rate at which resources have been developed to meet its demands. The evidence is provided by environmental crises such as the depletion of stocks, the degradation of soil, the pollution of water, contamination of land and loss of biodiversity.

Achieving sustainable development means balancing resource demand with supply. Birth control policies which reduce fertility are effective, provided they avoid coercion and are accompanied by improvements in primary healthcare. Sustained levels of resource production are only possible if techniques of production are environmentally sound.

Answers accessing markbands E/F will fully explore the population/resource relationship and recognize its complexity and dynamic nature. **[15]**

4 The answer should include a definition or show a clear understanding of the term fertility. The crude birth rate (the number of live births per thousand population per year) is acceptable, but more specific measures are preferable, such as either total fertility rate (the average number of births per thousand women of childbearing age) or the general fertility rate (the number of births per thousand women aged 15–49 years, sometimes 44).

There are four possible relationships for discussion:

- High fertility and weak economic growth, which causes poverty; this might be exemplified by sub-Saharan Africa, where children are regarded as an asset rather than an encumbrance. Reasons include the inability of the economy to keep pace with the demands of a growing population and their need for housing, health and education.
- High fertility and rapid economic growth is typical of South-East Asia, where over the last 30 years the population has been an important resource.
- Low fertility and affluence is typical of the richer countries, where children are costly to raise.
- Low fertility and poverty is typical of areas in both the rich and poor world suffering from out-migration and an ageing population.

A good answer accessing markbands E/F will mention at least three of these relationships. **[15]**

5 The global pattern may be described as rich North/poor South, but with recognition of intermediate rapidly advancing economies, particularly those of South-East Asia such as Taiwan and South Korea. Acceptable economic indicators include GDP/GNP/GNI per capita and the value of exports and AOVP. Several indicators should be evaluated, including one composite index such as HDI. An evaluative answer achieving markbands E/F is likely to promote composite indices and to indicate the shortfalls of single index such as GDP per capita, which disguises variation within one nation and ignores quality of life.

The allocation of marks to identifying the global pattern and evaluation need not be equally balanced to achieve markbands E/F. **[15]**

6 The goals are: 1 Eradicate extreme poverty and hunger; 2 Achieve universal primary education; 3 Promote gender equality and empower women; 4 Reduce child mortality; 5 Improve maternal health; 6 Combat HIV/AIDS, malaria and other diseases; 7 Ensure environmental sustainability; 8 Develop global partnership for development.

A good answer achieving markbands E/F will describe the general aim of the eight goals. Precise wording of the individual goals is not required, but they should be supported by statistics for target dates and achievement levels. **[15]**

7

- Environmental effects of temperature increase – changes in wind, pressure, precipitation and humidity; sea-level rise through thermal expansion and ice-melting causing coastal erosion and flooding; more frequent extreme events, such as storms, droughts, fire, erosion, landslides, sedimentation, avalanches, pests and diseases.
- Socio-economic effects – social disruption and economic losses in low-lying areas liable to coastal flooding and more frequent storms, e.g. in Egypt, the Netherlands and Bangladesh; reduced agricultural production due to drought in the US Grain Belt and many areas of southern Europe; expansion of tourism in areas of high latitude/altitude.
- Responses – These depend on the extent to which countries will be affected physically, socially and economically; their level of economic development and their ability to access new technologies for cutting emissions of greenhouse gases.

Further elaboration and specific detail on the following is expected – international initiatives, such as the Kyoto protocol; strategies such as planting forests to absorb carbon dioxide; keeping fewer cattle to reduce methane emissions; buying carbon credits from other countries that do not use their full quota; technological solutions to reduce CO_2, such as improved energy efficiency, fuel switching to renewable energy resources and nuclear power, and the capture and storage of CO_2. **[15]**

A good answer accessing markbands E/F will include a wide range of effects and responses.

8 Global climate change is most likely to be interpreted as global warming, which is currently its prominent feature. Answers which discuss historic changes that may have involved cooling are equally acceptable, but unlikely to be well supported by examples. The focus of the answer is on consequences alone; the causes are irrelevant.

The consequences of global warming are likely to include: sea-level rise leading to flooding in low-lying areas, increased storm activity, changes in agricultural patterns and a reduction in biodiversity. These consequences are likely to affect poor populations more than rich ones because they often live in marginal areas with a high risk, such the Ganges Delta. These populations are vulnerable because they lack coping mechanisms, they are less mobile, they lack insurance, they will take longer to recover and will suffer secondary effects of climate change and associated events such as drought, storm and epidemics. Wealthy countries with diversified and more robust economies can adopt strategies to protect themselves against some negative aspects of climate change. These include coastal defences, insurance and preparedness programmes. **[15]**

Answers accessing markbands E/F will consider a range of consequences and include at least two examples.

9

- Definition – Biodiversity involves plants, animals and micro-organisms. It refers to species diversity, genetic diversity and the interdependence of species within the ecosystem.
- Characteristics – The tropics are the richest area for biodiversity. Tropical forests contain over 50% of the world's species in just 7% of the world's land. They account for 80% of the world's insects and 90% of primates.
- Origins – Their biodiversity stems from a long history free from human disturbance when species evolved slowly (some have evolved over >50 million years). The optimum growing conditions and wide variety of habitats permit a wide range of niches and species.

The rainforest has nurtured this "pool" to become home for 170,000 of the world's 250,000 known plant species.

- Economic value – Fuelwood, charcoal, pulpwood, plywood, industrial chemicals, resins, rubber, medicinal plants.
- Ecological value – Integrity of a complex ecosystem with the potential for future genetic resources.

A good answer accessing markbands E/F should include a definition of biodiversity and most of the points listed, but the emphasis may vary. Credit relevant case studies. **[15]**

10 The answer should distinguish between physical and economic water scarcity. Physical water scarcity occurs where supply is limited, normally by inadequate rainfall and high rates of evaporation, and the demands may not be met by supply despite the application of technology. Economic scarcity is associated with poverty and occurs when human, institutional and financial capital limits access to water even though it might be physically available.

- Causes may include: drought; population growth leading to increasing domestic demand; economic development leading to increasing demand from industrial, domestic and especially agricultural sectors; international disputes jeopardizing supply, especially in the Middle East; pollution and poor quality resulting from effluents and poor infrastructure.
- The consequences include: food shortages and health crises; international conflict; decreasing river discharge and waste disposal problems; water pollution and ecosystem destruction.

An answer accessing markbands E/F should cover at least three causes and three consequences, although the attention given to each need not be balanced. **[15]**

11 The answer should cover the disadvantages of fossil fuels and the relative advantages of renewable energy resources

Disadvantages of fossil fuels include serious environmental impacts through mining/drilling, transport/piping, refining and emissions which are implicated with global warming and acid rain.

Geopolitical problems result from the uneven global pattern of oil production and consumption and the resource being used as an international political tool.

Renewable energy resources include HEP, geothermal, solar, wind and tidal.

Increasing use of renewables relates to their minimal environmental impact. However, their capacity to replace non-renewables and to generate large quantities of energy at peak times is still limited. Nuclear power would be accepted in the answer as a renewable type of energy.

Answers which focus only on renewable types of energy and make no comment on the disadvantages of fossil fuels should not move beyond markband D. **[15]**

12 The demand for some natural resources continues to rise at a pace that cannot be met by further exploitation of raw material, and alternative approaches are required. A reduction in consumption generally involves one or more of the following strategies: conservation, substitution, recycling and reuse. One or more resources must be chosen, but a detailed response that refers to one resource is acceptable, as is one that covers several resources in less detail.

Strategies used to reduce resource consumption may include the following:

- conservation – reduces the exploitation of raw material such as wood, oil and bauxite.
- substitution – involves replacing the non-renewable resource with an abundant alternative, e.g. the use of fibre-optic cables instead of copper.
- recycling – reduces consumption of the raw material, the energy used in secondary processing, and the amount of waste production. Paper is an ideal choice and both pre-consumer and post-consumer waste is now being reused. Recycling requires less energy than the original pulp mill. It also reduces the amount of landfill and the amount of methane generated. Aluminium recycling cuts the use of energy by 95% of that used in original smelter. It also conserves bauxite.
- reuse – involves only collection, cleaning and redistribution, e.g. glass bottles.

A good answer accessing markbands E/F will comment on conservation, substitution and recycling (which may include reuse). **[15]**

PAPER 2 – OPTIONAL THEMES FOR HL AND SL

Abbreviations:

AOVP – any other valid point
OWTTE – or words to that effect

Option A: Freshwater – issues and conflicts

A1 a i) These are freshwater resources which are unpolluted [1 mark] and from surface water such as rivers, reservoirs and aquifers [1 mark].

ii) Water dependence is the percentage of water that comes from sources outside a national boundary [1 mark]. Allow [1 mark] for an example of a country or AOVP. **[2 + 2]**

b Freshwater resources are related to climatic zones and reflect the relationship between precipitation and evaporation. Withdrawal is related to water demand, but is also dependent on availability, access, infrastructure and technology, all of which are linked to the level of economic development.

For each of the columns A and B, allow [2 marks] for explanation and [1 mark] for examples from the countries listed. **[6]**

c This issue may be addressed on either the national and/or the international scale. Competition arises from increasing demands from agriculture, industry and domestic usage, all of which are associated with economic development and increasing consumption.

International disputes may arise where river basins are shared. The problems of matching increasing demand with supply are particularly acute in areas of water shortage, such as the Middle East and North Africa.

Although examples are not a specific requirement of the question, they should be included to access markbands E/F. **[10]**

A2 a i) The boundary of one drainage basin [1 mark] which separates it from an adjoining basin [1 mark] or AOVP.

ii) A porous water-bearing rock [1 mark], from which groundwater can be abstracted to supplement surface supplies of fresh water. AOVP for [1 mark]. **[2 + 2]**

b The diagram should be clear and accurate [1 mark] with short explanations for the input, storages, transfers and outputs. Award [1 mark] for a description of the input (precipitation), [1 mark] for two storages (interception, groundwater storage, depression storage), [2 marks] for four transfers (infiltration, percolation, leaf drip, stem flow, overland flow/surface runoff, throughflow) and [1 mark] for two outputs (evapotranspiration, evaporation, channel runoff).

All annotations must be part of the diagram or linked to it by arrows or a key. Separate written description is unacceptable. **[6]**

c Common problems include: salinization resulting from irrigation, agrochemical runoff from pesticides and eutrophication from fertilizers.

Good answers accessing markband E and above may cover two out of these three aspects provided that they are detailed and make close reference to more than one example. Accept AOVP.

[10]

Option B: Oceans and their coastal margins

B3 a A – wave-cut platform [1 mark].

This is an erosional feature resulting from the interplay between marine and sub-aerial processes. Marine forces include hydraulic action and abrasion causing undercutting and overhang. Sub-aerial processes of weathering, such as freeze–thaw and salt crystallization, disintegrate the steep cliff face, resulting in mass movement such as rockfalls and eventual retreat. The stepped form of this wave-cut platform is a response to the dipped horizontal strata of this sedimentary rock (Jurassic limestone).

Award up to [3 marks] for three valid points, but these must refer to both cliff face and cliff foot processes. **[1 + 3]**

b Characteristics – The visible backshore is a storm beach consisting of coarse sediment such as cobbles, pebbles and shingle. The coarsest material (cobbles) is stranded at the back of the beach by high tides and relatively weak backwash. Speculation about the foreshore, which is currently obscured by high tide, is acceptable. The whole beach is likely to have a concave cross-profile with fine shingle or sand on the foreshore.

Award [3 marks] for three well-explained and realistic characteristics.

Sediment sources – Onshore movement from the seabed, cliff fall, littoral drift, fluvial deposits and possibly beach nourishment.

Award [3 marks] for a brief explanation of three different sources of sediment. **[6]**

c The degree of protection depends on the threat posed by a receding coastline, the rate of recession, the value of the land affected, the vulnerability of the population living in the coastal zone and national priorities in terms of expenditure.

High-risk and high-value coastal zones devoted to urban land uses and recreation have traditionally been protected by hard engineering structures. Coastlines with lower risk and lower land values have received less protection or none at all. Priorities change with coastal land values, but recently environmental concerns have assumed importance, resulting in the adoption of a more passive approach through schemes such as managed retreat. Some poor countries with vulnerable coastal populations are exposed to a high level of risk from hazards such as storm surges and tsunamis. Despite the necessity for protection, financial restraints restrict this.

A good answer accessing markbands E/F is expected to cover at least two contrasting approaches with at least two examples. **[10]**

B4 a Award [2 marks] for each correctly named and located abiotic resource.

Continental shelves are a source of oil, gas, diamonds, sand and gravel. E.g. oil from the Persian Gulf and diamonds off the coast of Indonesia.

The ocean floor is a source of gold and manganese. Ocean ridges and rift valleys are rich in sulphur deposits close to hydrothermal vents/"black smokers".

Where abiotic factors (salinity, temperature, water, oxygen, nutrients and energy) instead of resources are cited, award a maximum of [2 marks]. **[4]**

b Exclusive economic zones recognize the right of coastal states to have control over their ocean space and are designed to conserve resources and avoid international dispute. Coastal states are free to exploit, develop, manage and conserve all resources such as fish, oil, gas, gravel and sulphur found in the waters, on the ocean floor and in the subsoil of an area, extending almost 200 nautical miles from its shore. Almost 90% of all known sub-sea oil reserves fall under one country's EEZ and 98% of the world's fishing regions fall within an EEZ.

Award [6 marks] for an answer that explains the purpose, the resources involved and the extent of an EEZ. **[6]**

c Overfishing involves exceeding the maximum tonnage of fish which can be caught in any one year to maintain maximum sustainable yield indefinitely. It occurs because fishing technology has become too intensive and reproduction patterns providing new recruits of fish have been disrupted. The use of factory ships and new on-board technology has been a major cause of this increasing efficiency.

Remedies designed to conserve fish stocks include: preventing improvements in efficiency by increasing mesh size and discouraging the marketing of juvenile fish; reducing the fishing effort by restricting time spent at sea, number of boats; imposing fishing permits, quotas and import tariffs; satellite and logbook surveillance and penalties for illegal landings.

Answers accessing markbands E/F should include both the causes of overfishing and most of the remedies listed. Evalutation may be speculative but must be included. **[10]**

Option C: Extreme environments

C5 a The description should include impact zones and areas of disturbed cushion plants, which are found in the upper valleys. These areas attract large numbers of trekkers visiting glaciers and peaks at higher elevations, where vegetation is thin and the ground easily exposed to erosion. Other valid explanations may be given. **[4]**

b Exposure to frost cycles at high altitudes leads to frost shattering, and steep gradients encourage rock falls and slides. This is exacerbated by heavy rainfall during the summer monsoon and further by the activities of humans through trekking and vegetation removal, which destabilizes slopes. Award up to [3 marks] for an explanation of weathering and [3 marks] for mass movement processes. **[6]**

c All extreme environments are fragile, have a low carrying capacity and are easily damaged by human activities such as mining, tourism and agriculture. Examples include oil exploration in Alaska, trekking in Nepal and overgrazing in the Sahel. Increased levels of human activity and exploitation result from population pressure and/or increasing demand. However, degradation is not inevitable and conservation measures, such as sustainable agricultural practices and ecotourism, have been designed to protect these areas from further damage.

Answers accessing markbands E/F are expected to discuss (not just describe) the fragility of one extreme environment, the damaging effects of at least one type of human activity there, and the nature of sustainable management – i.e. strategies designed to conserve its resources and protect its environment for future generations. **[10]**

C6 a Continentality implies remoteness from the sea [1 mark], which results in most of the moisture being evaporated by the time winds reach the centre of a large land mass [1 mark]. Onshore winds are cooled by the current and evaporation reduced [1 mark], therefore dry air moves inland e.g. the Benguela Current and the Namib Desert. **[4]**

Award [1 mark] for AOVP.

b Mechanical weathering results from a high diurnal range of temperature [1 mark]. Rock is weathered by thermal expansion and contraction and can result in exfoliation or granular disintegration [1 mark], but the weathered material remains *in situ* [1 mark].

Erosion is caused by wind and/or water [1 mark] and involves the removal and transport of material [1 mark] away from the site of weathering [1 mark].

Maximum marks should be awarded only where there is an attempt to distinguish between these two processes. Allow [1 mark] per point. **[6]**

c These areas share some adverse physical conditions, such as climatic extremes, slope instability and thin soils with little agricultural potential. All extreme environments are fragile, have a low carrying capacity and are easily damaged by human activities such as agriculture, mining and tourism.

Although some generalization is acceptable, answers should refer to specific examples of damaging human activities in the three environments. E.g. oil exploration in Alaska, trekking in Nepal and overgrazing in the Sahel.

Answers also need to address the issue of non-sustainability. Increased levels of human activity and exploitation result from population pressure and/or increasing demand. However, degradation is not inevitable and conservation measures such as sustainable agricultural practices and ecotourism have been designed to protect these areas from further damage. This more optimistic viewpoint is equally acceptable.

An evaluative response which recognizes both the limitations and future potential of these extreme environments is likely to be credited at markbands E/F. **[10]**

Option D: Hazards and disasters – risk assessment and response

D7 a Each of the three trends should be accurately described to include positive and negative changes and rates [1 × 3 marks]. Allow [1 mark] overall for quantification. **[4]**

b Before 1980 (approximately) the number of reported disasters and the numbers affected were lower than in 1980–2007. This was due to limitations in transport and communication, and less human involvement due to lower global population levels. Thereafter, the lines on the graph diverge: the number of reported disasters increases due to improvements in transport and ICT, allowing for easier access and more extensive reporting. The increase in the number of people affected (requiring immediate assistance during a period of emergency, such as food, water, shelter, sanitation and medical help) is also explained by population growth and exposure. Urbanization and the occupation of marginal land have also increased vulnerability. Nevertheless, the number of people killed has continued to decline due to technology, better hazard mitigation, and improved medical assistance.

Allow [6 marks] only where explanation is given for the changes in trends and there is cross-referencing between reported disasters, population affected and population killed. There must be some quantification. **[6]**

c People continue to live in areas exposed to natural hazards for a number of possible reasons:

- The event is unpredictable and people believe that it will never happen to them.
- There may be a lack of alternative options due to social, political, economic and cultural restrictions.
- Many perceive the advantages to outweigh the disadvantages and risks of living in a hazardous area. E.g. soils in volcanic regions are particularly fertile.
- Some adopt a fatalistic approach and believe a hazard to be an "act of God" to be endured. Such an attitude is characteristic of societies which are poor and have few alternatives but to stay put.

Answers attaining markbands E/F should contain at least three of the reasons given above. **[10]**

D8 a i) A hazard is a threat (whether natural or human) that has the potential [1 mark] to cause loss of life, injury, property damage, socio-economic disruption and/or environmental degradation [1 mark]. OWTTE.

ii) A disaster is a major hazard event that causes widespread disruption to the community or region [1 mark]. The affected community is unable to deal with it adequately without outside help [1 mark]. OWTTE. **[2 + 2]**

b Answers must only refer to the hazards covered in this syllabus: earthquakes/volcanoes, drought, tropical cyclones and human-induced hazards. Some hazards, such as earthquakes and volcanoes, can be monitored and a small event usually heralds a large one, but the timing is still uncertain. Droughts have slow onset and are therefore more predictable. Tropical cyclones can be monitored and their landfall time estimated and prepared for. Human-induced hazards are seldom predictable.

The chosen hazards are likely to have different levels of predictability, but this point is not essential to achieve markbands E/F provided there is sufficient comparison. **[6]**

c The content may not be prescribed, but consideration needs to be given to human causes and the features of this event that turned it into a disaster. Accurate information is essential, such as the time and location of its occurrence, those responsible, the number of people affected/killed and the short- and long-term effects (social, economic and environmental). Response may be regarded as an effect and is therefore relevant, but not essential.

Answers which include a range of causes, as well as short- and long-term effects, are likely to be credited at markbands E/F. **[10]**

Option E: Leisure, sport and tourism

E9 a The pattern is that the majority of countries shown are highly industrialized, wealthy and concentrated in Europe [1 mark]. The USA, Japan and Australia are exceptions. South America and Africa are not represented at all. [1 mark]

The trends are that Olympic performance of rich countries (Australia, France, Italy and the Netherlands) has declined from 2000 to 2008 [1 mark]. The USA, Germany and Great Britain are the only exceptions. The representatives of the poor world (China and South Korea) have shown dramatic increases [1 mark]. **[4]**

Credit AOVP.

b The factors include: access to major international airports, political stability, minimal health risks, equable climate, educated workforce, cultural liberalism and AOVP.

Award up to [2 marks] for each factor that is fully explained, up to a maximum of [6 marks]. **[6]**

c The key factors affecting participation in world sporting events include the following:

- population size – usually the greater the population, the more potential to participate in a wide range of sporting events. However, this is not universally reliable
- income per capita – high-income countries are able to train and to afford the equipment and resources required by some sports. These costs may exclude poor countries
- home advantage – local support and media coverage encourage better performance on home ground
- politics – centrally planned economies encourage more specialization and divert national resources to enhance performance
- proximity to the hosting country is of less importance but is borne out by some statistics.

A good answer accessing markbands E/F is expected to cover most of these factors in detail. **[10]**

E10 a Leisure – any freely chosen activity or experience [1mark] that takes place in non-work time [1 mark].

Sport – any physical activity involving a set of rules or customs [1 mark]. The activity may be competitive [1 mark]. Accept AOVP. **[4]**

b The sketch may consist of a sector of the urban area or a general plan. A variety of leisure facilities should be shown and these might include the following (working from the city centre outwards):

- entertainment such as cinemas, concert halls or theatres, which are high-order facilities needing a central location in the CBD and access to a large regional or national population and possibly international tourists
- suburban leisure facilities are designed to serve local residents and, due to lower land prices, may occupy a large land area. Such facilities include leisure centres, sports stadia and school playing fields
- further out on the rural–urban fringe more selective and higher-priced facilities may exist.

These include golf clubs, nature reserves and amenities that require a large land area and threshold population and serve the whole region. The location of different facilities will depend on wealth and the location of the city. Alternative plans are equally acceptable.

Award [2 marks] for a well-drawn sketch map showing urban zones from the CBD to the outskirts. Award [1 mark] for each correctly located and logically explained leisure facility up to a maximum of [4 marks]. **[6]**

c Carrying capacity can be maintained through careful management by controlling the number of tourists in time and space. Problems of environmental damage can be minimized by limiting opening hours, imposing a charge, restricting access and parking, zoning of pedestrians and traffic, tourist education by signage and information boards, by protecting and reinforcing paths, and by surveillance and the employment of wardens.

A good answer accessing markbands E/F and above would be expected to cover most of these strategies and to include at least one developed example. **[10]**

Option F: The geography of food and health

F11a The areas which have been most enlarged and therefore have a disproportionate share of the world's unhealthy people tend to be where population numbers are high and living conditions are relatively poor. Examples include South and South-East Asia and China. Countries which have relatively few unhealthy people are Canada, the USA and Australia. **[4]**

b Population size – Countries with a large population such as India are likely to have a large number of unhealthy people, whereas with Canada the reverse is true.

Poor quality of life – Many people in less developed countries suffer from relatively poor living conditions, lack of access to sanitation and clean water, poor infrastructure and access to medical facilities. Therefore, illness, injury and disability are much more likely. **[6]**

c Example: malaria.

Factors causing the spread include: the increasing mobility of populations; accessibility to malarial areas; increasing tourism and trade; global warming; and poor irrigation practice, which expands the source area and breeding grounds.

Difficulties in containment include: ever-increasing population mobility; the environmental damage caused by DDT and organophosphates; and the resistance of the anopheles mosquito to such pesticides. Alternative methods such as breeding predators and screening mosquitoes by using bed nets have both been partially ineffective.

Strategies of containment, prevention and cure are dictated by the economic status of the country concerned. Poverty is still the main factor causing the spread of malaria and many other diseases.

Answers accessing markbands E/F should be discursive and identify the difficulties of disease containment from a variety of perspectives: social, economic and environmental. **[10]**

F12a Malnutrition is due to nutrient deficiency [1 mark], linked to inadequate food supply. Prevalent in sub-Saharan Africa [1 mark]. Obesity is another form [1 mark] found in some rich countries [1 mark]. **[4]**

b Technological innovations may include: drainage, irrigation, artificial fertilizers, pesticides, HYVs and GM crops, factory farming and others.

Award [1 mark] for each type of innovation and [1 mark] for explanation. **[6]**

c Famine is an acute shortage of food which may be triggered by climatic hazards such as droughts or floods, pestilence, civil war and other political upheavals. All these crises may be responsible for disrupting food production and cutting off market access. However, famine often results from long-term food insecurity. A population may experience this when it is unable to access food through lack of exchange entitlements, i.e. money or possessions that can be exchanged for food. Poverty is an underlying cause of food insecurity and overpopulation may be both its cause and its consequence.

Answers which include a recent and relevant example should access markbands E/F. **[10]**

Option G: Urban environments

G13a The left-hand side of the photograph shows a shanty town where dwellings are haphazardly constructed and made of temporary materials. Infrastructural provision such as roads and sewage systems are not in evidence and water appears to be running down the street. There is some suggestion of economic activity such as retailing and manufacturing amidst the dwellings.

In contrast, the architecturally designed high-rise apartments on the right are likely to accommodate affluent populations who can afford recreational facilities such as individual swimming pools and a communal tennis court. The contrasts in housing are explained by the sharp disparity in the wealth of urban populations in LEDCs, the pressure of population, competition for space and need for social segregation.

Award [2 marks] for description and [2 marks] for explanation that covers both these contrasting residential areas. **[4]**

b The informal sector has the following characteristics:

- Small-scale activity is prevalent and usually involves domestic premises.
- Activities are unregistered, unregulated and do not involve taxation.
- Activities have few employees, often family members, sometimes children.
- The activities may involve manufacturing or services. Examples are vehicle repairs, street vending and bar work.
- Illegal trading, petty theft, prostitution and drug dealing may be involved.

- It is typical of the less developed city, but is universal.
- Manufacturing tends to be located in squatter settlements and poor residential areas.
- Informal services tend to concentrate in areas of high pedestrian density, such as the CBD.

Award up to a maximum of [6 marks] where at least four characteristics are described in some detail. **[6]**

c Environmental problems are likely to include those of poor access to safe water and efficient sewerage systems, derelict land, traffic congestion, air pollution and noise. These problems are common to many cities but are often exacerbated by poverty.

The causes and remedies should both be addressed, but the answer need not be balanced to access markbands E/F. **[10]**

G14a Definition – The development of distinct urban zones which may be categorized as commercial, residential or open space. It may be applied to the CBD in more detail, identifying specific types of commercial function such as shops, offices, municipal services and entertainment. Award [4 marks] for four valid points. **[4]**

b Tall buildings – due to higher land values in the intense competition for space.

Retailing – specialist outlets selling goods requiring a high threshold population.

Offices and services – requiring access to clients and staff.

Administration – public buildings, municipal offices needing a central location.

Entertainment – cinemas, theatres and other centres of entertainment having a large threshold are located in the middle of the town for easy access.

Lack of green open space – due to higher land values. **[6]**

c Both processes involve the revival of the inner urban area and are typical of the rich world.

Re-urbanization involves the development of activities to increase residential population densities within the existing built-up area of the city. This may include the redevelopment of vacant land, refurbishment of housing and the development of new business enterprises. Housing is occupied by new and often wealthy immigrants.

Gentrification is closely associated with re-urbanization, but it has a more specific and narrow interpretation. It involves the refurbishment and reoccupation of housing previously owned by lower classes and upgraded to suit the tastes of wealthy immigrants. It therefore goes through the process of upward filtering.

The London Docklands illustrates both processes. **[10]**

PAPER 3 – HIGHER LEVEL EXTENSION – GLOBAL INTERACTIONS

1 a The answer should show an understanding of the meaning of culture, the process of diffusion and the agents involved. It does not require any comment on outcomes.

Culture provides a sense of identity and attachment to place. It has several modes of expression or traits. These include language, religion, customs, music, art, architecture, dress, food, technology and skills. The process of expansion diffusion involves the gradual spread of culture through person-to-person contact. It may result in a spatial pattern of distance decay and an S-shaped curve of adoption over time. An alternative process is location diffusion, which occurs where migrants (diasporas) have transferred aspects of their culture to a host society. The pace and extent of cultural diffusion have been accelerated recently by developments in communications technology.

TNCs have played a major role in the promotion of culture and consumption through their involvement in the media, music, TV and sport. Accusations of westernization, Americanization or, more seriously, cultural imperialism have been levelled at the "advanced" nations whose culture has been dominant.

The conclusion might observe that cultural diffusion has been occurring for centuries, but the speed and scale of the process is unprecedented and the impacts likely to be controversial.

A good answer accessing markbands D/E will identify cultural traits and the modes of diffusion. **[10]**

b Some have claimed that the uniformity created by global interactions, the freedom of movement and the breakdown of international boundaries have resulted in the loss of cultural diversity and even the death of geography.

The answer must address both sides of this question but it need not be balanced.

The argument supporting the notion of a reduction in global diversity might include some of the following points:

- Causes: the increasing mobility of people; widespread use of information technology; the domination of TNCs in the spreading of consumer culture.
- Consequences: the abandonment of moral codes and customs, usually in favour of westernized ones; the adoption of dominant cultural traits of dress, language, food and music; the spread of consumer culture; the homogenization of landscapes and architecture.

Valid alternative examples may be described and should be credited.

The counter-argument might put forward the following points:

- It is not inevitable that cultural traits will be accepted.
- There are many cases of adaptation rather than adoption where a commodity is modified to suit local tastes. This is known as glocalization.

- The process of cultural change is ongoing and eventually resistance may develop, supported by organizations and the nation states themselves. For example, after decades of cloning, some planners seek to diversify businesses and return to the original pattern of urban functions; dissatisfaction with the globalization of food production has resulted in promotion of farmers' markets and the revival of national culinary dishes in some parts of the world.
- In extreme cases there may be conflict resulting from religious fundamentalism and possibly terrorism.

Even though not a specific requirement of the question, examples of named places and TNCs are essential in a good answer accessing markbands D/E. **[15]**

2 a The term "global interaction" is closely linked to globalization. An acceptable description would include the transfer or exchange of goods, services, financial capital, people, technology, ideas and culture between nations. The first three are sufficient, provided the answer is detailed.

The pattern of globalization is uneven, with the majority of interactions occurring between nations of the rich "North". Some poor nations from the "South", such as those in sub-Saharan Africa, are excluded from most of the activity and financial gain. Nations which have become increasingly involved in global interactions include several of the Asia-Pacific region.

A good answer accessing markbands D/E requires a description of both characteristics and pattern, but some imbalance is allowed. Explanation is not required. **[10]**

b The aspects of the global economy should be identified as involving trade in goods and finance. The economic benefits may affect nation states and individual TNCs. They include the increased participation in world trading encouraged by the free-trade ethos of the WTO. This has allowed TNCs to operate globally and to benefit from access to cheap labour and raw materials with minimal regulation. The efficiency of operations has been facilitated by improvements in transport and ICT. The removal of protectionist barriers also allows for the free flow of finance, which permits instantaneous transactions.

These economic benefits of globalization are not evenly shared. Powerful countries and regions such as North America and north-west Europe control most of the trading activities, whereas parts of Africa are largely excluded. Trading blocs such as NAFTA, the EU and ASEAN protect the countries within them and limit the access of those outside the unions. In some cases the annual income of individual TNCs exceeds that of some smaller national economies.

Answers that include a range of benefits, citing examples, are likely to access markbands D/E. **[15]**

3 a The reduction in the friction of distance through transport improvements has led to time–space convergence. This has resulted in the improved efficiency of global economic operations by accelerating the speed of cross-border transfers of people, goods, services and ideas. For example, air freight permits perishable flowers to be flown in from Kenya to Europe and minerals can be shipped in bulk from areas of cheaper production in the less developed world. The use of ICT eliminates distance and allows for instantaneous business transactions to be made between organizations.

A good answer accessing markbands D/E will include an explanation of specific technological changes affecting transport and communications. **[10]**

b The globalization of economic activity has had profound environmental consequences through exploitative primary economic activities such as mining, forestry, fishing and agriculture. These activities may create local problems such as hydrological disturbance, scarring of the landscape through mining or quarrying, acidification and salinization of the soil.

International dispute arises where there is pressure on resources and this is often a consequence of their uneven global distribution. The geographical mismatch between areas of oil production and oil consumption requires long-distance transport and has resulted in pollution events and political conflict. The unnecessary food miles generated by rich consumers demanding out-of-season produce are another concern. The increasing demand for fresh water associated with growing populations and affluence becomes an acute problem where river basins are shared. Transboundary pollution involving shared resources of the ocean, the atmosphere and rivers are particularly contentious. Some localized environmental damage, such as deforestation of the Amazon basin and the loss of biodiversity, may have far-reaching long-term effects well beyond the area of damage. The legacy of past economic activity, such as species destruction, overfishing and water pollution, are persistent and difficult to solve.

It is expected that answers accessing markbands D/E will address most of these problems of resource exploitation and transfer and a variety of environmental impacts which result from global (and not local) interactions. **[15]**

4 a How is globalization measured?

There are several indices of globalization. Most commonly recognized are the Kearney and KOF indices. Both have the same measurable traits, but the method of calculation varies. Kearney's index tracks changes in the four key components of global integration: trade and investment flows; movement of people across borders; volumes of international telephone traffic and internet usage; participation in international organizations. Each one is broken down into subcategories but the four are equally weighted. KOF selects three broad categories: economic, social and political, with weightings of 36%, 38% and 23% respectively. KOF also takes into account negative values such as trading restrictions, which are part of the economic aspect. Each category is subdivided into a larger number of subcategories.

Why is globalization measured?

This part of the question is more difficult and likely to be less detailed, but an uneven balance is acceptable for access to markbands D/E.

Attempts to measure globalization are undertaken from an academic standpoint to gauge the pattern and possibly the rate. Meaningful correlations with other objective indicators such as GDP per capita or HDI may be made and used to support an argument for and against globalization. Further links may be made between globalization and economic development and the advantages and disadvantages assessed and attempts made to address the balance, in theory. **[10]**

b Answers should identify the principal aspects of globalization as: integration of culture and economy, political involvement and international migration.

The high levels of global involvement by rich countries originate from their initial advantages of abundant resources, colonialism, early industrialization and their long-standing economic domination. They trade mainly amongst themselves and this is facilitated by advanced infrastructures and communication systems and protected by trading blocs. The newly industrialized counties (NICs) such as China, India and Brazil have become increasingly integrated in the world economy and culture due mainly to the involvement of TNCs. Non-globalized countries suffer from inadequate infrastructure, lack of investment, debt and exclusion from the world market through disadvantageous terms of trade and protectionism. Better answers accessing markbands D/E may recognize the self-exclusion of some countries on religious and cultural grounds. **[15]**

5 a The statement suggests that globalization brings opportunities and wealth to rich and poor countries, through communications technology, creation of trading blocs and the relaxation of border controls. It allows greater access to a wide range of markets, labour and resources. Poor countries can gain economically through access to global markets, generating wealth and having a positive effect on the national economy. It benefits the labour force by offering them new opportunities elsewhere.

To access markbands D/E the answer should identify a range of opportunities and explain the new international division of labour. **[10]**

b The sovereignty of nation states has been taken for granted, but is now threatened by the growth of TNCs and the breakdown of national boundaries.

- TNCs which play a major role in the promotion of consumer goods and the ethics of consumption. TNCs may also meet with resistance when the profits of the company exceed that of annual GDP and the "trickle-down" effect is limited by repatriation, remittances and leakage of funds.
- The formation of trade blocs and increasing permeability of political boundaries which may lead towards the weakening and loss of the nation states – this is called "deterritorialization". If the sovereignty and national identity of the country is seriously threatened, resistance in the form of protest may occur.

Political outcomes include anti-globalist protests, national demonstrations, religious fundamentalism and possible terrorism. Such outcomes are not inevitable and with the support of NGOs the future of national sovereignty is deemed to be secure.

To access markbands D/E, the answer must show an understanding of the concept of national sovereignty, its changing state and the aspects of globalization that threaten it. **[15]**

6 a The pollution event must be major and have affected more than one nation. The chosen incident must be explained in terms of origins, global impacts and responses. Good answers accessing markband D and above are likely to consider these characteristics on a range of spatial and temporal scales. **[10]**

b Concern over environmental issues results from increasing international awareness of them. Improved communications technology and transport have revealed a number of environmental problems. These result from the increasing scale of activity and intensification of production by TNCs, lack of environmental regulations in some LEDCs and increasing volumes of international freight transfer.

Environmental damage may arise from the following activities:

- opencast and deep mining, causing landscape scarring, hydrological disruption, deforestation, soil erosion and water pollution
- agribusiness, causing loss of soil fertility through monoculture, contamination of land and water by excessive use of fertilizers and pesticides and air pollution through increased air freight
- logging on a large scale, adversely affecting local hydrology and climate and reducing biodiversity
- the expansion of unregulated manufacturing activities in LEDCs by TNCs, resulting in pollution
- the global transfer of toxic waste via the oceans.

Global environmental problems, such as climate change, ozone depletion and the tragedy of the commons (oceans and atmosphere), have also been revealed through scientific research and international awareness. The work of NGOs (civil societies) has been fundamental in raising awareness of these problems and the failure of some countries to comply with international agreements such as the Kyoto protocol.

Answers which identify a range of environmental issues and explain increasing global concern should access markbands D/E. **[15]**

INTERNAL ASSESSMENT: ADVICE TO STUDENTS AND TEACHERS

WHY FIELDWORK MATTERS

Fieldwork is an essential part of learning geography and is compulsory for both HL and SL students. It is referred to as "Internal Assessment" (IA), which means that it will be marked by your teacher and moderated by an external IB examiner.

Your fieldwork investigation is important because it will:

- help you make sense of some of the more difficult aspects of the subject
- improve your overall grade, especially if you don't perform so well in the external exams
- provide useful case study material when answering an external exam question
- provide research skills which will be useful in higher education or employment.

Internal assessment – the essentials

- IA counts for 20% of the total marks at HL and 25% at SL.
- It requires 20 hours of class time (including fieldwork).
- Group work is allowed for data collection.
- Fieldwork reports are written individually.
- Each report must be no more than 2500 words in length.
- It must be related to a topic on the syllabus.

Fieldwork research methods

Information must come from the student's own observations and measurements collected in the field. This **"primary information"** must form the basis of each investigation. Fieldwork should provide sufficient information to enable adequate interpretation and analysis.

Common errors

- The report exceeds the 2500 word limit.
- The chosen topic has no spatial element.
- The chosen topic is not geographical.
- The chosen topic does not relate to the syllabus.
- The fieldwork question is too simplistic.
- The information is collected only from the internet.
- The survey area is too large and covers the whole region.
- The fieldwork information is insufficient to answer the fieldwork question.
- The analysis is purely descriptive

STAGES IN UNDERTAKING FIELDWORK FOR INTERNAL ASSESSMENT

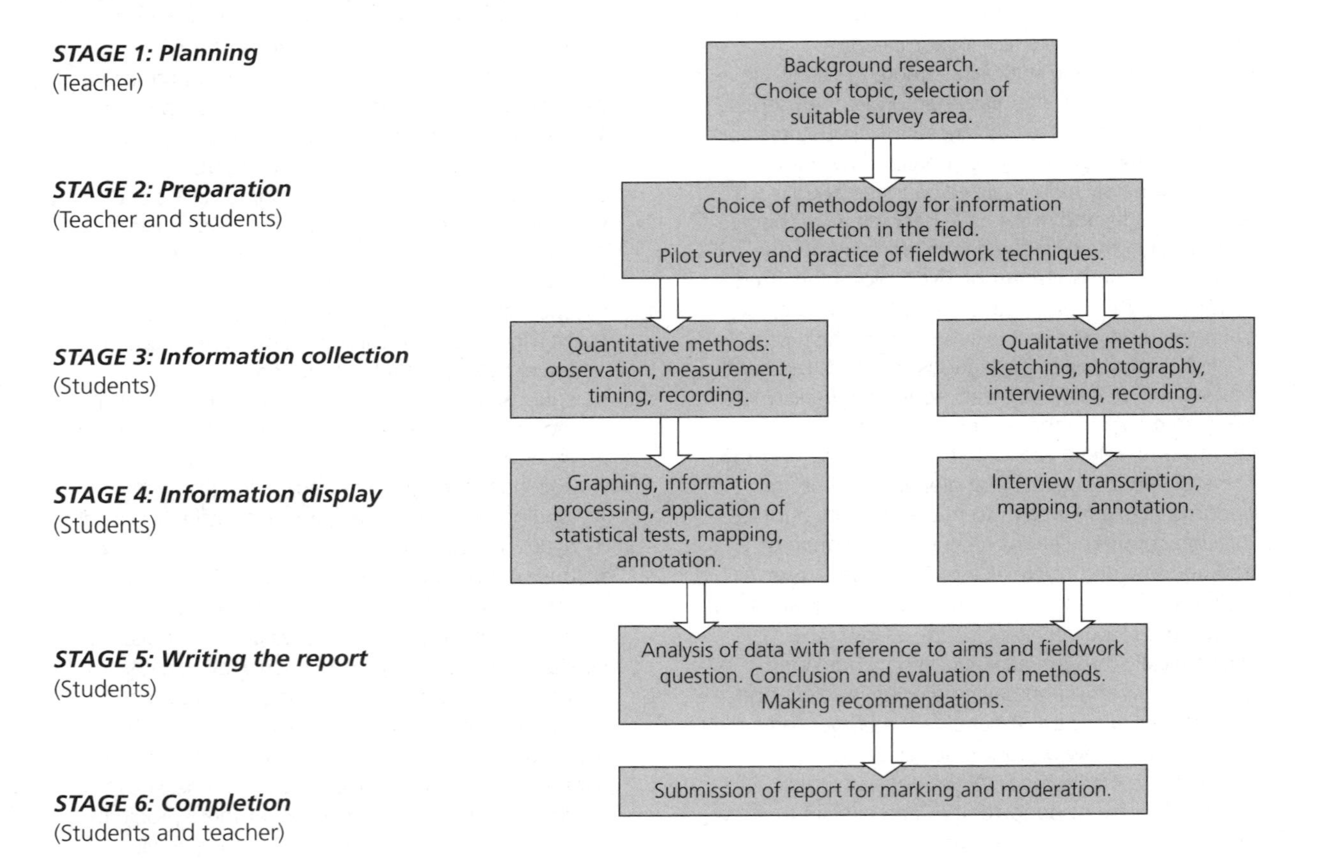

STAGE 1 (teacher only)

Background research

The success of your students' fieldwork will depend on your careful planning and preparation. The following resources are essential reading before you start.

- The Geography Subject Guide (2009 version with first exam in May 2011) can be found on the subject page of the online curriculum centre (OCC) at **http://occ.ibo.org**, a password-protected IB website designed to support IB teachers. The subject guide can also be purchased from the IB store at **http://store.ibo.org**.
- Additional publications such as teacher support materials, subject reports, internal assessment guidance and grade descriptors can also be found at the OCC.
- The Examiner's reports provide teachers with an overall review of investigations undertaken in a large range of schools and make recommendations for future investigations.

The IB Online Curriculum Centre is a discussion forum where geography teachers often exchange ideas on fieldwork.

Choosing the right topic

The fieldwork topic must be related to the syllabus, and the most suitable topics are found within the Optional Themes. The core and HL extension have very few topics which are suitable owing to their global scale.

The investigation must be:

- focused upon a clearly defined fieldwork question
- confined to a small area and on a local scale
- spatial
- based on the collection of primary information in the field
- manageable in terms of the area covered, the time allowed and the 2500 word limit
- able to fulfil the assessment criteria.

Choosing the right site

The viability and success of the fieldwork is determined by careful planning and preparation. It is essential that you select the survey area in advance of the fieldwork investigation to ensure that it fulfils the following criteria:

- It is on a local scale, but the area covered is large enough for sufficient information to be collected.
- The area can be covered by the students in the time allocated.
- All sites within the area are accessible at all times of day and at all seasons.
- The land is open to the public and research is permitted.

Where fieldwork is restricted to the school site, many successful investigations can be undertaken; for example, surveys of footpath erosion, microclimate, infiltration / ground compaction and waste management.

The role of the teacher, the group and the individual student

It is advisable for you to choose the fieldwork topic and test its viability before embarking on the class exercise. In general, the most successful undertakings are those involving groupwork, with the initial planning done by the teacher. The choice of topic, the scale of the investigation, the area covered and the time allowed will be determined by the number of students available to carry out the work.

Fieldwork methods used to collect information should be chosen by you, and the techniques and equipment should be practised prior to the investigation by the students. Once the fieldwork is over and the information made available to all members of the class, students should work individually and no further collaboration is allowed.

STAGE 2 (teacher and students)

Devising the fieldwork question

The fieldwork question forms a basis to the research, which should allow for an investigative rather than descriptive approach. The question should be clearly focused, unambiguous and answerable. If the question is simplistic and the answer obvious, it is unlikely to be worthy of execution. However, research topics which have uncertain outcome are still perfectly viable.

Collecting the right information

Fieldwork must involve the collection of primary information.

Primary informations may be qualitative or quantitative, or a combination of both (see diagram on page 202). In the case of a traffic survey, qualitative data might include photographs, interviews with pedestrians and the subjective assessment of perceived traffic hazard by the student. Quantitative information might include traffic counts, traffic delay times, length of tailback, noise levels in decibels or a survey of suspended particulate matter in the atmosphere.

Secondary or published information not collected by the students themselves may be used to supplement primary information but it must not form the basis of the report.

STAGE 3 (over to the student)

Once your teacher has done the initial planning and preparation, it is over to you to undertake the task of information collection. Remember that this is a one-off opportunity: the stormy conditions during which you collected your wave data cannot be repeated.

Collecting and justifying your fieldwork methods

You must be aware of all the techniques involved and be able to critically evaluate each of them. Before you start collecting information and before you leave the survey site, make sure you have:

- marked for the sites of information collection
- recorded the date and time of collection
- recorded the weather conditions or any special event occurring on the day that might affect the results
- recorded the technique of handling a particular instrument, where it is placed, the time interval between readings, the advantages and disadvantages of the technique
- justified the choice of survey sites and their number/frequency/location
- justified the choice of method used for information collection
- justified the sampling technique used.

STAGE 4 (students)

How to display your fieldwork information

Your fieldwork data should be displayed next to the text to which it refers and should not be confined to the end of the report. Use the table below as a guide.

Method	Do	Don't
Maps	✓ Include a map of the survey sites. ✓ Show your results at specific survey sites on this map. ✓ Annotate your map with brief comments.	✗ Include a national map; it is irrelevant. ✗ Include scruffy maps drawn with pencil.
Graphs	✓ Wherever possible, place a series of graphs on the same page for comparison. ✓ Use a variety of graphical techniques. ✓ Use transparent overlay maps to show spatial relationships	✗ Use a monotonous series of pie charts to represent your data page by page. ✗ Download maps from the internet without first modifying or adapting them for your purpose.
Photos	✓ Make sure that each photograph shows the time it was taken, its location and its orientation.	✗ Include photos of your friends and teacher unless they are strictly relevant to the investigation.
Sketches	✓ Make sure they are fully labelled/ annotated and dated.	✗ Include these unless relevant.
Method		
Generally	Make sure that all illustrations are properly referenced. Use a range of techniques, but make sure each is suitable. Map information wherever possible.	

STAGE 5 (students)

Writing your report

Your report should be structured using the assessment criteria shown below. Note that criterion C can be represented by illustrative material in any part of the report. Assessment of this criterion is not confined to one section.

The mark allocation, and the recommended and approximate number of words for each criterion, are both given right.

	Criteria		
A	**Fieldwork question**	**3 marks**	**300 words**
	This should be concise and clear to the reader. There should be one question only. You should comment briefly on the geographic context, explaining why that particular area of survey was chosen. It is essential to include a map showing the area under investigation. You should state the syllabus section to which the investigation relates.		
B	**Method(s) of investigation**	**3 marks**	**300 words**
	You should describe the method(s) used to collect information. The methods should be justified, which means explaining sampling techniques, the time chosen, the specific location and any other relevant information such as weather conditions.		
C	**Quality and treatment of information collected**	**5 marks**	***N/A**
	There is a range of possible techniques of information display that you might use in any investigation, but make sure that they are clear and effective. The type of method used will be determined by the nature of the particular investigation.		
D	**Written analysis**	**10 marks**	**C + D: 1350 words**
	In the written analysis you demonstrate your knowledge and understanding by interpreting and explaining the information collected in relation to the fieldwork question. This involves recognizing spatial patterns and trends found in the information collected. Where appropriate, you should attempt to explain anomalies.		
E	**Conclusion**	**2 marks**	**200 words**
	You should summarize the findings of your fieldwork investigation. There should be a clear, concise statement answering the fieldwork question. It is acceptable for the conclusion to state that the findings do not match any of your preliminary judgements or projections.		
F	**Evaluation**	**3 marks**	**300 words**
	You should review the methods you used to collect the information in the field. You should include any factors which threatened the validity of the data, such as an abnormal weather event. Suggest viable and realistic ways in which the study might be improved in the future.		
G	**Formal requirements**	**4 marks**	**N/A**
	The written report must meet the following five formal requirements of organization and presentation: • The work is within the 2500 word limit. • The report is neat and well structured. • The pages are numbered. • All sources are correctly referenced. • All illustrations are numbered, fully integrated into the body of the report and not placed in an appendix. General guidance on IB policy to referencing and sourcing can be found in the subject guide.		
	Finished report	**30 marks**	**< 2500 words**

*Criterion C assesses information display and does not include a word count (except for large sized annotations).

STAGE 6 (students and teacher)

Completion of the fieldwork report by the student

Complete this checklist before you submit your fieldwork report.

Tasks	Completed
The candidate name and number is stated on the front cover	
The report is bound or held together securely in a folder	
All plastic pockets have been removed	
There is a contents page	
All the pages are numbered	
All illustrations have figure numbers	
All illustrations are close to the relevant text	
All sources are referenced	
The appendix contains only raw information	
The report has a fieldwork question	
All methods of information collection are fully justified	
All maps have normal conventions of title, scale, north point and key	
The analysis refers to the fieldwork question and the information collected	
There is a conclusion	
The evaluation makes recommendations for improvements	

EXAMPLES OF DIFFERENT METHODS OF INFORMATION COLLECTION

Investigation using primary (qualitative and quantitative) and secondary methods of information collection

Title	An investigation of gentrification in area A in town X
Aim	To determine the effects of gentrification on area A in town X and to examine local attitude towards it
Fieldwork question	How has gentrification brought social, economic and environmental changes to town X?
Syllabus theme	Urban environments
Conceptual basis	See p. 132
Methods of information collection	• Socio-economic patterns and changes in area A – use secondary information from census or housing type and price from estate agents. Survey of local streets to record car type/age • Compare information with averages for town X or adjoining areas • In-migration – questionnaire survey of residents to discover occupation, length of residence and motives for moving • Local attitudes – questionnaire with long-term local residents to determine attitude to affluent newcomers and perception of long-term residents of local changes • Housing survey – evaluate housing condition and record signs of renovation and devise a housing quality index • Environmental quality survey – litter, vandalism, landscape quality, dereliction, noise pollution. In transect across area X and adjoining areas • Economic change – local facility survey – classify and map new shops and services, new bars, restaurants, good transport services
Methods of information display	• Annotated photos (qualitative) • Maps showing scores for survey sites using overlay (quantitative) • Graphical profiles for landscape quality/dereliction • Classification and mapping of shops and services
Analytical techniques	• Chi-squared test to investigate significance of socio-economic changes in one area over time or in space • Location quotients to identify above-average level of housing renovation scores

INDEX